Reaction–Diffusion Equations and Their Applications to Biology

Reaction–Diffusion Equations and Their Applications to Biology

N. F. Britton

School of Mathematics
Bath University
Bath, UK

1986

ACADEMIC PRESS
Harcourt Brace Jovanovich, Publishers
London Orlando New York San Diego Austin Boston
Toronto Sydney Tokyo

ACADEMIC PRESS INC. (LONDON) LTD
24/28 Oval Road
London NW1

United States Edition published by
ACADEMIC PRESS INC.
Orlando, Florida 32887

British Library Cataloguing in Publication Data

Britton, N. F.
Reaction–diffusion equations and their applications to biology.
1. Biomathematics 2. Differential equations
I. Title
574'.01'51535 QH323.5

ISBN 0-12-135140-8

Printed in Great Britain by J. W. Arrowsmith Ltd., Bristol

Preface

In the past few years there has been a great increase in the use of mathematics in the biological sciences. One of the areas of interest has been reaction-diffusion systems, which have applications in developmental biology, ecology, physiology and many other fields. With such a wide variety of applications the choice of what to include in this book was necessarily a personal one, and topics have been included both as interesting applications and as illustrations of the various mathematical techniques which may be used to analyse such systems. The emphasis throughout has been on practical methods, and on obtaining answers to the questions of biological importance, rather than on more abstract mathematical questions.

In Chapter 1 the equations are derived, and we justify the neglect of stochastic effects in certain circumstances. We shall always be concerned with applications in which such circumstances obtain. In Chapter 2 an introduction to the techniques of ordinary differential equations is given. In Chapter 3 a particular class of reaction–diffusion equations is considered, those whose kinetic equations are conservative. Such systems have limitations as models in ecology, as we show, but are widely and successfully used in epidemiology. We discuss a model for the spread of the rabies virus in Europe. In Chapter 4 scalar reaction–diffusion equations are considered; applications include the wave of advance of an advantageous gene in a population and the control of the spruce budworm of North America. In Chapter 5 general reaction–diffusion systems are discussed from a mathematical point of view; comparison theorems, which may be used to prove the existence and uniqueness of solutions and to find bounds for the solutions of intractable systems and which form the basis of numerical techniques for the computations of solutions, are stressed. Some results on the asymptotic behaviour of solutions as $t \to \infty$ are proved. In Chapter 6 steady and periodic solutions of such systems are discussed by using bifurcation theory, with applications to pattern formation, including the regeneration of hydranths in the marine hydroid *tubularia*, and to a prey-predator system. Techniques which allow construction of the solutions are employed.

In many applications small parameters appear in the equations, and much information may be obtained by the use of asymptotic methods; in the last three chapters such methods are discussed. In Chapter 7 oscillatory systems are considered. Applications include a model for target patterns in the well-known Belousov–Zhabotinskii chemical reaction. In Chapter 8 matching techniques for systems with fast and slow variables are considered, with applications in enzyme kinetics, nerve physiology and the control system in the slime mould *dictyostelium discoideum* and a different approach to the problem of target patterns in the Belousov–Zhabotinskii reaction. Finally, in Chapter 9 techniques for systems with small diffusion coefficients are considered, leading to boundary layer problems, and models for facilitated diffusion of oxygen into muscle and for carbon monoxide poisoning are analysed.

Although the book is largely self-contained, some knowledge of the mathematics of differential equations is necessary. Thus the book is intended for mathematicians who are interested in the application of their subject to the biological sciences and for biologists with some mathematical training. It is also suitable for postgraduate mathematics students and for undergraduate mathematicians taking a course in mathematical biology.

I should like to thank Dr J. D. Murray, Prof. J. F. Toland, and Dr J. E. Marshall, who read parts of an early draft of the book, and Dr S. M. Skevington for her support and encouragement.

April 1986 N. F. Britton

Contents

1. Introduction

1.1 Derivation of the Equations

Reaction–diffusion systems are coupled systems of parabolic partial differential equations. Applications are very wide; some of those discussed in this book are models for pattern formation in morphogenesis, for prey-predator and other ecological systems, for conduction in nerves, for epidemics, for carbon monoxide poisoning, and for oscillating chemical reactions. In their simplest form they may be written as

$$\partial \mathbf{u}/\partial t = \mathbf{u}_t = \mathbf{f}(\mathbf{u}) + D\nabla^2\mathbf{u} \tag{1.1}$$

where $\mathbf{u} = \mathbf{u}(\mathbf{x}, t)$ is the vector of dependent variables, $\mathbf{f}(\mathbf{u})$ is a nonlinear vector-valued function of $\mathbf{u}$ (the reaction terms) and D is the diffusion matrix. The reaction terms derive from any interaction between the components of $\mathbf{u}$; for example, $\mathbf{u}$ may be a vector of chemical concentrations, when $\mathbf{f}(\mathbf{u})$ represents the effect of chemical reaction on these concentrations; or the components of $\mathbf{u}$ may be population densities, when $\mathbf{f}(\mathbf{u})$ represents the effect of predator–prey relationships, competition or symbiosis. The diffusion terms may represent molecular diffusion or some random movement of individuals in a population. For applications of such equations to chemical reactor theory see Aris (1975). For applications to biological systems see for example Murray (1977), Fife (1979a), Sleeman (1981), Jones and Sleeman (1983) and Smoller (1983).

We shall derive the equations in the case of chemical reaction and molecular diffusion (Crank, 1975) and indicate the assumptions necessary for them to be applied in different contexts. We shall also include the effects of convection for completeness, although in most applications which we shall consider these effects are absent.

Let us consider a region (which may be the liquid in a test-tube or a living cell) in which chemical reactions are taking place. Let us take the concentration of the ith species taking part in the reactions to be u_i, and let its rate of formation through reaction be $f_i(\mathbf{u}, \mathbf{x}, t)$, where $\mathbf{u}$ is the vector of the concentrations u_i, $\mathbf{x}$ is position and t is time. Let the flux of this species be $\mathbf{J}_i = \mathbf{J}_i^{(d)} + \mathbf{J}_i^{(c)}$, where $\mathbf{J}_i^{(d)}$ is the flux due to diffusion and $\mathbf{J}_i^{(c)}$ that due to convection.

Consider an arbitrary fixed volume V of fluid inside the region, with surface S. Then the rate of increase of the amount of u_i inside the volume is equal to the amount formed by the reaction minus the flux through the surface. In mathematical terms

$$\frac{d}{dt}\int_V u_i\,dV = \int_V f_i\,dV - \int_S \mathbf{J}_i \cdot d\mathbf{S}$$

Using the divergence theorem, this becomes

$$\int_V \left(\frac{\partial u_i}{\partial t} - f_i + \nabla \cdot \mathbf{J}_i\right) dV = 0$$

or since the volume V is arbitrary,

$$\frac{\partial u_i}{\partial t} = f_i - \nabla \cdot \mathbf{J}_i$$

The function f_i comes from the application of the law of mass action to the reactions taking place. As an example, consider the reaction

$$m\mathrm{A} + n\mathrm{B} \underset{k_{-1}}{\overset{k_1}{\rightleftharpoons}} p\mathrm{C}$$

where A, B and C are reactants, k_1 and k_{-1} are rate constants and m, n and p represent the numbers of molecules of each reactant involved in the reaction and are known as stoichiometric parameters. Then the law of mass action states that

$$m\frac{d[\mathrm{A}]}{dt} = -k_1[\mathrm{A}]^m[\mathrm{B}]^n + k_{-1}[\mathrm{C}]^p = n\frac{d[\mathrm{B}]}{dt} = -p\frac{d[\mathrm{C}]}{dt}$$

where the square brackets denote concentrations. Here k_1 and k_{-1} are temperature-dependent (and hence possibly space- and time-dependent), but we shall very often consider reactions taking place at constant temperature and shall take such quantities to be constant. The convectional flux $\mathbf{J}_i^{(c)}$ is given by $\mathbf{J}_i^{(c)} = \mathbf{c}_i u_i$, where $\mathbf{c}_i$ is the velocity of convection. The diffusional flux is given by Fick's law as $\mathbf{J}_i^{(d)} = -D_i \nabla u_i$, where D_i is the diffusion coefficient. Thus the equations become

$$\frac{\partial u_i}{\partial t} = f_i - \nabla \cdot (\mathbf{c}_i u_i) + \nabla \cdot (D_i \nabla u_i)$$

We shall normally consider $\mathbf{c}_i = \mathbf{0}$ and D_i constant, although D_i may be a function of $\mathbf{x}$, t and $\mathbf{u}$. We shall also often consider f_i to be a function of $\mathbf{u}$ alone. In this case we have

$$\mathbf{u}_t = \mathbf{f}(\mathbf{u}) + D\nabla^2\mathbf{u}$$

where D is a diagonal matrix. It is also possible for diffusion of one species to affect the rate of production of another. In this case the diffusion matrix D is no longer diagonal.

In deriving these equations we have made various assumptions. First, we have assumed that the amount of each chemical present per unit volume may be described by a quantity known as its concentration which is twice continuously differentiable in space and once in time. This concentration is assumed to evolve deterministically. We have assumed the existence of functions f_i and of the convectional and diffusional fluxes in the forms given. Similar assumptions on quantities analogous to concentrations may be made in different contexts, and quite general assumptions on the interaction mechanism and on the redistribution processes analogous to chemical reaction and molecular diffusion lead to equations of similar form in many fields. For a discussion of the assumptions required see Fife (1979a). Thus reaction–diffusion systems are used as models in population dynamics, genetics, nerve conduction, epidemiology and combustion as well as in purely chemical systems. Care must be taken, however, in the use of these models, especially if some stochastic effect is suspected. In this case the equations *may* be a good model for the expected values of the quantities concerned, but this is not necessarily the case.

1.2 Stochastic Effects

As an example consider the spatially independent birth process $Y(t)$, where Y is the number of individuals in a certain population. Let the probability that Y increases to $Y+1$ in an infinitesimal time dt be proportional to Y and to dt, i.e.

$$P\{Y \to Y+1 \text{ in } (t, t+dt)\} = \alpha Y\,dt$$

Then it is easy to show that the expected value of Y, $y = E(Y)$, satisfies the differential equation

$$\dot{y} = \alpha y$$

However, if

$$P\{Y \to Y+1 \text{ in } (t, t+dt)\} = \alpha Y(1 - Y/\rho)\,dt$$

in other words there is some limiting process which reduces the rate of birth as Y approaches a value ρ, which may be thought of as the carrying capacity of the environment, then $y = E(Y)$ satisfies

$$\dot{y} = \alpha E\{Y(1 - Y/\rho)\} = \alpha y - \alpha E(Y^2)/\rho$$

Thus, unless $E(Y^2)=(E(Y))^2$, we cannot write

$$\dot{y}=\alpha y(1-y/\rho)$$

as we might have expected. Since the difference $E(Y^2)-(E(Y))^2$ is the variance of Y, the approximation $\dot{y}=\alpha y(1-y/\rho)$ may be used when Y has small variance. Similar results are true for higher order effects, and if more than one species is involved we shall require covariances etc., to be small. Normally, when dealing with chemical reaction we expect the concentrations of the reactants to have small variance, covariance, etc.—two chemical experiments set up with nearly identical initial concentrations and physical conditions produce nearly identical results. Thus we may in this case move directly from the stochastic to the deterministic model. The same may be true of the growth of a bacterial culture, but is unlikely to be true in, for example, the growth of a population of rabbits. In this case different experiments with identical initial conditions may produce entirely different results, and in this case the deterministic analogue of the stochastic model does not even give the development of the expected number of rabbits unless the model is linear.

A discussion of the relationships between deterministic and stochastic models is given by Mollison (1977).

2. Systems of Ordinary Differential Equations

2.1 Preliminaries

Systems of ordinary differential equations will turn up many times in this book. For example, spatially independent solutions of the reaction–diffusion system (1.1) satisfy the kinetic system

$$\dot{\mathbf{u}} = \mathbf{f}(\mathbf{u}) \tag{2.1}$$

Also, travelling wave solutions of reaction–diffusion equations and time-independent solutions in one space dimension satisfy similar systems. In this chapter we shall sketch some of the methods of analysis of these equations. For a fuller treatment there are many books available [e.g., Coddington and Levinson (1955), Sansone and Conti (1964), Jordan and Smith (1979), Cronin (1980) and Arrowsmith and Place (1982)].

We shall state some preliminary theorems, but since this chapter is only intended as an introduction to reaction–diffusion systems, we shall not in general prove them. The most fundamental is known as Picard's theorem, although it is also associated with the names of Cauchy and Lipschitz.

THEOREM 2.2 (Picard's Theorem): Consider the system of equations

$$\dot{\mathbf{u}} = \mathbf{f}(\mathbf{u}, t)$$

Let the functions $f_i(\mathbf{u}, t)$ satisfy Lipschitz conditions in their arguments. Then there exists a unique solution $\mathbf{u} = \mathbf{u}(t)$ in the neighbourhood of $t = 0$ satisfying initial conditions $\mathbf{u}(0) = \mathbf{u}_0$. Moreover, this solution is a continuous function of the initial conditions. If

$$\dot{\mathbf{u}} = \mathbf{f}(\mathbf{u}, t, \lambda)$$

where λ is a parameter, and each f_i also satisfies a Lipschitz condition uniformly in λ in a neighbourhood of λ_0 and is continuous in λ, then the same conclusions hold in a neighbourhood of λ_0. Moreover $\mathbf{u} = \mathbf{u}(t, \lambda)$ is a continuous function of λ in this neighbourhood.

In later chapters we shall be very concerned with stability considerations and we therefore state here some relevant definitions.

DEFINITION 2.3: Consider the system $\dot{\mathbf{u}}=\mathbf{f}(\mathbf{u}, t)$. The solution $\mathbf{u}=\boldsymbol{\phi}(t)$ defined in $[t_0, \infty)$ is *stable* if given any $\varepsilon>0$ there exists $\delta>0$ such that if $\boldsymbol{\psi}(t_0)$ is any vector satisfying

$$|\boldsymbol{\phi}(t_0)-\boldsymbol{\psi}(t_0)|<\delta$$

then the solution $\mathbf{u}=\boldsymbol{\psi}(t)$ with initial conditions $\mathbf{u}(t_0)=\boldsymbol{\psi}(t_0)$ exists in (t_0, ∞) and satisfies

$$|\boldsymbol{\phi}(t)-\boldsymbol{\psi}(t)|<\varepsilon$$

for all $t \geqslant t_0$.

DEFINITION 2.4: A solution $\boldsymbol{\phi}(t)$ is *asymptotically stable* if (a) it is stable and (b) there exists $\eta>0$ such that if

$$|\boldsymbol{\phi}(t_0)-\boldsymbol{\psi}(t_0)|<\eta$$

then

$$|\boldsymbol{\phi}(t)-\boldsymbol{\psi}(t)| \rightarrow 0 \quad \text{as} \quad t \rightarrow \infty$$

It is *globally asymptotically stable* if η may be chosen arbitrarily large.

DEFINITION 2.5: A solution $\boldsymbol{\phi}(t)$ is *unstable* if given any positive ε sufficiently small and any $\delta>0$ there is a solution $\boldsymbol{\psi}(t)$ such that (a)

$$|\boldsymbol{\phi}(t_0)-\boldsymbol{\psi}(t_0)|<\delta$$

and (b)

$$|\boldsymbol{\phi}(t)-\boldsymbol{\psi}(t)|>\varepsilon$$

for some $t>t_0$.

A system of equations is said to be *autonomous* if $\mathbf{f}(\mathbf{u}, t)=\mathbf{f}(\mathbf{u})$. We shall henceforth be concerned with autonomous systems unless otherwise stated.

DEFINITION 2.6: A solution $\boldsymbol{\phi}(t)$ of an autonomous system is *orbitally stable* if for any $\varepsilon>0$ there exists $\delta>0$ such that if

$$|\boldsymbol{\phi}(t_0)-\boldsymbol{\psi}(t_0)|<\delta$$

then

$$\inf_{\tau \geqslant t_0}|\boldsymbol{\phi}(t)-\boldsymbol{\psi}(\tau)|<\varepsilon$$

for each $t \geqslant t_0$.

A *critical point* (also called a *singular point* or an *equilibrium point*) of an autonomous system is a point $\mathbf{u}=\boldsymbol{\phi}$, where $\mathbf{f}(\boldsymbol{\phi})=\mathbf{0}$. In this case the equations have a solution $\mathbf{u}=\boldsymbol{\phi}$. The critical point is (asymptotically) stable if the solution $\mathbf{u}=\boldsymbol{\phi}$ is. The *order* of a system of equations is defined to be the dimension of the vector $\mathbf{u}$. In the remainder of this chapter we shall classify the types of critical point in second-order linear systems and consider the effect of nonlinearities on the nature of the critical points. We shall then discuss the asymptotic behaviour of autonomous systems as $t\to\infty$, and shall be particularly concerned with limit cycle solutions.

2.2 Systems of Linear Ordinary Differential Equations

In the general linear autonomous case

$$\dot{\mathbf{u}}=A\mathbf{u} \tag{2.7}$$

where A is a constant matrix. Then it can be shown that the unique critical point $\mathbf{u}=\mathbf{0}$ is stable if $\mathrm{Re}(\lambda)\leqslant 0$ for each eigenvalue λ of A and if λ is a simple eigenvalue whenever $\mathrm{Re}(\lambda)=0$, and asymptotically stable if and only if $\mathrm{Re}(\lambda)<0$ for each λ.

$$\begin{aligned}\dot{u}&=f(u,v)=au+bv\\ \dot{v}&=g(u,v)=cu+dv\end{aligned} \tag{2.8}$$

Since the system is linear, we look for solutions of the form $\mathbf{u}=\boldsymbol{\phi}\,e^{\lambda t}$, where $\boldsymbol{\phi}$ is a constant vector, to obtain

$$\lambda u=au+bv,\qquad \lambda v=cu+dv$$

These have solutions when $|A-\lambda I|=0$, where

$$A=\begin{bmatrix}a & b\\ c & d\end{bmatrix}$$

i.e. when λ is the eigenvalue of the matrix A and $\boldsymbol{\phi}$ the corresponding eigenvector. Since the eigenvalues λ satisfy a quadratic equation with roots λ_1 and λ_2 we can distinguish the following cases:

(i) Let λ_1 and λ_2 be real and distinct and of the same sign. Then

$$\mathbf{u}=k_1\boldsymbol{\phi}_1e^{\lambda_1 t}+k_2\boldsymbol{\phi}_2e^{\lambda_2 t}$$

for some constants k_1 and k_2. In the (u,v)-plane (the *phase plane*), the trajectories are of the form shown in Fig. 2.9(a). The critical point is stable if λ_1 and λ_2 are negative and unstable if they are positive. The critical point in this case is known as a stable or unstable *node*.

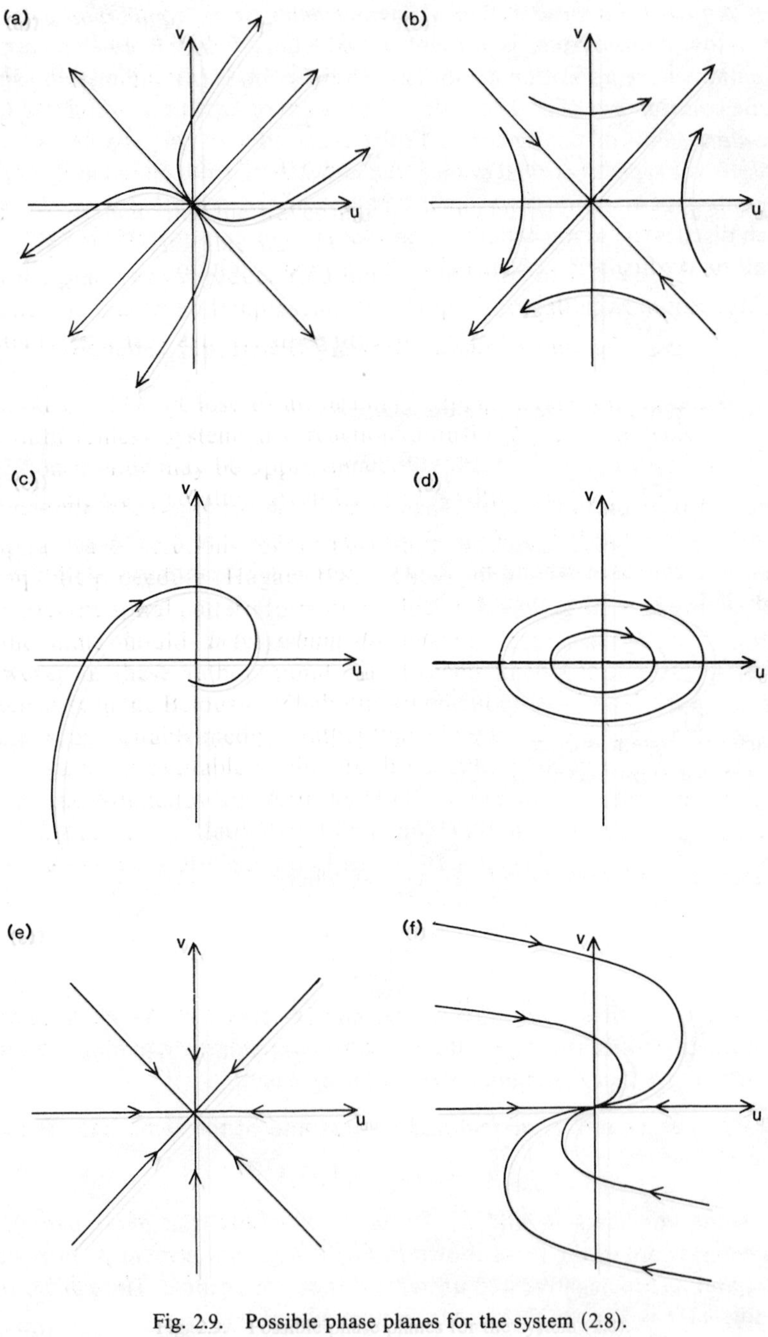

Fig. 2.9. Possible phase planes for the system (2.8).

(ii) Let λ_1 and λ_2 be real and of opposite sign. Then

$$\mathbf{u} = k_1\boldsymbol{\phi}_1 e^{\lambda_1 t} + k_2\boldsymbol{\phi}_2 e^{\lambda_2 t}$$

and the trajectories are shown in Fig. 2.9(b). The critical point is known as a *saddle point.*

(iii) Let λ_1 and λ_2 be conjugate complex. Then $\text{Re}(\lambda_1) = \text{Re}(\lambda_2) = \text{Re}(\lambda)$, say, and $\mathbf{u}$ is of the form

$$\mathbf{u} = e^{\text{Re}\,\lambda t}\{k_1\boldsymbol{\phi}_1 e^{i\,\text{Im}\,\lambda_1 t} + k_2\boldsymbol{\phi}_2 e^{i\,\text{Im}\,\lambda_2 t}\}.$$

Because of the imaginary part of λ_1 and λ_2, the trajectories in the phase plane will move around the critical point. If $\text{Re}(\lambda_1) = \text{Re}(\lambda_2) < 0$, they will spiral into the point, whereas if $\text{Re}(\lambda_1) = \text{Re}(\lambda_2) > 0$, they will spiral out, as shown in Fig. 2.9(c). The critical point is known as a stable or unstable *focus.* If $\text{Re}(\lambda_1) = \text{Re}(\lambda_2) = 0$, we have the limiting case of a *centre,* where the trajectories do not spiral in or out but are closed, as in Fig. 2.9(d).

(iv) Let us now consider the case of two equal (real) eigenvalues. We can, without loss of generality, take $a = d$ and $bc = 0$. If this does not hold, a simple linear transformation, or rotation of axes in the phase plane, ensures that it does. There are then two possibilities, the first being that $b = c = 0$. In this case $A = aI$ and the solution is $\mathbf{u} = \mathbf{u}_0 e^{at}$, where $\mathbf{u}_0$ is the initial condition. This is shown in Fig. 2.9(e). The second possibility is that one of b and c, say b, is non-zero. In this case we have

$$\dot{u} = au + bv, \qquad \dot{v} = av$$

which can be integrated to obtain

$$\mathbf{u} = \begin{bmatrix} u \\ v \end{bmatrix} = u_0 \begin{bmatrix} 1 \\ 0 \end{bmatrix} e^{at} + v_0 \begin{bmatrix} bt \\ 1 \end{bmatrix} e^{at}$$

This is shown in Fig. 2.9(f). Both these critical points are kinds of node.

(v) The last possibility is that one (or both) of the eigenvalues is zero. In this case A is singular, there is a non-trivial solution of $A\boldsymbol{\phi} = \mathbf{0}$, and all points $\alpha\boldsymbol{\phi}$ are critical points. Thus the origin is no longer an isolated singularity. If both eigenvalues are zero the critical point may be either stable or unstable, as may be seen by considering (iv) with $a = 0$ and either $b = 0$ or $b \neq 0$.

Linear systems of three or more ordinary differential equations may be dealt with in similar ways.

2.3 Systems of Nonlinear Ordinary Differential Equations

Consider the system

$$\dot{\mathbf{u}} = \mathbf{f}(\mathbf{u}) = A\mathbf{u} + \mathbf{N}(\mathbf{u}) \tag{2.10}$$

where A is a matrix and $\mathbf{N}(\mathbf{u}) = o(\mathbf{u})$ as $\mathbf{u} \to \mathbf{0}$ (i.e. $|\mathbf{N}(\mathbf{u})|/|\mathbf{u}| \to 0$ as $|\mathbf{u}| \to 0$), so that $\mathbf{N}(\mathbf{u})$ represents the nonlinear terms in the equations. We are of course assuming here that $\mathbf{f}(\mathbf{u})$ is sufficiently smooth to admit such a representation and without loss of generality that $\mathbf{f}(\mathbf{0}) = \mathbf{0}$. In most of the examples which we shall consider $\mathbf{f}(\mathbf{u})$ is a polynomial or a rational function without poles, so that such a representation is valid. Then the following theorem may be proved:

THEOREM 2.11 (Linearised Stability): The linearised system for (2.10) is

$$\dot{\mathbf{u}} = A\mathbf{u} \tag{2.12}$$

If the matrix A has (a) all eigenvalues with negative real part, or (b) at least one eigenvalue with positive real part, then the trivial critical point has the same character in the nonlinear system (2.10) as in the linearised system (2.12) in a sufficiently small neighbourhood of the origin. However, if the linearised system has one or more eigenvalues with zero real part and no eigenvalues with positive real part, then the nonlinear system may be stable, unstable or asymptotically stable depending on the nonlinear terms.

For a proof of this theorem we refer the reader to one of the works cited at the beginning of this chapter. In the cases where the nonlinear terms are important in determining stability, there are few general theorems stating which possibility occurs, and it is usually necessary to consider each new system separately. One method for proving (asymptotic) stability of a critical point is due to Lyapunov, and is known as Lyapunov's second, or direct, method. We require the following definitions:

DEFINITIONS 2.13: A function $V: \mathbb{R}^m \to \mathbb{R}$ is *positive definite* if (a) $V(\mathbf{0}) = 0$ and (b) $V > 0$ everywhere else in some open region $\Omega \subset \mathbb{R}^m$ containing the origin. For any solution $\mathbf{u} = \mathbf{u}(t)$ of $\dot{\mathbf{u}} = \mathbf{f}(\mathbf{u})$, the function $V(\mathbf{u}) = V(\mathbf{u}(t))$ is a function of t and its derivative is defined by

$$\frac{dV}{dt} = \dot{V} = \sum_{i=1}^{m} \frac{\partial V}{\partial u_i}\dot{u}_i = \sum_{i=1}^{m} \frac{\partial V}{\partial u_i} f_i(\mathbf{u})$$

if V is sufficiently smooth. A *Lyapunov function* $V: \mathbb{R}^m \to \mathbb{R}$ for the system $\dot{\mathbf{u}} = \mathbf{f}(\mathbf{u})$ is a function with continuous derivatives which is positive definite and such that $dV/dt \leq 0$ on Ω for any solution $\mathbf{u}$ of $\dot{\mathbf{u}} = \mathbf{f}(\mathbf{u})$.

THEOREM 2.14: If a Lyapunov function exists for the system (2.1) with $\mathbf{f}(\mathbf{0})=\mathbf{0}$, then the critical point at zero is stable.

THEOREM 2.15: If such a function exists and $-dV/dt$ is positive definite, then zero is asymptotically stable.

The proofs of these theorems will be omitted, but may be found in the references cited previously, or in LaSalle (1968).

For example, consider the system

$$\dot{u}=v+uh(u, v), \qquad \dot{v}=-u+vh(u, v)$$

where h is continuous near the origin and $h(0, 0)=0$. The linearised system has a centre at the origin. Consider the function $V(u, v)=u^2+v^2$. This is positive definite, and

$$\dot{V}=2u(v+uh(u, v))+2v(-u+vh(u, v))=2(u^2+v^2)h(u, v)$$

Hence the origin is asymptotically stable if h is negative definite in its neighbourhood and unstable if h is positive definite. Note that this is a non-local result, and if h is negative definite everywhere in the (u, v)-plane, then the origin is globally asymptotically stable, i.e., $(u, v)\to(0, 0)$ as $t\to\infty$ whatever the initial conditions. Lyapunov functions are therefore useful in giving a global picture of the solutions. In general, however, it is difficult to find a Lyapunov function for a specific system, and other techniques must be used to obtain global information.

2.4 Asymptotic Behaviour of Second-order Systems

Two-component autonomous systems

$$\dot{u}=f(u, v), \qquad \dot{v}=g(u, v) \tag{2.16}$$

are limited in their asymptotic behaviour as $t\to\infty$, as we see from the following theorem:

THEOREM 2.17 (Poincaré–Bendixson): If (u, v) remains bounded as $t\to\infty$ and neither is nor tends to any singular point, then the trajectory either is or tends to a periodic solution.

This is an example of a fixed point theorem, which we shall discuss in more detail later. For the present we shall consider periodic solutions. Systems with sufficiently smooth functions f and g and at least one periodic solution either have (a) a complete family of periodic solutions which follow the closed curves $G(u, v)=\text{const}$, for some function G, in the phase plane, or

(b) a number of isolated periodic solutions, or *limit cycles*. The former are known as conservative systems, and the function $G(u, v)$ is a first integral of the system. They are unsatisfactory as models for oscillatory phenomena in biology, because of their inherent structural instability. This will be discussed in Chapter 3. In fact, observable biological oscillations must be stable limit cycles. It is therefore important to be able to prove the existence or non-existence of a limit cycle solution of a given system. Before stating a theorem which is often useful in this respect, we need the following definition and lemma:

DEFINITION 2.18: Let Σ be a domain enclosed by a simple curve $\partial\Sigma$ in the phase plane. Then Σ is an *invariant set* for the system (2.16) if any solution of the system with initial conditions in Σ remains in Σ for all $t>0$. (This definition generalises to higher dimensional phase spaces in the obvious way.)

LEMMA 2.19: If

$$\mathbf{f}(\mathbf{u}) \cdot \mathbf{n}(\mathbf{u}) < 0$$

for all $\mathbf{n}(\mathbf{u}) \in \partial\Sigma$, where $\mathbf{n}(\mathbf{u})$ is the unit outward normal at $\mathbf{u} \in \partial\Sigma$, then Σ is an invariant set.

Proof: $\mathbf{f} \cdot \mathbf{n} = \dot{\mathbf{u}} \cdot \mathbf{n} < 0$. But $\dot{\mathbf{u}}$ points along the solution trajectory of (2.16) and therefore this trajectory points to the interior of Σ. Thus, no trajectory may leave Σ, and Σ is invariant. (If $\mathbf{f}$ and Σ are sufficiently smooth, this may be extended to $\mathbf{f} \cdot \mathbf{n} \leq 0$).

THEOREM 2.20: Let the system (2.16) have an invariant set Σ including a unique unstable critical point (which must be a focus). Then Σ contains a limit cycle.

Proof: Since any trajectory with initial conditions in Σ is bounded and does not tend to a critical point, it must either be or tend to a periodic solution by the Poincaré–Bendixson theorem (2.17). Since a system with an unstable critical point cannot be conservative, this periodic solution must be a limit cycle.

Two non-existence theorems follow:

THEOREM 2.21 (Negative Criterion of Bendixson): If the quantity $\nabla \cdot \mathbf{f} = f_u + g_v$ is of constant sign in an appropriate domain Ω (for example, in the positive quadrant if u and v are concentrations), then no limit cycle exists in Ω. Conversely, any limit cycle solution must cross the curve $\nabla \cdot \mathbf{f} = 0$.

Proof: Suppose for contradiction that there is a limit cycle C, where C has interior S, and $S \subset \Omega$. Then

$$\int_S \nabla \cdot \mathbf{f}\, dS = \int_S (f_u + g_v)\, dS = \oint_C (-g\, du + f\, dv)$$

$$= \oint_C (-\dot{v}\, du + \dot{u}\, dv) = 0$$

using Stokes' theorem. But this contradicts the assumption that $\nabla \cdot \mathbf{f}$ is of constant sign.

THEOREM 2.22 (Hanusse, 1972): No two-component reaction system containing only bi-molecular reactions has a limit cycle solution.

The proof is by classification of all such systems. Since there are very few tri-molecular reactions in biology this is a serious restriction on the use of such systems in modelling oscillations in biology. However, many systems with three or more components may be extremely closely approximated by two-component systems to which Hanusse's result does not apply, using methods such as those to be introduced in Chapter 8.

Finally, we state a positive result on the existence of limit cycles.

THEOREM 2.23 (Liénard Systems): A Liénard system has a unique stable limit cycle solution.

DEFINITION 2.24: A Liénard system is a system which can be written in the form

$$\ddot{x} + \phi(x)\dot{x} + \psi(x) = 0$$

where $\phi(x)$ and $\psi(x)$ satisfy the following conditions:

(i) $\phi(x)$ is even, $\psi(x)$ is odd, $x\psi(x) > 0$ for all $x \neq 0$ and $\phi(0) < 0$;

(ii) $\phi(x)$ and $\psi(x)$ are continuous, and $\psi(x)$ satisfies a Lipschitz condition;

(iii) $\Phi(x) = \int_0^x \phi(y)\, dy \to \pm\infty$ as $x \to \pm\infty$, $\Phi(x)$ has a single positive zero $x = a$ and for $x \geqslant a$ is a monotonic increasing function of x.

The proof and some extensions of this result are given in the books by Minorsky (1974) and Cesari (1963). Further results on limit cycles in two-component systems are given by Sansone and Conti (1964) and Minorsky (1974), and summarised by Murray (1977).

2.5 Limit Cycles in Systems of Order Greater Than Two: The Oregonator

If we are concerned with systems with more than two components then the well-developed theory of phase plane analysis is no longer available to us. In particular the Poincaré–Bendixson theorem 2.17 no longer holds, and we cannot show the existence of a limit cycle by proving that trajectories are bounded and do not tend to a critical point. Lyapunov techniques still apply, but the problem of finding the Lyapunov function becomes even greater. Bifurcation techniques and asymptotic analysis may be used; these are discussed in Chapters 6, 7 and 8. We are sometimes able to show the existence of a limit cycle solution by the use of Brouwer's fixed point theorem, which follows:

THEOREM 2.25 (Brouwer's Fixed Point Theorem): Let X be a compact convex subset of $\mathbb{R}^n$, and let T be a mapping of X into itself. In other words, T is a continuous transformation defined and such that $T(\mathbf{x}) \in X$ for all $\mathbf{x} \in X$. Then T has a fixed point, i.e. there exists $\mathbf{y} \in X$ such that $T(\mathbf{y}) = \mathbf{y}$.

For a proof this theorem see Lefschetz (1957). As an example of its use let us consider the following theorem relating to a model of the Belousov–Zhabotinskii reaction, an oscillating chemical reaction involving bromate and bromide ions, cerium ions or some other catalyst and citric or malonic acid. This reaction was first reported by Belousov (1959) who noted oscillations in an experiment involving citric acid, bromate and metal ions. Further work was done by Zhabotinskii (1964) who replaced the citric acid by malonic acid. A detailed reaction mechanism was proposed by Field *et al.* (1972), and a simpler version by Field and Noyes (1974). This model has come to be known as the Oregonator, after the University in which it was developed. Extensive reviews of the reaction and this mechanism have been written by Tyson (1976, 1984). A concise exposition also appears in Murray (1977). We shall therefore content ourselves with a short derivation of the equations from the five-step model set up by Field and Noyes. The five most important reactions are given as

$$Br^- + BrO_3^- + 2H^+ \xrightarrow{K_1} HBrO_2 + HOBr$$

$$Br^- + HBrO_2 + H^+ \xrightarrow{K_2} 2HOBR$$

$$2Ce^{iii} + BrO_3^- + HBrO_2 + 3H^+ \xrightarrow{K_3} 2Ce^{iv} + 2HBrO_2 + H_2O$$

$$2HBrO_2 \xrightarrow{K_4} HOBr + BrO_3^- + H^+$$

$$4Ce^{iv} + BrCH(COOH)_2 + 2H_2O \xrightarrow{K_5} 4Ce^{iii} + HCOOH + 2CO_2 + 5H^+ + Br^-$$

The oscillations occur as Ce^{iii} is oxidised to Ce^{iv} in reaction 3, and then reduced back to Ce^{iii} in reaction 5. Defining

$$A = BrO_3^-, \quad P = HOBr, \quad X = HBrO_2, \quad Y = Br^-, \quad Z = Ce^{iv}$$

we obtain the model reactions

$$A + Y \xrightarrow{k_1} X + P$$

$$X + Y \xrightarrow{k_2} 2P$$

$$A + X \xrightarrow{k_3} 2X + 2Z$$

$$2X \xrightarrow{k_4} P + A$$

$$Z \xrightarrow{k_5} fY$$

where f is a stoichiometric parameter to be determined, and the reaction rates $k_1, \ldots, k_5$, are related to $K_1, \ldots, K_5$. Noyes and Jwo (1975) have shown that the appropriate value of f is 0.5. On writing down the corresponding differential equations for A, P, X, Y and Z it can be seen that the equation for P is uncoupled from the system. In the model reactions A is assumed to be in excess, so that it can be taken to be constant on the time scale of the experiment. There are therefore three equations, for X, Y and Z, which when non-dimensionalised become

$$\varepsilon\dot{x} = x + y - xy - qx^2 = f(x, y, z)$$

$$\dot{y} = 2fz - y - xy = g(x, y, z)$$

$$p\dot{z} = x - z = h(x, y, z)$$

where typical values of the parameters are $p = 3.1 \times 10^2$, $q = 8.4 \times 10^{-6}$ and $\varepsilon = 2 \times 10^{-4}$. This system is essentially third order, and phase plane techniques cannot be used. However, three-dimensional phase space techniques such as the Brouwer fixed point theorem are applicable, and can be used to prove the existence of a periodic solution of the equations when the parameters are such that the single non-trivial steady state is unstable, as shown by Hastings and Murray (1975).

THEOREM 2.26 (Hastings and Murray): The Oregonator has a periodic solution.

The proof consists of showing that any trajectory which once passes through a certain plane subset of phase space, which does not include the only critical point of the system, must necessarily pass through it again. This area of phase space is the set X of Brouwer's theorem. Then for any $\mathbf{x} \in X$ let us define the trajectory $\mathbf{u}_{\mathbf{x}}(t)$ to be the solution of the equations satisfying $\mathbf{u}_\mathbf{x}(0) = \mathbf{x}$, and define $\tau = \tau(\mathbf{x})$ to be the smallest positive value of t such that $\mathbf{u}_\mathbf{x}(\tau) \in X$. Then the transformation T defined by $T(\mathbf{x}) = T(\mathbf{u}_\mathbf{x}(0)) = \mathbf{u}_\mathbf{x}(\tau)$, which takes any point to the point on the trajectory where it next passes through X, is a continuous mapping of X into itself. By Brouwer's theorem it therefore has a fixed point, i.e., a point $\mathbf{y}$ such that $\mathbf{y} = T(\mathbf{y})$, or $\mathbf{u}_\mathbf{y}(0) = T(\mathbf{u}_\mathbf{y}(0)) = \mathbf{u}_\mathbf{y}(\tau)$. But the trajectory starting at $\mathbf{y}$ is then a periodic trajectory. It thus follows that this model of the Belousov–Zhabotinskii reaction has a periodic solution, as required. This method is based on the ideas of Poincaré and T is sometimes known as a Poincaré map.

3. Conservative Systems

3.1 Introduction

In this chapter we wish to move on to systems of reaction–diffusion equations, but to restrict ourselves to a particular class of systems, those whose kinetic equations (that is, the equations without diffusional effects) are conservative. The fundamental property of a conservative system is the existence of a function of the dependent variables which is a constant of the motion and plays the role of energy (see Minorsky (1974)). That is, a system $\dot{\mathbf{u}} = \mathbf{f}(\mathbf{u})$ is conservative if there exists a function $G(\mathbf{u})$, known as a first integral, or simply an integral, of the system, such that

$$\dot{G}(\mathbf{u}) = \sum_{i=1}^{m} \frac{\partial G}{\partial u_i}\dot{u}_i = \sum_{i=1}^{m} \frac{\partial G}{\partial u_i} f_i(\mathbf{u}) = 0$$

Conservative systems often have oscillatory solutions and have therefore been widely used to model phenomena such as oscillations in prey and predator populations. They tend to be amenable to analysis, but they have some major disadvantages as models for real systems related to their behaviour as $t \to \infty$, as we shall see. The most widely studied oscillatory conservative system is that of Lotka and Volterra, and we shall consider this system with and without diffusion effects. Other conservative systems are often used as models in epidemiology, and since they are only considered over a finite length of time, they are not subject to the same disadvantages as the Lotka–Volterra system. We shall consider some examples at the end of this chapter.

3.2 The Lotka–Volterra System

In 1920 Lotka proposed a theoretical chemical reaction scheme which exhibited undamped oscillations. In 1926 Volterra obtained the same mathematical equations by modelling a prey–predator system which was observed to oscillate. We shall derive the so-called Lotka–Volterra system from the latter point of view.

The populations considered by Volterra were phytoplankton with biomass P, the prey, and a second type of plankton, herbivorous copepods with biomass H, the predators. The model was intended to suggest an explanation for the oscillations in the numbers of these plankton which had been observed to occur in the Adriatic Sea. The populations were assumed to evolve according to the following laws:

$$\begin{array}{rcccccc} dP/dt = & \alpha_1 P & - & \beta_1 P & - & \gamma PH \\ & \uparrow & & \uparrow & & \uparrow \\ & \text{rate of birth} & & \text{rate of natural mortality} & & \text{rate of mortality due to predation} \end{array}$$

$$\begin{array}{rcccccc} dH/dt = & e\gamma PH & + & \alpha_2 H & - & \beta_2 H \\ & \uparrow & & \uparrow & & \uparrow \\ & \text{rate of increase due to predation} & & \text{rate of birth} & & \text{rate of mortality} \end{array}$$

We thus assume birth and mortality rates to be proportional to the biomass of plankton present. α, P is in fact the rate of increase of P in the absence of predation and natural mortality and is due to photosynthesis. $\alpha_2 H$ is the corresponding term for H and is included for generality; normally we take $\alpha_2 = 0$. We assume that $\alpha_1 > \beta_1$ and $\alpha_2 < \beta_2$ so that dP/dt and dH/dt are not perpetually negative and positive, respectively. We assume that the biomass P is lost through predation at a rate which is proportional to both P and H, and that this is converted to biomass of H with an efficiency e.

The steady states of the system are given by $P = H = 0$ and $P = P_s = (\beta_2 - \alpha_2)/e\gamma$, $H = H_s = (\alpha_1 - \beta_1)/\gamma$. It is convenient to non-dimensionalise the system by defining $u = P/P_s$, $v = H/H_s$ and $\tau = (\alpha_1 - \beta_1)t$. We obtain

$$\dot{u} = u(1 - v), \qquad \dot{v} = av(u - 1) \tag{3.1}$$

where $a = (\beta_2 - \alpha_2)/(\alpha_1 - \beta_1)$ is a non-dimensional parameter and the dot denotes differentiation with respect to τ. The singular points are at $(0, 0)$ and $(1, 1)$. If the system is linearised about $(0, 0)$, it has a saddle point there, and hence by Theorem 2.11 the full nonlinear system has the same character in the region of the origin. If it is linearised about $(1, 1)$, it has a centre there (two purely imaginary eigenvalues), and we cannot deduce the character of the singular point at $(1, 1)$ in the full nonlinear system. However, this system is conservative and is therefore amenable to analysis in the following way.

It is easily seen from (3.1) that

$$av(u - 1)\, du = u(1 - v)\, dv$$

This is a differential equation with separable variables, and can be integrated to obtain

$$G(u, v) = a(u - \log u) + v - \log v = A \tag{3.2}$$

where A is a constant. Hence the quantity $G(u, v)$ is conserved and can be thought of as the energy of the system. The Lotka–Volterra system is therefore conservative.

Since $G(u, v)$ does not change as we move along a trajectory or solution curve of the equations, these trajectories are defined by the curves $G(u, v) = A$ for different values of the constant A. It follows from this that the singular point $(1, 1)$ cannot be a stable focus. For if it were, then all curves in a neighbourhood of it would tend to it, and hence would have $G(u, v) = G(1, 1)$, since G is a continuous function. But this implies that G is constant in a neighbourhood of $(1, 1)$, which contradicts its definition. Similarly, the singularity cannot be an unstable focus (just take the limit as $t \to -\infty$). It also follows by similar arguments that there are no stable or unstable limit cycles surrounding the point $(1, 1)$. It is easily seen that all trajectories starting in the positive quadrant are bounded, so the only possiblity is that the phase plane consists of closed trajectories around the singular point $(1, 1)$, each with a different value of the energy $G(u, v)$, as in Fig. 3.3(a). For a proof of this using a Poincaré map see Robinson (1985). The corresponding time dependence of the populations is shown in Fig. 3.3(b).

Hence Volterra's model for oscillating plankton populations in the Adriatic certainly has oscillatory solutions. However, there are other questions which must be asked before the model can be accepted as realistic. It must be ascertained whether the solutions are stable. If they are unstable, they will not be observed in practice, because any small disturbance from the solution, such as may be brought about by external means or because of the inherent stochastic nature of a prey–predator system, will be magnified, and the observed trajectory will diverge from the unstable solution. A stable solution on the other hand will be observed, assuming that the model accurately reflects the true situation. In our case the solutions are orbitally stable but not stable, in the sense of Definitions 2.6 and 2.3. This means that two solutions which start close together produce trajectories which remain close together for all time. However, a solution on one trajectory will complete a revolution more quickly or slowly than one on an adjacent trajectory and will gain or lag more and more. Assuming that the model accurately reflects reality, we must ask whether oscillatory behaviour of the populations will actually be observed.

We are thus concerned with the cumulative effect of small random perturbations on such a system. It can be seen that such perturbations will cause the solution to wander about between trajectories until it finally meets

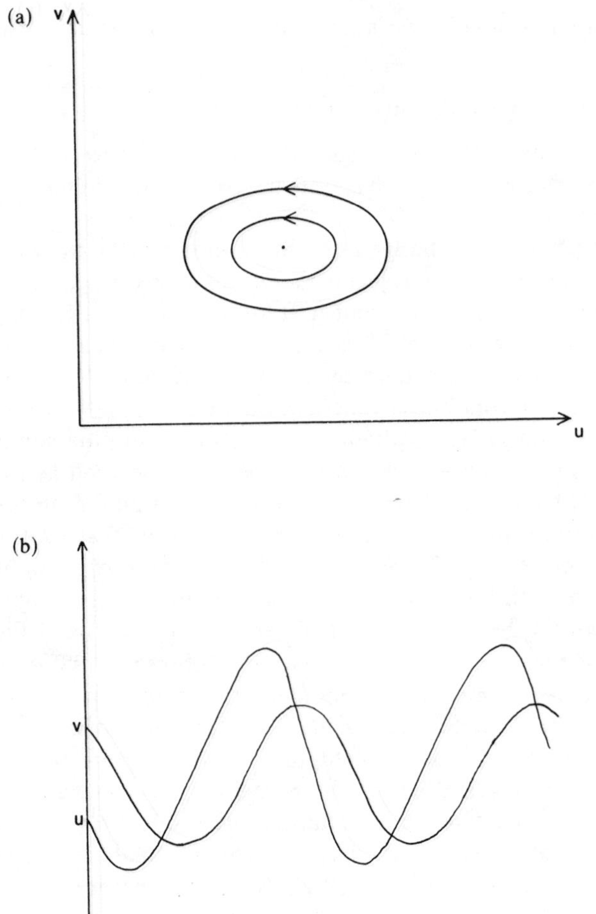

Fig. 3.3. (a) The phase plane for the system (3.1). (b) Time dependence of a solution of (3.1).

one of the axes $u=0$ or $v=0$. Physically, this corresponds to extinction of prey or predator, so that the model cannot be considered adequate in describing long-term oscillatory behaviour in prey–predator systems.

There is another way in which small perturbations in the system may arise. In any attempt at modelling, certain effects are neglected, and for the model to be a good one it is necessary that such neglect does not affect the final solution markedly. In Volterra's model, for example, it can be seen that if there were no predator, the prey population would increase exponentially. This is obviously unrealistic; the population must be self-limiting. A simple way to achieve that is to add a term $-\varepsilon u^2$ to the first of Eqs. (3.1) to obtain

$$\dot{u}=u(1-v)-\varepsilon u^2, \qquad \dot{v}=av(u-1) \tag{3.4}$$

It may appear that if ε is extremely small, the new term will not have a significant effect on the solutions of the system. However, an eigenvalue analysis shows that the critical point which is now at $(1, 1-\varepsilon)$ is a stable focus in the linearised system and is therefore a stable focus in the nonlinear system, however small ε may be. Alternatively, $V(u, v)=a(u-\log u)+v-(1-\varepsilon)\log v$ is a global (in the positive quadrant) Lyapunov function for the system. The solutions now spiral in to the critical point, and the system can no longer be put forward as a model for the sustained oscillations observed by Volterra. It is essential that models of real biological systems with periodic solutions do not possess this kind of behaviour, which is known as *structural instability*. In mathematical terms, a system $\dot{\mathbf{u}}=\mathbf{f}(\mathbf{u})$ is *structurally stable* if there exists a homeomorphism from the orbits of $\dot{\mathbf{u}}=\mathbf{f}(\mathbf{u})$ to the orbits of $\dot{\mathbf{u}}=\mathbf{f}(\mathbf{u})+\mathbf{p}(\mathbf{u})$ for sufficiently small perturbations $\mathbf{p}(\mathbf{u})$. All conservative systems are structurally unstable, and hence should be used with great care.

Conservative systems are, however, useful in suggesting avenues for further investigation. The Lotka–Volterra system may not be satisfactory as it stands, but it does suggest that oscillations may be produced by a simple interaction between one species of predator and one species of prey. It is possible that the addition of further essential, but small, effects into the equations will produce a limit cycle solution, i.e. a stable periodic solution, which may be the basis of observed oscillations. This has been discussed from the mathematical, rather than the ecological, point of view by Freedman and Waltman (1975a, b).

3.3 Conservative Reaction–Diffusion Systems

By conservative reaction–diffusion systems we mean systems whose kinetic equations (that is, the equations without diffusion) are conservative. In this case the existence of a first integral for the diffusionless case may lead to a method of analysis for the full reaction–diffusion system. As an example, we may prove the following theorem:

THEOREM 3.5 (Murray, 1975): The diffusional Lotka–Volterra system with equal diffusion coefficients and zero-flux boundary conditions tends to a spatially uniform state as $t\to\infty$.

Proof: The equations may be written as

$$u_t=u(1-v)+D\,\nabla^2 u, \qquad v_t=av(u-1)+D\,\nabla^2 v \tag{3.6}$$

in $\Omega \times (0, T]$ with boundary conditions

$$\frac{\partial u}{\partial n}(\mathbf{x}, t) = \frac{\partial v}{\partial n}(\mathbf{x}, t) = 0 \tag{3.7}$$

for $\mathbf{x} \in \partial\Omega$, where n is the outward normal, and initial conditions

$$u(\mathbf{x}, 0) = u_0(\mathbf{x}), \qquad v(\mathbf{x}, 0) = v_0(\mathbf{x}) \tag{3.8}$$

for $\mathbf{x} \in \Omega$; Ω is a domain in n- (usually two- or three-) dimensional space. Define $s(\mathbf{x}, t)$ by

$$s = a(u - \log u) + v - \log v \tag{3.9}$$

so that s is the energy of the diffusionless system, and $s_t = 0$ if $D = 0$. We wish to obtain a corresponding differential equation for s in the diffusional case. Differentiating (3.9), we have

$$s_t = a(u_t - u_t/u) + v_t - v_t/v$$

$$\nabla^2 s = a(\nabla^2 u - \nabla^2 u/u + |\nabla u|^2/u^2) + \nabla^2 v - \nabla^2 v/v + |\nabla v|^2/v^2$$

so that

$$\begin{aligned} s_t - D\,\nabla^2 s &= a(u_t - D\,\nabla^2 u)(u-1)/u + (v_t - D\,\nabla^2 v)(v-1)/v \\ &\quad - aD|\nabla u|^2/u^2 - D|\nabla v|^2/v^2 \\ &= -aD|\nabla u|^2/u^2 - D|\nabla v|^2/v^2 \leqslant 0 \end{aligned} \tag{3.10}$$

using equations (3.6). Physically, the energy of the system is dissipated by the diffusional terms. The boundary and initial conditions for s are

$$\frac{\partial s}{\partial n}(\mathbf{x}, t) = 0 \tag{3.11}$$

for $\mathbf{x} \in \partial\Omega$, and

$$s(\mathbf{x}, 0) = a(u_0 - \log u_0) + v_0 - \log v_0 = s_0(\mathbf{x}) \tag{3.12}$$

say, for $\mathbf{x} \in \Omega$. Let us now define the total amount of energy in the system at time t by

$$S(t) = \int_\Omega s(\mathbf{x}, t)\, d\mathbf{x}$$

where $d\mathbf{x}$ denotes the element of integration in Ω. Then, integrating (3.10) and using the boundary condition (3.11), we obtain

$$\dot{S}(t) = \int_\Omega s_t(\mathbf{x}, t)\, d\mathbf{x} = -D \int_\Omega \left(\frac{a|\nabla u|^2}{u^2} + \frac{|\nabla v|^2}{v^2} \right) d\mathbf{x}$$

Thus S is monotone non-increasing and either tends to a finite limit or to $-\infty$ as $t\to\infty$. But the definition of $s(\mathbf{x}, t)$ implies $s(\mathbf{x}, t)\geq a+1$ and so

$$S(t)=\int_\Omega s(\mathbf{x}, t)\, d\mathbf{x}\geq (a+1)|\Omega|$$

where

$$|\Omega|=\int_\Omega d\mathbf{x}$$

Hence S tends to a finite limit, so that the system

$$\dot{S}=-D\int_\Omega\left(\frac{a|\nabla u|^2}{u^2}+\frac{|\nabla|^2}{v^2}\right) d\mathbf{x}\to 0.$$

It follows that ∇u and $\nabla v\to 0$ and tends to a spatially uniform state as $t\to\infty$.

This result is easily extended to the case of non-equal diffusion coefficients and implies that the Lotka–Volterra system cannot be used as a model for ecological patchiness, the observed spatial distribution of species in an environment. The method is based on the idea of Lyapunov, since $V=S-(a+1)|\Omega|$ satisfies

(i) $V\geq 0$, $V=0$ if and only if $u\equiv 1$, $v\equiv 1$ in Ω,
(ii) $\dot{V}\leq 0$,

and is therefore essentially a Lyapunov functional (cf. Definition 2.13). The method may also be extended to more general systems of Lotka–Volterra type; see section 5.7, de Mottoni and Rothe (1979), Mimura (1979), Nalleswamy (1983), Redheffer and Walter (1983), Shukḷa and Das (1982) and Voorhees (1982). For a review of pattern formation in such systems see Conway (1984).

3.4 Conservative Systems in Epidemiology

The defects of conservative systems as models of real systems are all concerned with the behaviour as $t\to\infty$. If we are concerned only with finite time intervals, then continuity of the solutions of systems of ordinary differential equations with respect to the initial conditions and the form of the equations themselves (Theorem 2.2) ensures that conservative systems are not structurally unstable over such intervals and may be acceptable as mathematical models. An important class of systems of this sort arises in epidemiology. We consider here the classic epidemic model of Kermack and McKendrick (1927); see also Bailey (1975) and Hethcote (1976). Let a population of N individuals contain S who are susceptible to some infection, I who are infective and R who are either immune or have been removed from the community through isolation or death. A susceptible is assumed to become infective through contact with another infective, and then leaves the class of susceptibles. Thus,

$$dS/dt = -\beta IS$$

where β is a constant parameter. Infectives are assumed to leave the infective class at a rate γ and to enter the removed class. Thus,

$$dI/dt = \beta IS - \gamma I, \qquad dR/dt = \gamma I$$

The R equation is uncoupled from the system, that is, R may be found by quadrature, or straightforward integration, if S and I are known. We thus have the two-component system

$$\dot{S} = -\beta IS, \qquad \dot{I} = \beta IS - \gamma I \tag{3.13}$$

so that

$$\dot{S} + \dot{I} = -\gamma I = (\gamma/\beta)\dot{S}/S = (1/\sigma)\dot{S}/S$$

where the parameter $\sigma = \beta/\gamma$ is known as the infectious contact number and describes the average number of infectious contacts per susceptible which might be made by an infective. Integrating, we obtain

$$G(S, I) = \sigma S - \log S + \sigma I = A \tag{3.14}$$

where A is a constant. The phase plane of the system may be drawn from this relationship and is shown in Fig. 3.15. The energy G of the system is constant on any trajectory, so that the Kermack–McKendrick system is conservative.

Note that $(S, 0)$ is an equilibrium point for all S, and that the introduction of a small number of infectives into a population of susceptibles greater than $1/\sigma$ leads to a rise and subsequent fall to zero in the number of infectives as the number of susceptible falls to some value below $1/\sigma$. The model therefore seems to reflect the behaviour of certain infectious immunity-conferring or lethal diseases. It is, of course, structurally unstable over an infinite time interval and should therefore be used only to study epidemics of finite duration.

It is of interest to study the spatial behaviour of such epidemics, which often travel at steady rates across continents. Examples are the black death of the Middle Ages and rabies at the present time. For general information

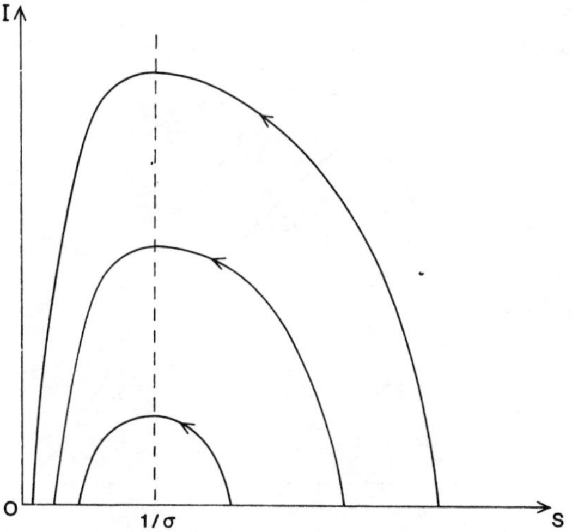

Fig. 3.15. The phase plane for the Kermack–McKendrick system (3.13).

on rabies see Kaplan (1977) and MacDonald and Bacon (1982). The rabies virus is moving across Western Europe at a rate of about 40 km per year and is approaching the English Channel. The main vectors of this virus are foxes, which in the healthy state tend to stay in their own territory, but when rabid will travel large distances and attack other foxes. For this reason Källén *et al.* (1985) have proposed a rabies model which includes diffusive effects for infective but not for susceptible foxes. It may be written as

$$S_t = -\beta IS, \qquad I_t = \beta IS - \gamma I + D\nabla^2 I \tag{3.16}$$

where D is the diffusion coefficient of the infectious foxes, and where S and I should now be thought of as the population *densities* of susceptible and infectious foxes, respectively. This model has been analysed by Källén *et al.* (1985) and Källén (1984). For other work on rabies see Mollison (1983) and Mollison and Kuulasmaa (1984). An epidemic is a wave of infectives propagating into a population of susceptibles. Let us take the initial density of susceptibles, before the passage of the wave, to be S_0. The initial and final densities of infectives, before and after the passage of the epidemic, will be zero. The final density of susceptibles is unknown. A diagrammatic representation of the epidemic is shown in Fig. 3.17. Non-dimensionalising equations (3.16) by

$$u = S/S_0, \qquad v = I/S_0,$$

and re-scaling the space and time variables appropriately, we obtain

$$u_t = -uv, \qquad v_t = uv - rv + \nabla^2 v \tag{3.18}$$

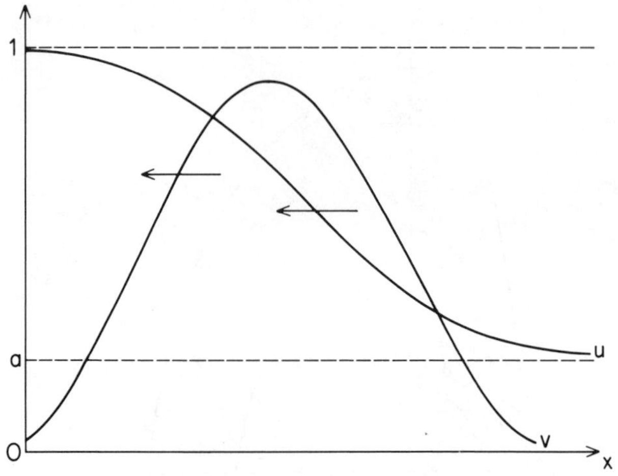

Fig. 3.17. A travelling wave solution of the rabies model (3.16).

where $r = \gamma/\beta S_0$; $1/r$ is the non-dimensional version of the infectious contact number σ. Seeking travelling wave solutions of the form $u = u(x+ct)$ and $v = v(x+ct)$, where c is some constant wave speed which we take to be positive without loss of generality, we obtain

$$cu' = -uv, \qquad cv' = uv - rv + v''$$

where primes denote differentiation with respect to $\xi = x + ct$. Defining $v' = w$, this becomes

$$u' = -uv/c, \qquad v' = w, \qquad w' = -uv + rv + cw \tag{3.19}$$

The initial and final conditions become

$$u(-\infty) = 1, \qquad v(\pm\infty) = 0, \qquad w(\pm\infty) = 0 \tag{3.20}$$

the last of these since $w = v'$ and therefore tends to zero as v tends to zero at $\pm\infty$. We denote the unknown final density of susceptibles by $u(\infty) = a$.

This system is third-order and is therefore difficult to handle using phase space analysis. However, just as in the Lotka–Volterra system, the existence of an energy function $G(S, I)$, given by (3.14), of the spatially uniform system gives a clue for the analysis of the spatially non-uniform system. The non-dimensional form of the energy is given by $g(u, v) = u - r \log u + v$. This is conserved in the diffusionless system. In this case the diffusional system still admits an invariant quantity

$$h(u, v, w) = u - r \log u + v - w/c$$

It is easy to check that $h' = 0$. The system may therefore be written in the form

$$h' = 0, \qquad u' = -uv/c, \qquad v' = -c(h - u + r \log u - v)$$

Since h is constant this is essentially a second-order system for u and v. The value of h may be obtained by applying the conditions (3.20) at $-\infty$; $h = h(-\infty) = 1$ and the system is

$$u' = -uv/c, \qquad v' = -c(1 - u + r \log u - v) \tag{3.21}$$

We also obtain an equation for $u(\infty) = a$ by using the facts that $h = 1$ and $v = w = 0$ at ∞, in the form

$$a - r \log a = 1 \tag{3.22}$$

This has a solution a satisfying $0 < a < 1$ if and only if $r < 1$, and therefore a travelling wave of the form considered cannot exist if $r \geq 1$. We shall assume henceforth that $0 < r < 1$; we wish to show that a travelling wave front exists in this case. The system (3.21) is two-dimensional and amenable to phase plane analysis. It has two critical points at $(a, 0)$ and $(1, 0)$; and a travelling wave front is a trajectory in the phase plane connecting these. From physical considerations (this also follows from the equations), the trajectory must remain in the positive quadrant. Linearising about the critical point at $(b, 0)$ (where b is either a or 1), we obtain eigenvalues and eigenvectors

$$\lambda_{\pm} = c/2 \pm \sqrt{c^2/4 - (b - r)}, \qquad \phi_{\pm} = (1, -b/c\lambda_{\pm})^T.$$

Hence if $c^2 < 4(1 - r)$, the eigenvalues at $(1, 0)$ are complex and we have oscillatory solutions near $(1, 0)$, which contradicts the requirement that the trajectory remain in the positive quadrant. We therefore take $c \geq 2\sqrt{1 - r}$. Then $(1, 0)$ is an unstable node and $(a, 0)$ is a saddle point, and the phase plane is shown in Fig. 3.23. A travelling wave front from $(1, 0)$ to $(a, 0)$ exists if the trajectory T, which enters $(a, 0)$ as $\xi = x + ct \to \infty$, tends to $(1, 0)$ as $\xi \to -\infty$. Consider the region D defined by

$$D = \{(u, v) \,|\, a < u < 1, 0 < v < m(r \log u + 1 - u)\}$$

It is left as an exercise to show that if $c \geq 2\sqrt{1 - r}$ then m may be chosen such that a trajectory at any non-stationary point of the boundary of D immediately leaves D. It follows that T could not have entered D at any finite value of ξ and must therefore be inside D for all $\xi \in (-\infty, \infty)$, and tend to $(1, 0)$ as $\xi \to -\infty$. This proves the existence of a travelling wave front from $(1, 0)$ to $(a, 0)$ for any $c \geq 2\sqrt{1 - r}$. In fact Källén (1984) has shown that the stable wave speed is $c_0 = 2\sqrt{1 - r}$.

The fact that a travelling wave front cannot exist if $r \geq 1$ has possible implications for the prevention of the spread of rabies; if a barrier to the

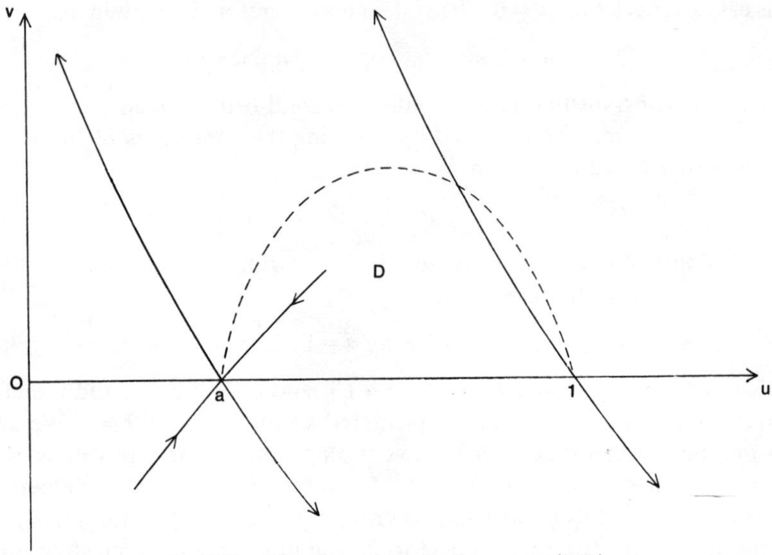

Fig. 3.23. The phase plane for the system (3.21).

epidemic is set up where $r = \gamma/\beta S_0$ is increased beyond unity by reducing S_0 sufficiently (that is, by reducing the numbers of susceptible foxes), then the wave cannot travel further. The required width of the barrier may be estimated by further analysis of the system and is found to be around 14 km (Källén *et al.* (1985)). Such a barrier has been set up to prevent rabies from spreading into Denmark, and seems to be effective.

4. The Scalar Reaction–Diffusion Equation

4.1 Introduction

So far we have mainly considered cases where the reaction–diffusion system reduces to a system of ordinary differential equations. The simplest form of reaction–diffusion system which cannot be so reduced is the special case of (1.1) where the vector $\mathbf{u}$ has only one component; the system then reduces to the single equation

$$u_t = f(u) + D\,\nabla^2 u \tag{4.1}$$

where $D > 0$ is a constant diffusion coefficient. This kind of equation occurs frequently, either as an approximation to a more complicated system or in cases in which there is only one species. A simple example is Fisher's (1937) equation, where $f(u) = su(1-u)$. This equation was also considered by Kolmogorov *et al.* (1937) and has been much studied since then. Some general references are Fife (1979a), Henry (1976), Smoller (1983), Diekman and Temme (1976) and Fitzgibbon and Walker (1977). Mollison (1977) gave a comprehensive account of the relationship between deterministic and stochastic models of this type, and this aspect has also been discussed by Fife (1979a).

In this chapter we wish to summarise some of the more important techniques for analysing the single reaction–diffusion equation (4.1) in one or more spatial dimensions. The theorems we shall state also apply to more general nonlinear parabolic equations. The function f may depend on $\mathbf{x}$, t and ∇u, as well as u, as long as it is sufficiently smooth, and operators other than $D\,\nabla^2$ may be considered; for details see Friedman (1964). Examples in which f depends on values of u at previous times may also be considered; see for example Cohen *et al.* (1979a) and Cohen *et al.* (1979b). In the exposition we shall restrict ourselves to the simplest case, so that the equation may be written, taking $D = 1$ without loss of generality, as

$$Nu \equiv Lu - f(u) \equiv u_t - \nabla^2 u - f(u) = 0 \tag{4.2}$$

N is thus a nonlinear operator and L the linear operator $\partial_t - \nabla^2$.

The various mathematical problems consist of the differential equation (4.2) in a given spatio-temporal domain together with conditions on the boundaries of this domain. Problems are usually classified as follows:

DEFINITION 4.3: (a) Let $Q_T = \mathbb{R}^n \times (0, T)$, $S_0 = \mathbb{R}^n \times \{0\}$ and $S_T = \mathbb{R}^n \times \{T\}$. Then the *Cauchy*, or *initial value, problem* consists of the differential equation $Nu = 0$ in $Q_T \cup S_T$ with initial conditions $u(\mathbf{x}, 0) = u_0(\mathbf{x})$ in S_0, and possibly some boundedness conditions on u as $|\mathbf{x}| \to \infty$.

(b) Let $Q_T = \Omega \times (0, T)$, where Ω is a given bounded time-independent n-dimensional spatial domain, and define the boundaries of Q_T by $S_0 = \Omega \times \{0\}$, $S_T = \Omega \times \{T\}$ and $\Gamma = \partial\Omega \times (0, T)$. (Problems where the spatial domain depends on t may also arise, but we do not consider these here.) Then the *initial-boundary-value problems* consist of the differential equation $Nu = 0$ in $Q_T \cup S_T$ with initial conditions $u(\mathbf{x}, 0) = u_0(\mathbf{x})$ specified on S_0 and boundary conditions of one of the following types on Γ:

(i) $u(\mathbf{x}, t) = b(\mathbf{x}, t)$ (Dirichlet conditions);
(ii) $\partial u/\partial \nu(\mathbf{x}, t) = b(\mathbf{x}, t)$ (Neumann conditions); and
(iii) $Bu \equiv c(\mathbf{x}, t)u(\mathbf{x}, t) + d(\mathbf{x}, t)\, \partial u/\partial \nu(\mathbf{x}, t) = b(\mathbf{x}, t)$, where $c \geq 0$, $d \geq 0$ and $c^2 + d^2 > 0$ (Robin conditions).

Here $\partial/\partial \nu$ denotes any outward derivative. Note that (i) and (ii) are, in fact, special cases of (iii), and we may take the general boundary condition to be $Bu = b$.

(c) Let Ω be as in (b), $Q = \Omega \times (-\infty, \infty)$, and $\Gamma = \partial\Omega \times (-\infty, \infty)$. Then the *boundary-value problems* consist of the differential equation $Nu = 0$ in Q with the given boundary conditions $Bu = b$ on Γ.

(d) *Stationary* (*or steady*) *problems* are as for (c), with the additional requirement that u be time-independent, when B and b are time-independent, or with $\Omega = \mathbb{R}^n$, and no boundary conditions.

(e) *Periodic problems* are as for (c), with additional requirement that u is T-periodic, when B and b are T-periodic, or with $\Omega = \mathbb{R}^n$, and no boundary conditions.

Having set up a mathematical problem, we wish to know whether it has a solution, and if so whether this solution is unique.

DEFINITION 4.4: We define a *solution of the initial-boundary-value problem* 4.3(b) to be a function u continuous in $\bar{Q}_T$, with u_t and the derivatives occurring in $\nabla^2 u$ continuous in $Q_T \cup S_T$, which satisfies $Nu = 0$ in $Q_T \cup S_T$, $u(\mathbf{x}, 0) = u_0(\mathbf{x})$ in S_0 and $Bu = b$ on Γ. Solutions of the other problems are defined analogously.

REMARK 4.5: The definitions may be extended to cope with the case of discontinuous initial or boundary data (Friedman, 1964), but we shall not consider this increased generality.

In proving the existence of solutions of parabolic equations such as (4.2) it is often necessary to find *a priori* bounds on the set of all solutions u and their derivatives. An important tool here is the maximum principle for parabolic equations, and related comparison theorems, which will be considered in the next section. Maximum principles are discussed in general in Gidas *et al.* (1979), Protter and Weinberger (1967) and Sperb (1981), and for parabolic equations in Friedman (1964). For scalar reaction–diffusion equations in particular see Aronson (1976), Diekmann and Temme (1976) and Fife (1979a).

We shall also be concerned with the stability of the solutions of the various mathematical problems. Comparison theorems are useful tools here. We state some definitions, which mirror those of Section 2.1.

DEFINITION 4.6: Consider the solution $u=\phi(\mathbf{x}, t)$ of $Nu=0$ in $\Omega\times(0,\infty)$ satisfying $\phi(\mathbf{x}, 0)=\phi_0(\mathbf{x})$ for $\mathbf{x}\in\Omega$, and $B\phi=b$ on $\partial\Omega\times(0,\infty)$, and let $u=\psi(\mathbf{x}, t)$ satisfy $N\psi=0$ in $\Omega\times(0, T)$ for some $T>0$ and $B\psi=b$ on $\partial\Omega\times(0, T)$. Then ϕ is a *stable solution of the initial-boundary-value problem* 4.3(b) if, given any $\varepsilon>0$, there exists δ such that whenever $\psi(\mathbf{x}, 0)=\psi_0(\mathbf{x})$ satisfies

$$\|\phi_0(\cdot)-\psi_0(\cdot)\|<\delta$$

then (i) ψ may be continued to be a solution of $N\psi=0$ in $\Omega\times(0,\infty)$ with $B\psi=b$ on $\partial\Omega\times(0,\infty)$ and (ii)

$$\|\phi(\cdot, t)-\psi(\cdot, t)\|<\varepsilon$$

for all $t>0$; ϕ is *asymptotically stable* if, in addition, δ can be chosen so that

$$\|\phi(\cdot, t)-\psi(\cdot, t)\|\to 0 \qquad \text{as} \quad t\to\infty$$

Stability and asymptotic stability for the Cauchy problem may be similarly defined; a solution is unstable if it is not stable.

REMARK 4.7: We have deliberately not restricted the definition to any particular norm. When considering questions of stability in a particular problem the norm must, of course, be specified. The definitions may easily be extended to cases in which u is a vector by choosing the norm appropriately.

A simple example of a scalar reaction–diffusion equation was considered by Fisher (1937). He showed that under various assumptions the rate of increase in frequency u of an advantageous gene in a population is given by $su(1-u)$ for some constant s (a measure of intensity of selection), so that in the spatially homogeneous case the equation is $u_t = su(1-u)$. He was interested in the effect of spatial inhomogeneity on the behaviour of such systems. On the assumption of a flux of u proportional to $-u_x$, as in the derivation of Fick's law in Chapter 1, he added to the spatially homogeneous equation a term Du_{xx} to represent the effect of spatial variations in the frequency. If the total population density is constant then the flux of frequency is proportional to a flux of individuals, and the term Du_{xx} is then equivalent to an assumption that these individuals move at random. Justification of the diffusion term in selection–migration problems has been given since then by many authors, and the subject is reviewed in Fife (1979a). With the initial condition $u(0) = u_0$, the spatially uniform equation has solution

$$u = u_o e^{st}/(1-u_0+u_0 e^{st})$$

Thus $u \equiv 0$ is an unstable steady state of the equation, whereas $u \equiv 1$ is asymptotically stable, and any positive initial frequency u_0 of the advantageous gene, however small, eventually leads to a frequency of 1. How does the introduction of the diffusion term affect these conclusions? The answer to this question depends on the domain and the boundary conditions. We shall show that in a finite domain with zero-flux boundary conditions, the conclusions still hold. However, if the boundary conditions are $u = 0$ on the boundary, then $u \equiv 1$ is no longer a solution of the problem. In this case the behaviour depends on the size of the domain. In a small domain, $u \equiv 0$ is asymptotically stable, but if the domain exceeds a certain size this solution loses its stability and a non-trivial steady-state solution becomes asymptotically stable. This bifurcation phenomenon is discussed in Section 4.5. In an infinite domain there is the possibility of a travelling wave solution, that is, a solution of the form $u(x, t) = U(x+ct)$, where c is the constant wave speed. For this solution $u(x, t) \to 1$ for any fixed value of x, but at any time t there are points in the domain where u may be as small as we like, so that the convergence to $u \equiv 1$ is not uniform in x. This solution corresponds to a wave of the advantageous gene sweeping through the population after being introduced at some part of the spatial domain. Such solutions will be discussed in Sections 4.6 and 4.7.

4.2 The Maximum Principle and Comparison Theorems

DEFINITION 4.8: A *regular subsolution* $\underset{\sim}{u}$ of a differential equation $Nu = 0$ in $Q_T \cup S_T$ is a function $\underset{\sim}{u}$ continuous in $\bar{Q}_T$ and with the relevant derivatives continuous in $Q_T \cup S_T$ satisfying $N\underset{\sim}{u} \leq 0$. It is a *regular subsolution of the initial-boundary-value problem* 4.3(b) if also $\underset{\sim}{u}(\mathbf{x}, 0) \leq u_0(\mathbf{x})$ and $B\underset{\sim}{u} \leq b$. Analogous definitions for the Cauchy and boundary-value problems may be introduced. We define $\underset{\sim}{v}(\mathbf{x},t)$ to be a *subsolution* of $Nu = 0$ in $Q_T \cup S_T$ if

$$\underset{\sim}{v}(\mathbf{x}, t) = \max_{1 \leq i \leq p} \underset{\sim}{u}_i(\mathbf{x}, t)$$

for some collection of regular subsolutions $\underset{\sim}{u}_i$, $i = 1, 2, \ldots, p$. Subsolutions for the various initial- and boundary-value problems are defined in the obvious way. Supersolutions are defined similarly, with the inequalities reversed. Note that any subsolution $\underset{\sim}{v}$ is a continuous function in $\bar{Q}_T$ and satisfies $N\underset{\sim}{v} \leq 0$ wherever the relevant derivatives are defined.

The maximum principle for linear equations may be stated as follows:

THEOREM 4.9 (Maximum Principle): Let u be a function continuous in $\bar{Q}_T$ with the relevant derivatives continuous in $Q_T \cup S_T$ satisfying

$$(L-h)u \equiv u_t - \nabla^2 u - h(\mathbf{x}, t)u \leq 0 \quad \text{in} \quad Q_T \cup S_T, \qquad u \leq 0 \quad \text{on} \quad S_0$$

where h is any function which is bounded above.

Dirichlet problem: If also

$$u \leq 0 \quad \text{on} \quad \Gamma$$

then

$$u \leq 0 \quad \text{in} \quad \bar{Q}_T$$

This is the weak maximum principle for the Dirichlet problem. Moreover, the corresponding strong maximum principle states that either

(a) $u < 0$ in $Q_T \cup S_T$ or
(b) $u \equiv 0$ in $Q_{t^*} \cup S_{t^*}$ for some $t^* \leq T$.

Finally, the boundary point lemma states that if $u = 0$ at some point $P^* = (\mathbf{x}^*, t^*) \in \Gamma$, and if the domain Q_T is sufficiently smooth at P^*, then either

(a) $\partial u / \partial n(\mathbf{x}^*, t^*) > 0$ or
(b) $u \equiv 0$ in $Q_{t^*} \cup S_{t^*}$.

The requirement on the domain Q_T is that it satisfies the interior sphere property at P^*, that is there exists a closed ball B such that $B \subset Q_T$ and $B \cap \Gamma = \{P^*\}$.

Neumann and Robin problems: If instead

$$Bu \leq 0 \quad \text{in} \quad \Gamma$$

and also the domain Q_T satisfies the interior sphere property at each point P of Γ, then the same conclusions hold.

Cauchy problem: Let $Q_T = \mathbb{R}^n \times (0, T)$ and assume that u is bounded. Then the weak and strong maximum principles still hold, with $t^* = T$.

The proof of this fundamental classical theorem due to Nirenberg (1953) may be found in Protter and Weinberger (1967). It has an important corollary for nonlinear problems, which is most conveniently stated as a comparison theorem.

THEOREM 4.10 (Comparison Theorem): If (i) $\underset{\sim}{u}$ and $\tilde{u}$ are regular sub- and supersolutions for the Cauchy problem 4.3(a) (in which case we must assume that $\underset{\sim}{u} - \tilde{u}$ is bounded) or the initial-boundary-value problem 4.3(b), (ii) f is uniformly Lipschitz continuous, and (iii) (if we are concerned with the Neumann or the Robin problem) Q_T satisfies the interior sphere property, then

$$\underset{\sim}{u} \leqslant \tilde{u} \quad \text{in} \quad \bar{Q}_T$$

This is the weak comparison theorem. Moreover, the strong comparison theorem states that either

(a) $\underset{\sim}{u} < \tilde{u}$ in $Q_T \cup S_T$ or
(b) $\underset{\sim}{u} \equiv \tilde{u}$ in $Q_{t^*} \cup S_{t^*}$ for some $t^* \leqslant T$.

Finally, for the initial-boundary-value problems, the boundary point lemma states that if $\underset{\sim}{u} = \tilde{u}$ at some point $P^* = (\mathbf{x}^*, t^*) \in \Gamma$ and if Q_T satisfies the interior sphere property at P^*, then either

(a) $\partial \underset{\sim}{u}/\partial n > \partial \tilde{u}/\partial n$ at P^* or
(b) $\underset{\sim}{u} \equiv \tilde{u}$ in $Q_{t^*} \cup S_{t^*}$.

REMARK 4.11: Note that either $\underset{\sim}{u}$ or $\tilde{u}$ may be a solution and that if alternative (b) holds, they both are (in $Q_{t^*} \cup S_{t^*}$).

Proof: Define $v = \underset{\sim}{u} - \tilde{u}$. Then

$$\begin{aligned} 0 \geqslant N\underset{\sim}{u} - N\tilde{u} &= L\underset{\sim}{u} - f(\underset{\sim}{u}) - L\tilde{u} + f(\tilde{u}) \\ &= Lv - \frac{f(\underset{\sim}{u}) - f(\tilde{u})}{\underset{\sim}{u} - \tilde{u}} v = (L - h)v \end{aligned}$$

where

$$h = \frac{f(\underset{\sim}{u}) - f(\tilde{u})}{\underset{\sim}{u} - \tilde{u}}$$

But since f is uniformly Lipschitz continuous by hypothesis, then h is bounded, and $(L-h)v \leq 0$. Moreover,

$$v(\mathbf{x}, 0) = \underset{\sim}{u}(\mathbf{x}, 0) - \tilde{u}(\mathbf{x}, 0) \leq 0 \quad \text{in} \quad S_0$$

and

$$Bv = B\underset{\sim}{u} - B\tilde{u} \leq 0 \quad \text{in} \quad \Gamma$$

if we are considering an initial-boundary-value problem. Hence the maximum principle applies to v, so that the conclusions of the theorem follow.

REMARK 4.12: If $\underset{\sim}{u}$ and $\tilde{u}$ are sub- and supersolutions of the appropriate problem, but are no longer required to be regular, then the same theorem holds. The proof follows from consideration of each pair of regular sub- and supersolutions obtained from the collections defining $\underset{\sim}{u}$ and $\tilde{u}$.

As an application consider the initial-boundary-value problem for Fisher's equation, given in non-dimensional form by

$$\begin{aligned} u_t &= u(1-u) + u_{xx} && \text{in} \quad Q_T \cup S_T = (x_1, x_2) \times (0, T] \\ u(x, 0) &= u_0(x) && \text{in} \quad S_0, \quad 0 \leq u_0(x) \leq 1 \\ Bu(x, t) &= b(x, t) && \text{in} \quad \Gamma, \quad 0 \leq b(x, t) \leq c(x, t) \end{aligned} \tag{4.13}$$

for continuous data $u_0(x)$ and $b(x, t)$. Let $u(x, t)$ be any solution of this problem. Then an application of the comparison theorem (Theorem 4.10) with $Nu \equiv u_t - u(1-u) - u_{xx}$, $\tilde{u} = u$ and $\underset{\sim}{u} = 0$, noting that $N0 = 0 = Nu$, $0 \leq u_0$ and $B0 = 0 \leq Bu = b$, leads to the result that $0 \leq u(x, t)$, and with $\underset{\sim}{u} = u$ and $\tilde{u} = 1$, noting that $N1 = 0 = Nu$, $u_0 \leq 1$ and $b = Bu \leq B1 = c$, gives $u(x, t) \leq 1$, so that $0 \leq u(x, t) \leq 1$.

4.3 Existence and Uniqueness of Solutions

Uniqueness follows immediately from the comparison theorem.

COROLLARY 4.14 (Uniqueness): If hypotheses (ii) and (iii) of Theorem 4.10 hold, then there is at most one solution of the initial-boundary problem 4.3(b), and at most one bounded solution of the Cauchy problem 4.3(a).

Proof: Assume that there are two solutions, v and w. Then v and w are both subsolutions and both supersolutions, so $v \leq w$ and $w \leq v$, forcing $v = w$.

REMARK 4.15: The requirement that the solution of the Cauchy problem be bounded is essential. The equation

$$u_t - u_{xx} = 0 \qquad \text{for} \quad (x, t) \in \mathbb{R} \times (0, \infty)$$

with zero initial conditions admits both the trivial solution and the solution

$$u(x, t) = \sum_{k=0}^{\infty} \frac{1}{(2k)!} x^{2k} \frac{d^k}{dt^k} \left\{ \exp\left(-\frac{1}{t^2} \right) \right\}$$

REMARK 4.16: The requirement that f be Lipschitz continuous is also, of course, essential. As a simple example (essentially from ordinary differential equations), the equation

$$u_t = u^{1/2} + u_{xx} \qquad \text{in} \quad (a, b) \times (0, T]$$

with zero-flux boundary conditions and zero initial conditions admits both the trivial solution and the solution $u = t^2/4$.

The comparison theorem, by supplying *a priori* bounds on the set of all solutions of certain problems, also leads to theorems on the existence of solutions for all time. We first introduce the concept of an invariant set for parabolic equations.

DEFINITION 4.17: Recall (cf. Definition 2.18) that an *invariant set* of the kinetic (spatially uniform) system $\mathbf{u}_t = \mathbf{f}(\mathbf{u})$ is a set $\Sigma \subset \mathbb{R}^m$, where m is the dimension of $\mathbf{u}$, such that if $\mathbf{u}$ is a solution of $\mathbf{u}_t = \mathbf{f}(\mathbf{u})$ in $0 < t < T$ for some $T \leq \infty$, with initial conditions $\mathbf{u}(0) \in \Sigma$, then $\mathbf{u}(t) \in \Sigma$ for $0 \leq t \leq T$. Similarly, an invariant set for the initial-boundary-value problem 4.3(b) is a set $\Sigma \subset \mathbb{R}$ such that if u is a solution of

$$u_t = f(u) + \nabla^2 u \qquad \text{in} \quad Q_T \cup S_T$$

with

$$u(\mathbf{x}, 0) \in \Sigma \qquad \text{for each point } (\mathbf{x}, 0) \in S_0$$

and

$$Bu(\mathbf{x}, t) \in c(\mathbf{x}, t)\Sigma \equiv \{c(\mathbf{x}, t)v \mid v \in \Sigma\} \qquad \text{for each point } (\mathbf{x}, t) \in \Gamma$$

then

$$u(\mathbf{x}, t) \in \Sigma \qquad \text{for each point } (\mathbf{x}, t) \in \bar{Q}_T$$

An invariant set for the Cauchy problem is similarly defined, with the stipulation that u be a bounded solution. We may now state

THEOREM 4.18 (Invariant Sets): Consider the initial-boundary-value problem 4.3(b), or the Cauchy problem 4.3(a), where f is Lipschitz continuous and Q_T satisfies the interior sphere property (if we are concerned with the Neumann or the Robin problem). Let a and b be such that $f(a) \geq 0$ and $f(b) \leq 0$. Then $\Sigma = \{v \mid a \leq v \leq b\}$ is an invariant set for the initial-boundary-value problem or the Cauchy problem.

Proof: Any solution of the initial-boundary-value problem is a continuous function on the bounded domain $\bar{Q}_T$ and so is bounded. Thus f is uniformly Lipschitz continuous in the relevant domain. The conditions on a and b ensure that they are respectively sub- and supersolutions of the initial-boundary-value problem, so that $a \leq u(\mathbf{x}, t) \leq b$, i.e., $u(\mathbf{x}, t) \in \Sigma$, for all $(\mathbf{x}, t) \in \bar{Q}_T$, as required. The proof for the Cauchy problem is analogous, since boundedness of the solution is required by the definition.

As an example, the set $\Sigma = \{u \mid 0 \leq u \leq 1\}$ is an invariant set for the initial-boundary-value problem for the Fisher equation, as we have already shown.

We may now prove a global existence theorem. We point out that our proof is *constructive*, and may be used as the basis of a numerical computation of the solution of any given problem.

THEOREM 4.19 (Global Existence): Let f be Lipschitz continuous, and let Q_∞ satisfy the interior sphere property (if we are concerned with the Neumann or the Robin problem). Let $f(a) \geq 0$, $f(b) \leq 0$, $-\infty < a < b < \infty$, and $u(\mathbf{x}, 0) \in \Sigma = \{u \mid a \leq u \leq b\}$ on S_0, and for the initial-boundary-value problem, $Bu \in c\Sigma$ on Γ. Then there exists a (unique, by Corollary 4.14) solution of the initial-boundary-value problem 4.3(b), or a unique bounded solution of the Cauchy problem 4.3(a), for all time.

Proof: Clearly Σ is an invariant set for all time so that $a \leq u \leq b$ whenever the solution u exists, and a is a subsolution and b a supersolution of the initial-boundary-value problem. Thus $f(u)$ is *uniformly* Lipschitz continuous in the required domain, with constant K, say. We define sequences $\{v^n\}$ and $\{w^n\}$ by

$$v^0 = a, \qquad Lv^n + Kv^n = f(v^{n-1}) + Kv^{n-1}$$

$$w^0 = b, \qquad Lw^n + Kw^n = f(w^{n-1}) + Kw^{n-1}$$

where v^n and w^n satisfy the initial and boundary conditions of (4.13). We claim that the sequence $\{v^n\}$ is monotonic increasing, and make the inductive hypothesis $v^{n+1} - v^n \geq 0$. This is clearly true by the maximum principle for $n = 0$, since

$$L(v^1 - v^0) + K(v^1 - v^0) = f(a) \geq 0$$

and the necessary initial and boundary inequalities hold. Assume it to be true up to $n-1$. Then

$$\begin{aligned} L(v^{n+1}-v^n)+K(v^{n+1}-v^n) &= f(v^n)+Kv^n-f(v^{n-1})-Kv^{n-1} \\ &\geq -K(v^n-v^{n-1})+K(v^n-v^{n-1})=0 \end{aligned}$$

by the Lipschitz continuity of f. Again the initial and boundary inequalities hold, so that $v^{n+1}-v^n \geq 0$. Similarly, it may be shown that $\{w^n\}$ is decreasing, so that $v=\lim_{n\to\infty} v^n$ and $w=\lim_{n\to\infty} w^n$ exist. Then it is easy to show that v and w are solutions, so that by Corollary 4.14 $v=w$ is the unique solution of the initial-boundary problem 4.3(b), which exists for all $t\geq 0$. The proof for the Cauchy problem is analogous.

If no invariant set exists then for global existence f must not be allowed to grow too fast with respect to u. For example, consider the equation

$$u_t = u^{1+\alpha}+u_{xx}, \qquad x\in(-1,1), \qquad t\in(0,\infty)$$

where α is a positive constant, with boundary conditions

$$u_x=0 \qquad \text{at} \quad x=\pm 1$$

and initial conditions

$$u=u_0 \qquad \text{at} \quad t=0,$$

where u_0 is a positive constant. Looking for a spatially independent solution, we obtain

$$u^\alpha = u_0^\alpha/(1-\alpha u_0^\alpha t)$$

which becomes infinite at $t=u_0^{-\alpha}/\alpha$.

If the function f in (4.2) also depends on ∇u, then it may be possible to prove existence and uniqueness theorems if we can obtain some *a priori* estimates on the derivatives of u. Such estimates may be found, such as those of Schauder type (Friedman, 1964). We shall not state these explicitly but refer the reader to Friedman (1964) or to Fife (1979a). For global existence we again require that f does not grow too fast, but now with respect to ∇u. For example, consider the equation

$$u_t = u_x^2+u_{xx}$$

It is easy to check that

$$u=(x-x_0)^2/(t_0-t)-\log(t_0-t)/2$$

is a solution of this equation which becomes infinite at $t=t_0$.

4.4 Stationary Solutions and Asymptotic Behaviour for the Neumann and Cauchy Problems

Assuming that a solution of the initial-boundary value problem 4.3(b) or the Cauchy problem 4.3(a) does exist for all time, we may wish to know its asymptotic behaviour as $t \to \infty$. In this section we shall first concentrate on the Neumann problem, that is

$$u_t = f(u) + u_{xx} \text{ in } \Omega \times (0, \infty), \qquad \frac{\partial u}{\partial n} = 0 \text{ on } \partial\Omega \times (0, \infty) \tag{4.20}$$

with given initial conditions, where Ω is a bounded interval, and ask if this tends to a stationary (time-independent) solution; f will always be taken to be a smooth function. Many of the results of this section are similar to those of Aronson and Weinberger (1975, 1978) reviewed by Aronson (1978). We first characterise the stationary solutions $u(x, t) = v(x)$ of (4.20). They satisfy

$$v'' + f(v) = 0 \text{ in } \Omega, \qquad v' = 0 \text{ on } \partial\Omega$$

where prime denotes differentiation. This problem may be analysed in the phase plane by defining $w = v'$, so that

$$v' = w, \qquad w' = -f(v) \tag{4.21}$$

Defining F to be the integral of f, this has integral curves given by

$$w^2/2 + F(v) = v'^2/2 + F(v) = \text{constant.} \tag{4.22}$$

The phase plane diagram depends on the function f. Let us first consider f satisfying the conditions

$$f(0) = f(1) = 0, \qquad f > 0 \text{ in } (0, 1), \qquad f'(0) > 0, \qquad f'(1) < 0 \tag{4.23}$$

This includes the Fisher equation, where $f(u) = u(1-u)$, and is sometimes called the (generalised) Fisher equation. It is easy to show that the critical points are at $(v, w) = (0, 0)$ (a centre) and $(1, 0)$ (a saddle point). The trajectories close to $(0, 0)$ must be closed by the symmetry of the system about the v-axis, and the phase plane can be drawn as in Fig 4.24. Solutions of the Neumann problem start and end on the v-axis. The following theorem enumerates the solutions of the stationary problem and describes the asymptotic behaviour of solutions of the corresponding initial-boundary value problem for the generalised Fisher equation:

THEOREM 4.25: The stationary problem 4.3(d) for the generalised Fisher equation (4.20) in $\Omega \times (-\infty, \infty)$ with zero-flux boundary conditions, where Ω is a bounded interval and f satisfies (4.23), has precisely two solutions with $0 \leq u \leq 1$, namely $u \equiv 0$ and $u \equiv 1$. Any solution u of the corresponding initial-boundary-value problem with continuous intial conditions

$$u(x, 0) = u_0(x), \qquad 0 \leq u_0 \leq 1$$

satisfies

$$\lim_{t \to \infty} u(x, t) = 1$$

uniformly in x in $\bar{\Omega}$ unless $u_0 \equiv 0$, in which case $u \equiv 0$.

Proof: The first statement is easily seen to be true from Fig 4.24, and it is clear that if $u_0 \equiv 0$, then $u \equiv 0$. It remains to show that if u_0 is not identically zero, then $u \to 1$ as $t \to \infty$, or equivalently that $u(x, t+k) \to 1$ as $t \to \infty$ for some fixed $k > 0$. Since the comparison theorem shows (as for the Fisher equation) that $0 \leq u \leq 1$, this is accomplished, again using the comparison

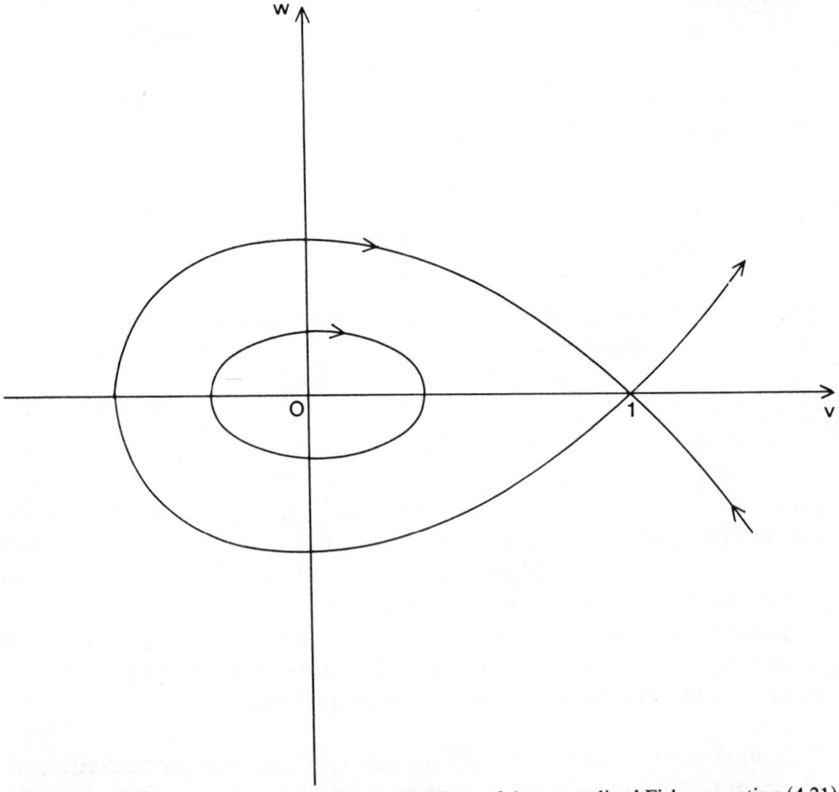

Fig. 4.24. The phase plane for stationary solutions of the generalised Fisher equation (4.21) with (4.23).

theorem, by constructing a subsolution $\underset{\sim}{u}$ for $u_k(x, t) \equiv u(x, t+k)$ which tends to unity as $t \to \infty$. Let $\underset{\sim}{u} = \underset{\sim}{u}(x, t)$ satisfy $N\underset{\sim}{u} = 0$ with initial conditions $\underset{\sim}{u}(x, 0) = \underset{\sim}{u}_0(x)$ and zero-flux boundary conditions. For any fixed $k > 0$, we claim that there exists a number ε, $0 < \varepsilon \leq 1$, such that $u_k(x, 0) = u(x, k) \geq \varepsilon > 0$ for all $x \in \bar{\Omega}$. Since the comparison theorem ensures that $u(x, k) > 0$ in Ω, it is only necessary to show that $u(x, k) > 0$ in $\partial\Omega$. Assuming otherwise, $u(x^*, k) = 0$ for some $x^* \in \partial\Omega$. But then, by the boundary point lemma of Theorem 4.10, $\partial u/\partial n\ (x^*, k) < 0$, which contradicts the boundary condition $\partial u/\partial n = 0$. Thus $u(x, k) > 0$ in $\bar{\Omega}$, and since $\bar{\Omega}$ is compact, $u(x, k) \geq \varepsilon > 0$ in $\bar{\Omega}$, as required. We may now take $\underset{\sim}{u}_0(x) \equiv \varepsilon$. Clearly $\underset{\sim}{u}$ is spatially independent and satisfies $\underset{\sim}{u}_t = f(\underset{\sim}{u})$, $\underset{\sim}{u}(x, 0) \equiv \varepsilon$, so that it is given implicitly by

$$t = \int_{\varepsilon}^{\underset{\sim}{u}} \frac{dz}{f(z)}$$

and the conditions (4.23) on f imply that the spatially independent subsolution $\underset{\sim}{u} \to 1$ as $t \to \infty$, so that $u(x, t) \to 1$ as $t \to \infty$ uniformly in x.

REMARK 4.26: Construction of a *spatially independent* subsolution depends crucially on the fact that the problem is a homogeneous Neumann problem. The corresponding theorems for the Dirichlet and Cauchy problems are not so easy to prove, as we shall see.

Interpreting this result in terms of the original formulation, where u is the frequency of an advantageous gene, this theorem states that if the situation may be modelled by Fisher's equation in a bounded interval with zero-flux boundary conditions, then the introduction of even a very small amount of advantageous genetic material will finally result in the frequency of the gene becoming unity throughout the population.

A similar theorem may be proved in the degenerate case in which $f'(0) = 0$.

THEOREM 4.27: Theorem 4.25 also holds if f satisfies

$$f(0) = f(1) = 0, \qquad f > 0 \text{ in } (0, 1); \qquad f'(0) = 0, \qquad f'(1) < 0 \tag{4.28}$$

In this case the phase plane is as shown in Fig. 4.29. The proof follows the same lines as those of Theorem 4.25.

In the next example let f satisfy, for some $a \in (0, 1)$,

$$\begin{gathered} f(0) = f(a) = f(1) = 0, \qquad f < 0 \text{ in } (0, a), \qquad f > 0 \text{ in } (a, 1) \\ f'(0) < 0, \qquad f'(1) < 0, \qquad F(1) = \int_0^1 f(z)\, dz > 0 \end{gathered} \tag{4.30}$$

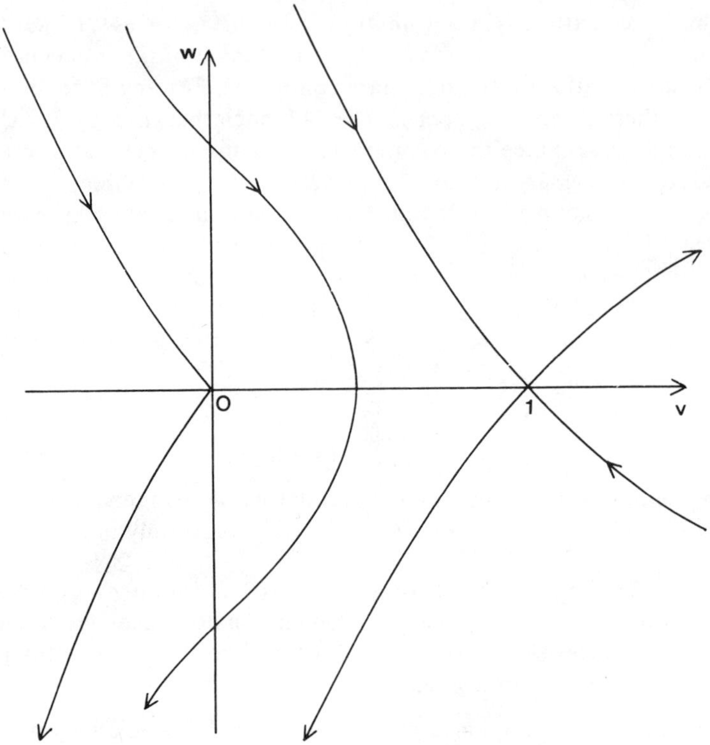

Fig. 4.29. The phase plane for stationary solutions of the degenerate equation (4.21) with (4.28).

The prototype for this is Nagumo's equation (Nagumo *et al.*, 1962), which has applications to models of nerve conduction and is given by (4.20) with $f(u)=u(1-u)(u-a)$, where $0<a<\frac{1}{2}$. The equation (4.20), where f satisfies (4.30), is often referred to as the (generalised) Nagumo equation or bistable equation (see Theorem 4.32). In this case there are three steady states in the phase plane for the stationary problem at $(v, w)=(0,0)$, $(a,0)$ and $(1,0)$. The phase plane diagram is shown in Fig. 4.31. Again solutions of the Neumann problem start and end on the v-axis. In this case the behaviour of the solutions of the stationary Neumann problem and the corresponding initial-boundary-value problem are given by the following theorem.

THEOREM 4.32: The stationary problem 4.3(d) for the generalised Nagumo equation (4.20) in $\Omega\times(-\infty,\infty)$ with zero flux boundary conditions, where Ω is a bounded interval and f satisfies (4.30), has three or more solutions with $0\leq u\leq 1$, depending on the length of Ω. The two stationary solutions

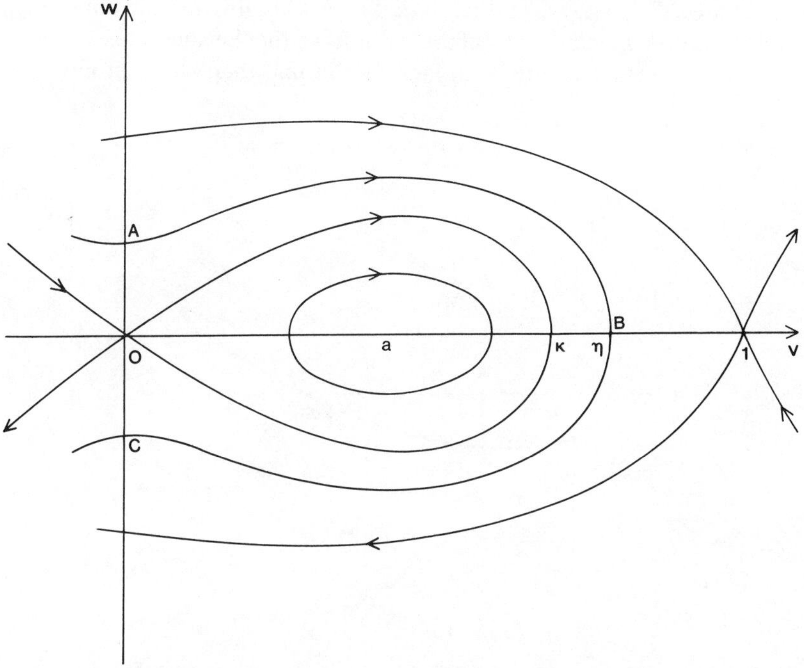

Fig. 4.31. The phase plane for stationary solutions of the generalised Nagumo or bistable equation (4.21) with (4.30).

$u \equiv 0$ and $u \equiv 1$ are both asymptotically stable as solutions of the corresponding initial-boundary-value problem, whereas $u \equiv a$ is unstable.

Proof: Clearly, $u \equiv 0$, $u \equiv a$ and $u \equiv 1$ are solutions of the stationary problem. Other solutions may be obtained, depending on the size of the domain, by taking portions of the solutions oscillating about the centre in the phase plane at $u = a$. The asymptotic stability of $u \equiv 0$ and $u \equiv 1$ may be proved in exactly the same way as was the asymptotic stability of $u \equiv 1$ for the Fisher equation, by showing that any solution of Nagumo's equation with zero-flux boundary conditions and initial conditions satisfying $u(x, 0) \geq a$ and $u(x, 0) \not\equiv a$ tends to 1, and any solution with initial conditions satisfying $u(x, 0) \leq a$ and $u(x, 0) \not\equiv a$ tends to 0. This also proves the instability of $u \equiv a$.

The non-constant stationary solutions $v(x)$ of the Neumann problem may be analysed by considering the integral curves (4.22), or

$$v'^2 + 2F(v) = \text{constant}$$

The graph of F is shown in Fig. 4.33. Since v attains its maximum v_{max} and its minimum v_{min} and $v'=0$ at these points (either because the extremum is internal or because of the boundary conditions) then we must have

$$v'^2+2F(v)=2F(v_{min})=2F(v_{max})$$

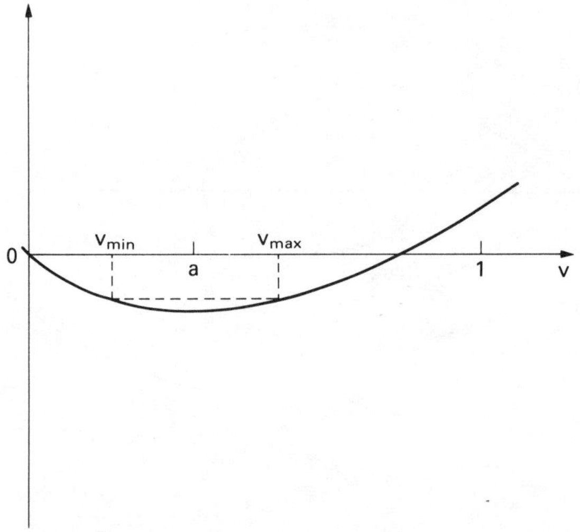

Fig. 4.33. The graph of the function $F(v)=\int_0^v f(z)\,dz$, where f satisfies (4.30).

The length of the domain for a solution oscillating between v_{max} and v_{min} depends on the number of oscillations. In the simplest case the solution is monotonic, with $v=v_{min}$ at one boundary and $v=v_{max}$ at the other, and the length of the domain is given by

$$l=\int_{v_{min}}^{v_{max}} \frac{dz}{\sqrt{2F(v_{max})-2F(z)}}$$

This integral represents the "time" taken for the solution to pass from v_{min} to v_{max}, or vice versa, and is known as the time map (Conley and Smoller, 1980). A solution with $m-1$ internal extrema is possible if the length of the domain is given by

$$l=m\int_{v_{min}}^{v_{max}} \frac{dz}{\sqrt{2F(v_{max})-2F(z)}}$$

A diagrammatic representation of these solutions is given in Fig. 4.34. Here $\Omega=(0,l)$: the solutions v_{m1} and v_{m2} have m internal zeros, and the solutions v_{01} and v_{02} are monotonic decreasing and increasing respectively. It may be shown that all these non-constant solutions are unstable; see Chafee (1975), who proves this for arbitrary functions f. Conley and Smoller

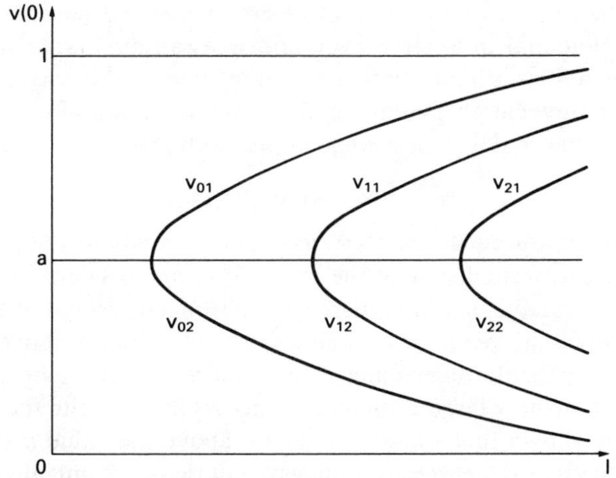

Fig. 4.34. The bifurcation diagram for the Neumann problem for the generalised Nagumo equation.

(1980; see also the references therein) find the dimension of the unstable manifold in the case of the Nagumo equation.

Let us now consider the corresponding problems in infinite domains. The analogue of Theorem 4.25 for the Fisher equation still holds if $\Omega = \mathbb{R}$, with the usual requirement that u must be a *bounded* solution of the Cauchy problem. The non-compactness of $\bar{\Omega}$ leads to complications in constructing the subsolution tending to unity, but these may be overcome using techniques similar to those to be introduced in the next section. It also leads to a weaker result in that the convergence is no longer uniform in x, but pointwise. The theorem for Nagumo's equation differs from Theorem 4.32 in that there is now an infinite number of solutions of the stationary problem corresponding to each of the closed trajectories around the centre in the phase plane at $(a, 0)$. The asymptotic stability of $u \equiv 0$ and $u \equiv 1$ and the instability of $u \equiv a$ still holds.

Nagumo's equation arises from the system of equations which he proposed, following Fitzhugh (1961), as a simplification of the Hodgkin–Huxley (1952) equations for nerve conduction. A further useful simplification, the piecewise linear approximation, was proposed by McKean (1970) and has since been studied by Rinzel and Keller (1973), McKean (1983, 1984), Terman (1983) and Sleeman and Tuma (1984). The Fitzhugh–Nagumo system may be written as

$$u_t = u(u-a)(1-u) - w + u_{xx}$$

$$w_t = \varepsilon(u - bw)$$

where $b \geq 0$, $0 < a < \frac{1}{2}$ and $0 < \varepsilon \ll 1$. Here u may be considered as the membrane potential in a nerve axon and w as an auxiliary variable. The steady state $u = w = 0$ represents the resting state of the nerve. Since ε is small, w is a slow variable compared to u, and in an initial time period we may assume that w does not change appreciably, i.e. $w = 0$. Thus

$$u_t = u(u-a)(1-u) + u_{xx}$$

which is Nagumo's equation. Here $u = 0$ corresponds to the resting state and $u = 1$ to the excited state of the nerve. Theorem 4.34 ensures that both these states are stable, and that there is therefore a threshold phenomenon. In other words the nerve is excitable, but only stimuli above a certain threshold are capable of exciting it, as required by the biology. It is of interest to know how large a stimulus is necessary to excite the nerve. We have already shown that stimuli which are above the value a everywhere along the length of the nerve are sufficient, but these are unrealistic biologically, since a stimulus is a localised phenomenon. The next theorem shows that a localised stimulus can excite a nerve of indefinite length. Before stating it, let us define the stationary solution $v_\eta(x)$ of Nagumo's equation to be that solution which passes from A to $B = (\eta, 0)$ to C in the phase plane (Fig. 4.31) which takes its maximum at $x = 0$; it increases from zero to η and decreases to zero again in a *finite* interval.

THEOREM 4.35: The bounded solution of the Cauchy problem for the generalised Nagumo equation $u_t = f(u) + u_{xx}$ in $\mathbb{R} \times (0, \infty)$, where f satisfies (4.30), with initial conditions

$$u(x, 0) = u_0(x) \geq \max\{v_\eta(x), 0\}.$$

satisfies

$$u(x, t) \to 1 \qquad \text{as} \quad t \to \infty$$

for each $x \in \mathbb{R}$.

Proof: The proof follows by showing that $\underset{\sim}{u}(x, t)$ satisfying $N\underset{\sim}{u} = 0$ in $\mathbb{R} \times (0, \infty)$ with initial conditions $\underset{\sim}{u}(x, 0) = \max\{v_\eta(x), 0\}$ is a subsolution of the problem for u which tends to unity as $t \to \infty$. We leave the details to the reader.

It should be noted that $u(x, t)$ does not converge uniformly to unity as $t \to \infty$. In fact it may be shown that if u_0 has bounded support then it converges to a pair of wave fronts travelling with constant speed in either direction, where u is close to zero before the passage of the front and close to unity afterwards. It follows that the problem on the semi-infinite domain $(0, \infty)$ with initial conditions above $\max\{v_\eta(x), 0\}$ tends to a single travelling wave front as $t \to \infty$. Such a travelling wave front corresponds to a nerve impulse in the Nagumo model of nerve conduction, and will be discussed further in Section 4.6. The model with $w = 0$ thus predicts that a finite localised stimulus can be sufficient to trigger a wave front such that the nerve is in its rest state before its passage and its excited state afterwards. Thus information is passed along the nerve. For a model of nerve conduction to be realistic there must be a mechanism for return to the rest state, so that the nerve may again be excited by a stimulus. This is the role of the slow variable w, and will be discussed in Chapter 8.

In this section we have always considered Ω to be a one-dimensional domain, but all the results on the stability and instability of the constant solutions of the problems carry over in a straightforward way to problems in higher dimensions. This includes the global asymptotic stability of $u \equiv 1$ for the Fisher equation, so that there are again no non-constant stationary solutions of the Neumann problem in this case. The non-constant stationary solutions of the Neumann problem for the Nagumo equation can no longer be characterised by trajectories in a phase plane, of course. However, analysis of the Laplacian operator with zero flux boundary conditions gives a similar bifurcation diagram to Fig. 4.34, at least locally. For convex domains, Casten and Holland (1978) have shown the non-existence of *stable* non-constant stationary solutions of the Neumann problem for arbitrary functions f; in non-convex domains such solutions are possible (Matano, 1979, Keyfitz and Kuiper, 1983). Finally, theorem 4.35 may easily be extended to give a similar result in n dimensions, replacing $v_\eta(x)$ by the principal eigenfunction of the Dirichlet problem for the Laplacian in a suitable domain.

4.5 Stationary Solutions and Asymptotic Behaviour for the Dirichlet Problem: The Spruce Budworm Model

As an example of the application of comparison methods to Dirichlet problems in nonlinear diffusion, we consider an ecological model. The spruce budworm is the larva of a moth which lives on balsam fir and spruce trees in northeastern North America, and is a particular problem to forest managers in the Canadian province of New Brunswick. Here the budworm population is normally controlled by natural predators and diseases, but epidemics occur every 30 to 70 years which defoliate and kill as many as 80% of the mature balsam firs and cause severe damage to the spruce trees. After seven to fourteen years of an epidemic, the damage is so great that the spruce budworm population can no longer sustain its high level, and falls rapidly to its scarce but endemic level. In an attempt to prevent the outbreak of such an epidemic, insecticides were used to spray trees in the area in 1952. The budworm population was reduced, but not to its endemic level, so that the danger of an outbreak was still present. It has been found necessary to spray the forests every year since then (except 1959, when budworm populations seemed to be in decline), so that although an epidemic has not occurred, the cost of keeping the budworm under control has been very large.

A mathematical model for the spruce budworm system was set up by Ludwig *et al.* (1978). In its simplest form it may be written as an equation for the budworm density B, namely

$$\frac{\partial B}{\partial T} = rB\left(1 - \frac{B}{\kappa S}\right) + \beta \frac{B^2}{\alpha^2 S^2 + B^2} + D\frac{\partial^2 B}{\partial X^2}$$

where all the parameters are positive. Defining non-dimensional variables by

$$t = rT, \qquad x = X(r/D)^{1/2}, \qquad u = B/\kappa S$$

and non-dimensional parameters by

$$\varepsilon = \alpha/\kappa, \qquad R = r\alpha S/\beta,$$

we obtain

$$u_t = u(1-u) - \frac{1}{R}\frac{u^2}{\varepsilon^2 + u^2} + u_{xx} \tag{4.36}$$

Here ε is a constant small parameter ($\varepsilon \simeq 1/300$) and R is a parameter which measures the amount of foliage available. R is assumed to be constant over the time-scales we are interested in. This may be justified because the budworm density changes much more quickly than the amount of foliage available (Ludwig *et al.*, 1978).

The equation contains the logistic term $u(1-u)$, reflecting the fact that the budworm population when small tends to increase at a rate proportional to its value, but that it levels off at a saturation value, the carrying capacity of the environment. The term $-u^2/(\varepsilon^2+u^2)R$ represents the effects of predation and parasitism on the budworm population. Predation grows faster than linearly for small u, since some predators such as birds shift their attention to the new source of food as it becomes more plentiful, and levels off at large u since predators have finite appetites. Also it is smaller at larger values of R since the budworm is more difficult to find when foliage is plentiful. A simple function satisfying these conditions is chosen for the model. More complicated but qualitatively similar functions will produce qualitatively similar results.

For ε fixed, the number of steady-state solutions of the budworm equation depends on the value of R. The zero function is always a solution, and other steady states are given by

$$R(1-u) = u/(\varepsilon^2 + u^2)$$

This equation may be treated graphically. The solutions are the intersections of the graph of $u/(u^2+\varepsilon^2)$ with $R(1-u)$. It can be seen (Fig. 4.37) that there are two critical values R_1 and R_2 with the following properties: (i) If $R < R_2$, there is only one non-trivial steady state for u which is small (the endemic state); it is easy to show that it is stable. (ii) If $R = R_2$, a second (large) steady state for u appears, which is semi-stable (i.e. stable for initial

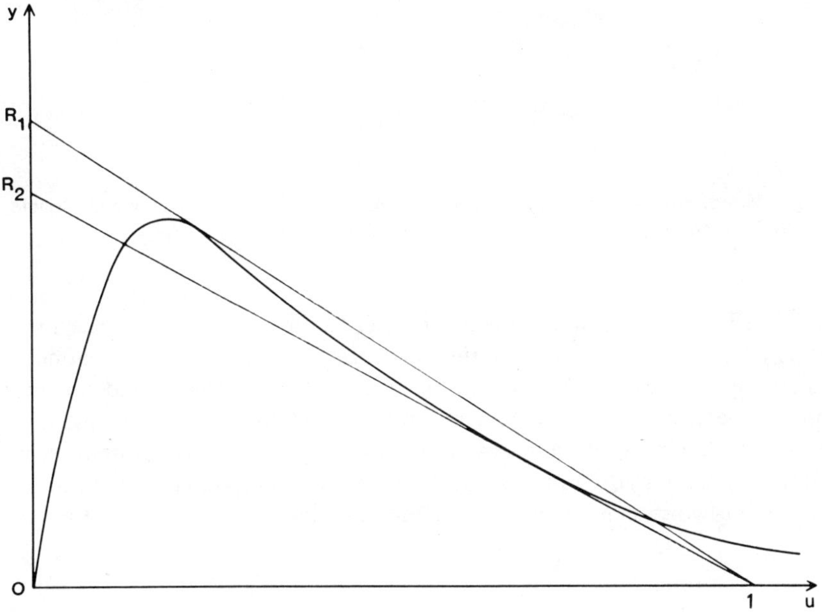

Fig. 4.37. Plots of $y = u/(u^2+\varepsilon^2)$ and $y = R(1-u)$ for fixed ε and various values of R.

conditions above the steady state, but unstable if they are below it). (iii) If $R_2 < R < R_1$, there are three steady states, the lower (endemic) and upper (epidemic) ones being stable whereas the middle one is unstable. (iv) If $R = R_1$, the lower and middle steady states coalesce and become semi-stable (i.e. stable for initial conditions below the steady state, but unstable if they are above it). (v) If $R > R_1$, there is only one large steady state (the epidemic state). These results are summarised in Fig. 4.38.

Ludwig *et al.* (1979) consider the possibility of controlling spruce budworm by the construction of barriers, or strips of forest in which the foliage density R has been reduced from its normal level. They address two questions: (i) How wide must a patch of forest be to support an epidemic?

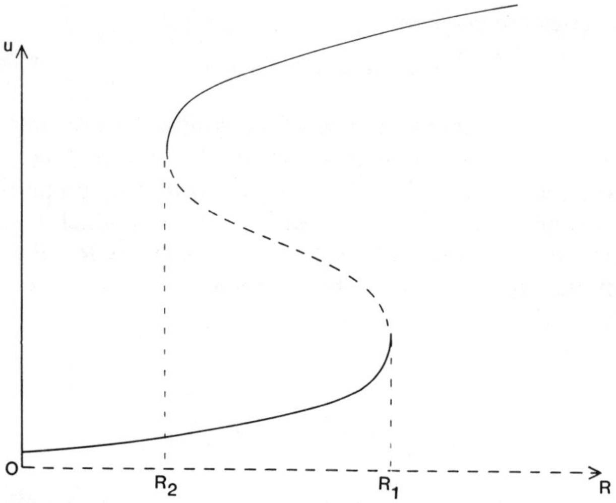

Fig. 4.38. Steady state solutions of the budworm equation (4.36) for fixed ε and variable R. Bold lines represent stable and dotted lines unstable steady states.

(ii) What is the width of an effective barrier to the spread of an epidemic? We confine our attention here to the first question and follow the treatment of Ludwig *et al.* (1979). A variation of this problem including wind convection has been discussed by Murray and Sperb (1983). To fix ideas, and because both are important cases in their own right, we consider first linear and then logistic budworm growth. We take the boundary conditions to be that the budworm cannot survive outside the region $-l/2 < x < l/2$, that is,

$$u(-l/2, t) = u(l/2, t) = 0 \tag{4.39}$$

THEOREM 4.40 (Linear Case): The linear equation

$$u_t = u + u_{xx} \tag{4.41}$$

with boundary conditions (4.39) has critical patch width $l = \pi$, where the critical patch width is defined to be that value of l above which the trivial solution of the problem becomes unstable.

Proof: The stability of the trivial solution of (4.41) with (4.39) depends on the eigenvalues of the problem. These are given by $\lambda_n = 1 - n^2\pi^2/l^2$, where n is a positive integer. It can be seen that if $l < \pi$, then all the eigenvalues are negative, so that the trivial solution is stable, but if $l > \pi$, then there is at least one positive eigenvalue, so that instability results. Thus $l = \pi$ is the critical patch size.

THEOREM 4.42 (Logistic Case): The diffusional logistic equation

$$u_t = u(1-u) + u_{xx} \tag{4.43}$$

with boundary conditions (4.39) has no non-trivial non-negative steady-state solution if $l < \pi$, and a unique such solution if $l > \pi$.

Proof: Steady-state solutions $v(x)$ of this problem satisfy

$$v'' + v(1-v) = 0 \quad \text{in} \quad |x| < l/2 \tag{4.44}$$

with

$$v(-l/2) = v(l/2) = 0 \tag{4.45}$$

The phase plane for (4.44) is shown in Fig. 4.46. Since $u \geq 0$ by the comparison theorem, we are only interested in non-negative solutions $v(x)$, and because of the boundary conditions (4.45), non-trivial solutions correspond to trajectories from points such as A on the $v = 0$ axis at $x = -l/2$ to B on the v-axis to C on the $v = 0$ axis at $x = l/2$. It is clear by symmetry that $x = 0$ at $B = (\mu, 0)$, and since no trajectory passes *through* the saddle point at $(1, 0)$ then $\mu < 1$. Moreover, $v' > 0$ in $(-l/2, 0)$ and <0 in $(0, l/2)$.

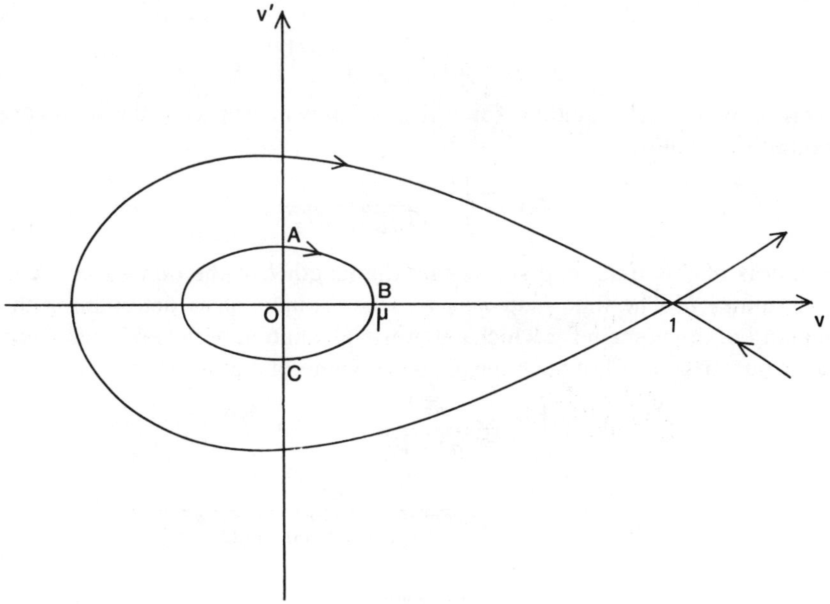

Fig. 4.46. The phase plane for equation (4.44).

The equation for v may be integrated by first multiplying through by v' to obtain

$$v'^2/2+v^2/2-v^3/3=A$$

where A is a constant, or

$$v'^2/2+F(v)=F(\mu)$$

which defines F, by applying conditions at $x=0$, where $v=\mu$ and $v'=0$. Thus

$$v'=\begin{cases}\sqrt{2(F(\mu)-F(v))}, & -l/2\leqslant x\leqslant 0\\ -\sqrt{2(F(\mu)-F(v))}, & 0\leqslant x\leqslant l/2\end{cases}$$

using our knowledge of the sign of v'. Integrating from x to 0 for $-l/2<x<0$,

$$\int_{v(x)}^{\mu}\frac{dz}{\sqrt{2(F(\mu)-F(z))}}=-x$$

and for $0<x<l/2$,

$$\int_{v(x)}^{\mu}\frac{dz}{\sqrt{2(F(\mu)-F(z))}}=x$$

or combining these,

$$\int_{v(x)}^{\mu}\frac{dz}{\sqrt{2(F(\mu)-F(z))}}=|x|$$

This is an implicit equation for v if μ is known. Applying the boundary conditions (4.45),

$$l=\sqrt{2}\int_{0}^{\mu}\frac{dz}{\sqrt{F(\mu)-F(z)}}$$

Analysis of this time map shows that the length l of the domain *uniquely* determines μ. The time map may also be thought of as determining the length l of the domain for which a non-trivial solution with maximum value μ occurs. The smallest such length is the value of l as $\mu\to 0$, that is,

$$\begin{aligned}l_c&=\lim_{\mu\to 0}\sqrt{2}\int_{0}^{\mu}\frac{dz}{\sqrt{F(\mu)-F(z)}}\\&=\lim_{\mu\to 0}\sqrt{2}\int_{0}^{\mu}\frac{dz}{\sqrt{\frac{1}{2}(\mu-z)(\mu+z+\frac{2}{3}(\mu^2+\mu z+z^2))}}\\&=\lim_{\mu\to 0}2\int_{0}^{\mu}\frac{dz}{\sqrt{\mu^2-z^2}}=\pi\end{aligned}$$

as required. The graph of μ against l is plotted in Fig. 4.47.

Thus there is a unique non-trivial stationary solution $v(x)$ of the logistic equation (4.43) with zero boundary conditions (4.39) as long as $l > \pi$. We are interested in the stability of this solution.

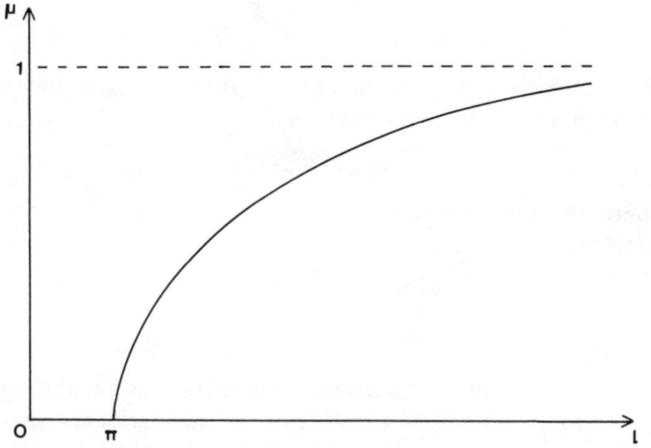

Fig. 4.47. The maximum value of v, the solution of (4.44) with zero boundary conditions (4.45), as a function of the domain width l.

THEOREM 4.48: Consider (4.43) with zero boundary conditions (4.39) and initial conditions

$$u(x, 0) = u_0(x) > 0 \tag{4.49}$$

in $(-l/2, l/2)$. Then (i) if $0 < l < \pi$, $u(x, t) \to 0$ as $t \to \infty$ and (ii) if $l > \pi$, $u(x, t) \to v(x)$ as $t \to \infty$.

REMARK 4.50: Just as in the proof of Theorem 4.25, (4.49) may be relaxed to

$$u(x, 0) = u_0(x) \geq 0, \qquad u_0(x) \not\equiv 0.$$

Proof of (i): By the comparison theorem (Theorem 4.10) with $Nu = u_t - u(1-u) - u_{xx}$, $\tilde{u} = u$ and $\underset{\sim}{u} = 0$, we have $u(x, t) \geq 0$, so that 0 is a subsolution. To complete the proof we require a supersolution $\tilde{u}$ which tends to 0 as $t \to \infty$. Define $\tilde{u}(x, t)$ to be the solution of

$$\tilde{u}_t = \tilde{u} + \tilde{u}_{xx}$$

in $|x|<l/2$, $t>0$, with boundary conditions

$$\tilde{u}(\pm l/2, t)=0$$

and initial conditions

$$\tilde{u}(x, 0)=u_0(x)$$

Note that

$$N\tilde{u}=\tilde{u}_t-\tilde{u}(1-\tilde{u})-\tilde{u}_{xx}=\tilde{u}^2\geqslant 0=Nu$$

Since $\tilde{u}=u$ initially and on the boundaries, it follows from the comparison theorem, with $\tilde{u}=\tilde{u}$ and $\underset{\sim}{u}=u$, that

$$u(x, t)\leqslant \tilde{u}(x, t)$$

everywhere. But the linear analysis above showed that $\tilde{u}(x, t)\to 0$ as $t\to\infty$ if $l<\pi$, so that

$$u(x, t)\to 0 \qquad \text{as} \quad t\to\infty$$

as required.

Proof of (ii): We construct sub- and supersolutions for the initial-boundary-value problem, each of which tends to the stationary solution $v(x)$. The nonlinear operator N is defined as before by $Nu=u_t-u(1-u)-u_{xx}$. Define $M_0=\sup_{|x|\leqslant l/2} u_0(x)$, $M=\max\{1, M_0\}$, and let $\tilde{u}$ be the solution of

$$N\tilde{u}=0, \qquad |x|<l/2, \qquad t>0$$

$$\tilde{u}(x, 0)=M, \qquad |x|<l/2$$

$$\tilde{u}(\pm l/2, t)=0, \qquad t>0$$

Then $N\tilde{u}=0\geqslant 0=Nu$, $\tilde{u}(\pm l/2, t)=0\geqslant 0=u(\pm l/2, t)$ and $\tilde{u}(x, 0)=M\geqslant M_0\geqslant u(x, 0)$ by the definitions of M and M_0, so that $\tilde{u}$ is a supersolution of the problem for u, and $\tilde{u}(x, t)\geqslant u(x, t)$ for all $|x|\leqslant l/2$ and $t\geqslant 0$, by the comparison theorem. It is easily shown that 0 is a subsolution and M is a supersolution of the problem for $\tilde{u}$, so that $0\leqslant\tilde{u}(x, t)\leqslant M$. The next step is to show that $\tilde{u}$ is monotone decreasing and hence tends to a limit, i.e. that for arbitrary positive h, $\tilde{u}(x, t)\geqslant\tilde{u}(x, t+h)$. This is again accomplished by using the comparison theorem, defining $U_h(x, t)=\tilde{u}(x, t+h)$ for any fixed h. We need to show that U_h is a subsolution of the problem for $\tilde{u}$. But U_h satisfies

$$NU_h=0, \qquad |x|<l/2, \qquad t>0$$

$$U_h(x, 0)=\tilde{u}(x, h), \qquad |x|<l/2$$

$$U_h(\pm l/2, t)=0, \qquad t>0$$

Thus U_h is a subsolution of the problem for $\tilde{u}$ if $U_h(x, 0) \leq \tilde{u}(x, 0)$. But $U_h(x, 0) = \tilde{u}(x, h) \leq M = \tilde{u}(x, 0)$, since we have already shown that $\tilde{u}(x, t) \leq M$ for all t. It follows that $\tilde{u}(x, t+h) = U_h(x, t) \leq \tilde{u}(x, t)$ for all x and t, so that u is monotone decreasing. It is also bounded below by zero so that $\lim_{t\to\infty} \tilde{u}(x, t) = w(x)$ exists and satisfies $0 \leq w(x) \leq M$. It can be shown (Aronson and Weinberger, 1975, 1978) that $w(x)$ is a non-negative solution of (4.43) with (4.39) in $|x| \leq l/2$. We need to show that it is non-zero. But since $v(x)$ satisfies $Nv = 0$ with zero boundary conditions and initial conditions $v(x) \leq 1 \leq M = \tilde{u}(x, 0)$, we may again use the comparison theorem to show that $\tilde{u}(x, t) \geq v(x)$, so $w(x) \geq v(x)$. Since $v(x)$ is the unique positive solution of (4.43) with (4.39) in $|x| \leq l/2$, it follows that $w(x) = v(x)$, that is

$$\lim_{t\to\infty} \tilde{u}(x, t) = v(x) \tag{4.51}$$

The solution $\underset{\sim}{u}(x, t)$ bounding $u(x, t)$ from below is more difficult to construct. The aim is to take initial conditions $\underset{\sim}{u}_0(x)$ for $\underset{\sim}{u}$ which are below those for u and which ensure that $\underset{\sim}{u}(x, t)$ is an increasing function of t. In particular, we want $\underset{\sim}{u}(x, t) \geq \underset{\sim}{u}_0(x)$, which suggests application of the comparison theorem to the function $\underset{\sim}{u}_0(x)$, and therefore that $\underset{\sim}{u}_0$ should perhaps be a solution of $N\underset{\sim}{u}_0 = 0$. In fact $\underset{\sim}{u}_0$ is constructed to satisfy $N\underset{\sim}{u}_0 = 0$ almost everywhere. It may be helpful to refer to Fig. 4.52.

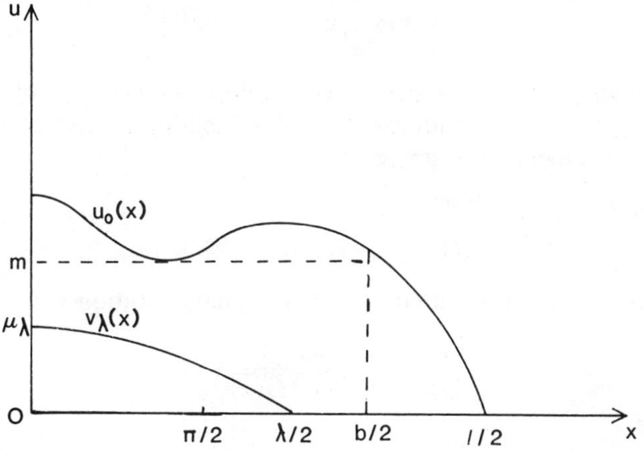

Fig. 4.52. The construction of a subsolution for the solution of the diffusional logistic equation (4.43) with zero boundary conditions (4.39) and initial conditions (4.49).

Define $v_\lambda(x)$ to be the solution of $v'' + v(1-v) = 0$ on $(-\lambda/2, \lambda/2)$ with $v(\pm\lambda/2) = 0$. Then for any fixed b, $\pi < b < l$, we may choose $\lambda > \pi$ sufficiently close to π that $v_\lambda(x) < u_0(x)$ in $(-\lambda/2, \lambda/2)$, since $\mu_\lambda = \sup_x v_\lambda(x) \to 0$ as $\lambda \to \pi$, and $u_0 > 0$ in $[-b/2, b/2]$ and is thus bounded below by some constant $m > 0$. Define $\underset{\sim}{u}(x, t)$ to be the solution of

$$N\underset{\sim}{u} = 0$$

$$\underset{\sim}{u}(x, 0) = \underset{\sim}{u}_0(x) = \begin{cases} v_\lambda(x), & |x| < \frac{1}{2}\lambda \\ 0, & \frac{1}{2}\lambda < |x| < \frac{1}{2}l \end{cases}$$

$$\underset{\sim}{u}(\pm l/2, t) = 0$$

Then $\underset{\sim}{u}(x, t) \leq u(x, t)$, $0 \leq \underset{\sim}{u}(x, t) \leq 1$, and for arbitrary h, $\underset{\sim}{u}(x, t) \leq \underset{\sim}{u}(x, t+h)$. This last follows directly from the comparison theorem for the two functions $\underset{\sim}{u}(x, t)$ and $\underset{\sim}{u}(x, t+h)$, as in the proof of monotonicity for the supersolution, if we can prove the requirement on the initial conditions, i.e. $\underset{\sim}{u}(x, 0) \leq \underset{\sim}{u}(x, h)$. This must hold for any h, so we require $\underset{\sim}{u}(x, 0) \leq \underset{\sim}{u}(x, t)$. If $|x| \geq \lambda/2$, then $\underset{\sim}{u}(x, 0) = 0 \leq \underset{\sim}{u}(x, t)$. If $|x| \leq \lambda/2$, then the comparison theorem on $\underset{\sim}{u}(x, t)$ and $\underset{\sim}{u}(x, 0) = \underset{\sim}{u}_0(x) = v_\lambda(x)$ give the required result, since $N\underset{\sim}{u} = Nv_\lambda = 0$, $\underset{\sim}{u}(x, 0) = v_\lambda(x)$ and $\underset{\sim}{u}(\pm\lambda/2, t) \geq 0 = v_\lambda(\pm\lambda/2)$. Thus $\underset{\sim}{u}$ is monotone increasing in t, and hence by a similar argument to that for $\tilde{u}$, it can be shown that

$$\lim_{t\to\infty} \underset{\sim}{u}(x, t) = v(x)$$

Since $\underset{\sim}{u} \leq u \leq \tilde{u}$, it follows from this and (4.51) that

$$\lim_{t\to\infty} u(x, t) = v(x)$$

In other words if $l > \pi$, the stationary solution $u(x, t) = v(x)$ of the logistic equation in $(-l/2, l/2)$ with zero boundary conditions and positive initial conditions is globally attracting.

The budworm equations

$$u_t = u(1-u) - (1/R)u^2/(\varepsilon^2 + u^2) + u_{xx}$$

may be dealt with in a similar way. A stationary solution $v(x)$ satisfies

$$\int_{v(x)}^{\mu} \frac{dz}{\sqrt{G(\mu) - G(z)}} = |x|$$

where

$$\int_0^{\mu} \frac{dz}{\sqrt{G(\mu) - G(z)}} = \frac{1}{2}l \tag{4.53}$$

and

$$G(v) = v^2 - 2v^3/3 - 2v/R + 2(\varepsilon/R)\arctan(v/\varepsilon)$$

In this case, $G(v)$ depends on R (and ε). For any value of R the dependence of μ on l can be calculated from the time map (4.53). The results are summarised qualitatively in the bifurcation diagram (Fig. 4.54). The stability of the various solutions may also be considered in a way similar to that of the logistic equation, although the algebra is more difficult, or by using bifurcation theory (see Chapter 6). The results are as follows:

THEOREM 4.55 (Ludwig *et al.*, 1979): The budworm equation

$$u_t = u(1-u) - (1/R)u^2/(\varepsilon^2 + u^2) + u_{xx}$$

for $x \in (-l/2, l/2)$, with boundary conditions

$$u(-l/2, t) = u(l/2, t) = 0$$

has no stationary solutions if $l < \pi$. If $l > \pi$, it has the following stationary solutions: (i) For R small, $R < R_1$, there is one non-trivial solution. It is at endemic level and is stable. (ii) As R increases ($R_1 < R < R_2$), the stable solution at endemic level still exists for all $l > \pi$; but for $l > l_1(R)$ there is also a stable solution at epidemic level and an unstable one at an intermediate

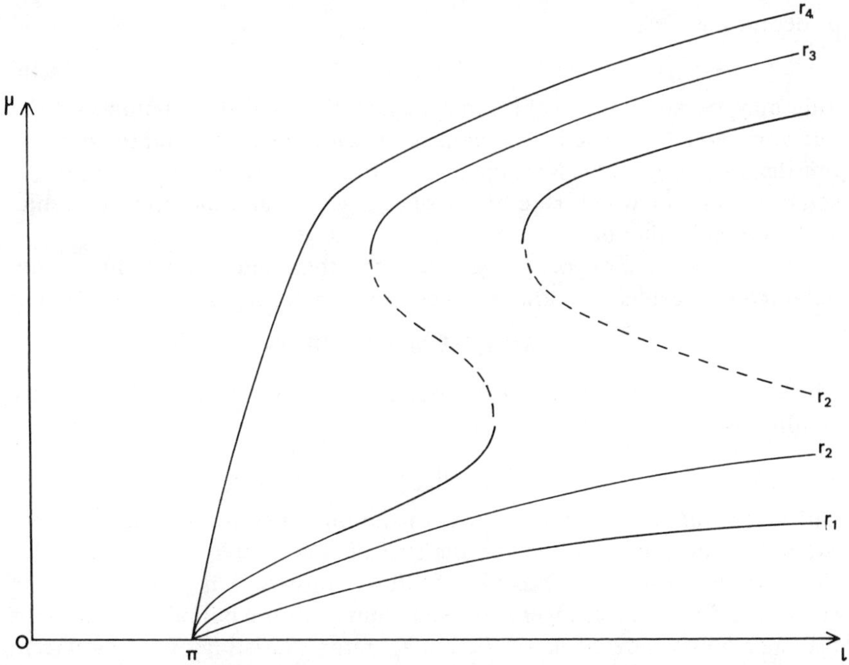

Fig. 4.54. The bifurcation diagram for the budworm equation (4.36) with zero boundary conditions (4.39), in the cases (i) $R = r_1 < R_1$, (ii) $R = r_2$, $R_1 < r_2 < R_2$, (iii) $R = r_3$, $R_2 < r_3 < R_3$, (iv) $R = r_4 > R_3$.

level. (iii) This is also the case for $R_2 < R < R_3$, but here the solution at an endemic level disappears for $l > l_2(R)$. When there are two stable solutions, the one reached depends on the initial conditions. Finally, (iv) for $R > R_3$, the solution rises directly to a stable epidemic level as l increases through π.

Thus to prevent the outbreak of an epidemic if the trees are healthy and foliage is abundant ($R > R_3$), the width of the strips of forest must be less than π. In terms of the original dimensional variables this is $\pi\sqrt{D/r}$, where r is the per capita rate of growth of budworm at small budworm densities and D is its diffusion coefficient. If R is smaller ($R_1 < R < R_3$), then the critical length below which epidemics cannot occur is increased to $l_1(R)$, or in dimensional terms $l_1(R)\sqrt{D/r}$.

Note that in the logistic equation and the budworm equation, the trivial and the nontrivial solution branches coincide at a critical value of l. This is not invariably true. For example, consider steady-state solutions $v(x)$ of the Nagumo equation $u_t = u(u-a)(1-u) + u_{xx}$ with Dirichlet boundary conditions, where $0 < a < \frac{1}{2}$. These satisfy

$$0 = v(v-a)(1-v) + v''$$

or defining $w = v'$,

$$v' = w, \qquad w' = -v(v-a)(1-v) \tag{4.56}$$

This may be analysed in the same way as the logistic equation and the budworm equation. The phase plane is shown in Fig. 4.57, and the bifurcation diagram in Fig. 4.58. Note the sudden appearance of two large amplitude solutions, one of which (the upper one) is stable and the other unstable, at the critical value of l.

Similar results may be proved in more than one spatial dimension, although the results here are not as comprehensive. Consider the equation

$$u_t = kf(u) + \nabla^2 u \text{ in } \Omega \times (0, \infty)$$

where k is a parameter and Ω is a bounded domain in $\mathbb{R}^n$, with boundary conditions

$$u = 0 \qquad \text{on} \quad \partial\Omega \times (0, \infty)$$

and given initial conditions on S_0. In the one-dimensional case we considered k constant and allowed the size of the domain to vary; here we shall, equivalently, keep Ω fixed and allow k to vary; k is a measure of the size of the domain. Homogeneous Neumann or Robin boundary conditions may also be considered. Let L be the operator consisting of the Laplacian on Ω with homogeneous Dirichlet boundary conditions. We shall need the following classical results on the spectral theory of L. Recall that the spectrum $\sigma(L)$ of L is the set of $\lambda \in \mathbb{C}$ such that $L - \lambda I$ is not invertible.

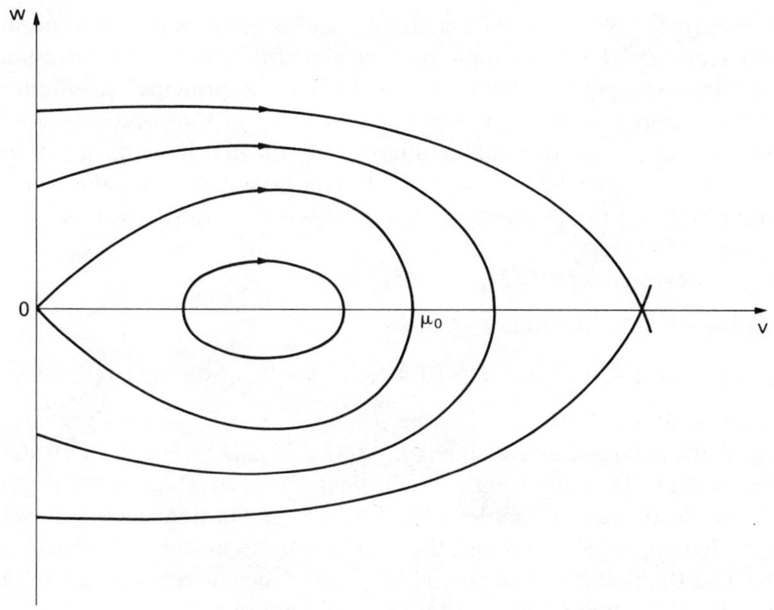

Fig. 4.57. The phase plane for the equations (4.56).

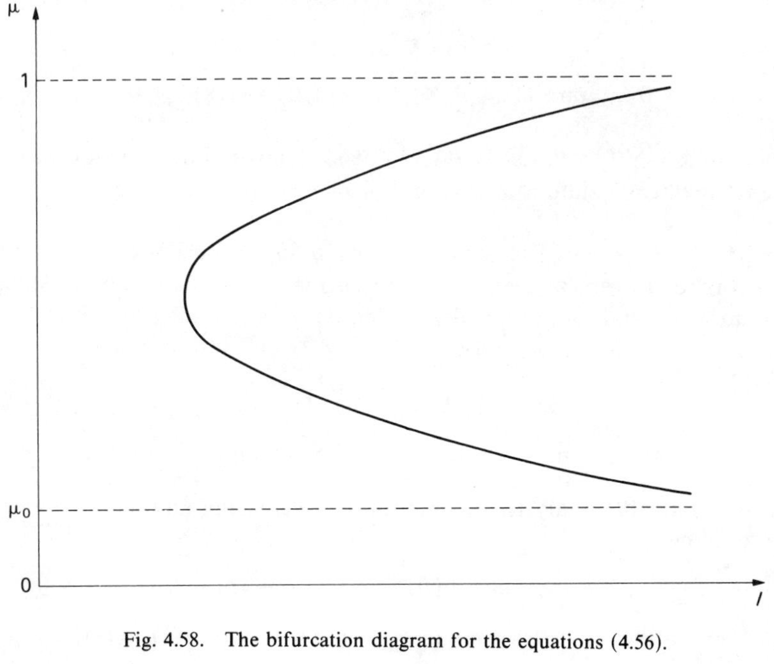

Fig. 4.58. The bifurcation diagram for the equations (4.56).

PROPOSITION 4.59: (a) $\sigma(L)$ is discrete and consists only of real negative eigenvalues. (b) L has a unique positive eigenfunction, which corresponds to the largest eigenvalue. We shall call these the principal eigenfunction and eigenvalue and denote them by ϕ and $\lambda_0 = \lambda_0(k)$, respectively. (c) The principal eigenvalue depends monotonically on the domain, in the sense that if $k_1 > k_2$, then $\lambda_0(k_1) > \lambda_0(k_2)$. (d) The principal eigenvalue depends continuously on the domain, i.e. λ_0 is a continuous function of k.

Proof: See Smoller (1983).

Let us consider the linear problem

$$u_t = ku + \nabla^2 u \quad \text{in} \quad \Omega \times (0, \infty), \qquad u = 0 \quad \text{on} \quad \partial\Omega \times (0, \infty)$$

The eigenvalues of this are given by $\mu = k + \lambda$, where λ is an eigenvalue of L, so that the largest eigenvalue is $k + \lambda_0(k)$. If $\lambda_0(k) < -k$, or equivalently if the domain Ω is sufficiently small, then all eigenvalues μ are negative and the steady state is stable. If $\lambda_0(k) > -k$, or the domain is sufficiently large, then $\mu_0 = k + \lambda_0 > 0$ and the steady state is unstable. We shall now show that the stability of the trivial solution of the corresponding nonlinear problem is determined by a linear stability analysis.

THEOREM 4.60 (Principle of Linearised Stability): Consider the problem

$$u_t = ku + h(u) + \nabla^2 u \quad \text{in} \quad \Omega \times (0, \infty),$$

$$u = 0 \quad \text{on} \quad \partial\Omega \times (0, \infty), \qquad u(\mathbf{x}, 0) = u_0(\mathbf{x}) \quad \text{in} \quad S_0,$$

where $h(0) = h'(0) = 0$. Let $\lambda_0 = \lambda_0(k)$ be as defined above. Then if $\lambda_0 < -k$, the trivial steady state is stable, and if $\lambda_0 > -k$, it is unstable.

Proof: Let $\lambda_0 < -k$. The idea is to find a supersolution of the problem which tends to zero as $t \to \infty$. Since the principal eigenvalue of L depends continuously and monotonically on the domain, we may find $k' > k$ such that $\lambda_0(k) < \lambda_0(k') < -k$, and consider $\tilde{u}(\mathbf{x}, t)$, the solution of

$$\tilde{u}_t = k'\tilde{u} + \nabla^2 \tilde{u} \quad \text{in} \quad \Omega \times (0, \infty),$$

$$\tilde{u} = 0 \quad \text{on} \quad \partial\Omega \times (0, \infty), \qquad \tilde{u}(\mathbf{x}, 0) = u_0(\mathbf{x}) \quad \text{in} \quad S_0$$

By an eigenvalue analysis, $\tilde{u}$ is monotone decreasing and tends to zero as $t \to \infty$. Now

$$N\tilde{u} \equiv \tilde{u}_t - k\tilde{u} - h(\tilde{u}) - \nabla^2 \tilde{u} = (k' - k)\tilde{u} - h(\tilde{u})$$

Since $k' > k$, there exists $\varepsilon > 0$ such that $(k' - k)u - h(u) \geqslant 0$ for all $u \in (0, \varepsilon)$.

Hence if u_0 satisfies $u_0(\mathbf{x}) \leq \varepsilon$ for all $\mathbf{x} \in \Omega$ then $\tilde{u}(\mathbf{x}, t) \leq u_0(\mathbf{x}) \leq \varepsilon$ and $(k'-k)\tilde{u} - h(\tilde{u}) \geq 0$, i.e.

$$N\tilde{u} \geq 0$$

Thus $\tilde{u}$ is a supersolution of the problem for u and the trivial solution is asymptotically stable.

Instability of the trivial solution for $\lambda_0 > -k$ may be proved similarly. We leave the details to the reader.

We now prove the existence of non-trivial steady-state solutions for the logistic equation. A similar proof holds for the budworm equations.

THEOREM 4.61: If $\lambda_0 > -k$, then there exists at least one non-trivial non-negative steady-state solution of

$$u_t = ku(1-u) + \nabla^2 u \qquad \text{in} \quad \Omega \times (0, \infty), \qquad u = 0 \qquad \text{on} \quad \partial\Omega \times (0, \infty) \tag{4.62}$$

Proof: We exhibit sub- and supersolutions for the problem with suitable initial conditions. Let ϕ be the principal eigenfunction of L, as before, and let $\delta > 0$. Then $\nabla^2 \phi = \lambda_0 \phi$, where $\lambda_0 + k > 0$, $\phi > 0$ in Ω and $\phi = 0$ on $\partial\Omega$. Thus if $\underset{\sim}{u} = \delta\phi$, then

$$\underset{\sim}{u}_t - k\underset{\sim}{u}(1-\underset{\sim}{u}) - \nabla^2 \underset{\sim}{u} = -(\lambda_0 + k)\delta\phi + \delta^2\phi^2 \leq 0 \qquad \text{in} \quad \Omega \times (0, \infty)$$

if δ is sufficiently small, and $\underset{\sim}{u} = 0$ on $\partial\Omega \times (0, \infty)$.

For a supersolution, take $\tilde{u} \equiv 1$. Clearly $N\tilde{u} = 0$ in Ω and $\tilde{u} \geq 0$ on $\partial\Omega \times (0, \infty)$. It follows, in a similar way to the one-dimensional case, that there exists a steady-state solution $v(x)$ of (4.62) satisfying $\underset{\sim}{u}(x) \leq v(x) \leq \tilde{u}(x)$, as required.

4.6 Travelling Wave Fronts

The Fisher equation was proposed to model the wave of advance of an advantageous gene, and the Fitzhugh–Nagumo equation to model impulses travelling along nerve axons. It is therefore natural to ask whether scalar reaction–diffusion equations have travelling wave solutions and whether these are stable. A travelling wave is a function of the form

$$u(x, t) = U(x + ct) \tag{4.63}$$

for some constant c, the wave speed. We shall restrict our attention to the equation

$$u_t = f(u) + u_{xx} \tag{4.64}$$

For examples with convective terms see Pauwelussen and Peletier (1981) and Segel (1984). Consider a smooth function f satisfying

$$f(0)=f(1)=0$$

and look for a solution of (4.64) of the form (4.63) which tends to 0 as $x\to-\infty$ and to 1 as $x\to\infty$. Any solution tending to 1 as $x\to-\infty$ and to 0 $x\to\infty$ may then be obtained by changing the signs of x and c. Since we shall ultimately consider the travelling wave front as the asymptotic form of the solution of an initial value problem with initial conditions $u(x,0)=u_0(x)$ where $0\leqslant u_0\leqslant 1$, we also impose the condition $0\leqslant u\leqslant 1$. By substituting (4.63) into (4.64), we obtain

$$U''-cU'+f(U)=0$$

or defining $V=U'$,

$$U'=V, \qquad V'=cV-f(U) \tag{4.65}$$

where primes denote differentiation with respect to $\xi=x+ct$. We require a trajectory from $(0,0)$ to $(1,0)$ in the phase plane which remains in the strip $0\leqslant U\leqslant 1$. Such a trajectory must remain in $V>0$ except at its end-points; in other words any such wave front must be *monotonic*. This is easily seen from phase plane considerations, noting that a trajectory is directed towards the right if V is positive and to the left if it is negative. It follows that the critical points $(0,0)$ and $(1,0)$ cannot be centres or foci, since solutions close to such points must oscillate. Trajectories which pass from one critical point to another are known as *heteroclinic* orbits. Certain systems also admit *homoclinic* orbits, which start and end at the same critical point, but we shall not consider these in this section.

The nature of the critical points is determined by the linearisation of the system (4.65) about $(0,0)$ and $(1,0)$. Linearisation about $(0,0)$ gives the eigenvalue equation

$$\begin{vmatrix} -\lambda & 1 \\ -f'(0) & c-\lambda \end{vmatrix} = \lambda^2 - c\lambda + f'(0) = 0$$

The origin is a saddle point if $f'(0)<0$. If $f'(0)>0$, then $c^2\geqslant 4f'(0)$ for real eigenvalues (complex eigenvalues lead to oscillatory solutions about the origin which have been excluded by the fact that $U\geqslant 0$). The origin is then a node, unstable if c is positive and stable if c is negative. By stability of the origin in this context we mean that all trajectories tend to the origin as $\xi\to\infty$, not as $t\to\infty$. It is easy to see that whether c is positive or negative, the origin is *unstable* in terms of the time variable. If $f'(0)=0$, then the eigenvalues are $\lambda=0$ and $\lambda=c$; the critical point is degenerate and its nature depends on the nonlinear term $f(U)$.

A similar analysis of the critical point at $(1,0)$ shows that it is a saddle point if $f'(1)<0$ and a node if $f'(1)>0$ and $c^2 \geq 4f'(1)$; this node is unstable if $c>0$ and stable if $c<0$ (again it is unstable in terms of the time variable). We again have a degenerate critical point if $f'(1)=0$.

Note that there is no trajectory from a node at $(0,0)$ to a node at $(1,0)$, or *vice versa*; for if this were so, then one of these nodes would be stable and the other unstable, which is impossible since they are both unstable if $c>0$ and both stable if $c<0$. The remaining possibilities are node–saddle (or equivalently, saddle–node) or saddle–saddle transitions, or cases involving degenerate critical points. We shall consider each of these possibilities.

(i) *Node–Saddle Orbits*: Consider the generalised Fisher equation, where f satisfies (4.23), i.e. $f'(0)>0$, $f'(1)<0$ and $f(u)>0$ in $(0,1)$. In this case the origin is a node and the point $(1,0)$ is a saddle point. We wish to find the values of c for which a trajectory from $(0,0)$ to $(1,0)$, i.e. a travelling front, exists. Then the origin must be unstable, i.e. $c>0$. Since $(1,0)$ is a saddle point, there are exactly two trajectories which tend to it as $\xi \to \infty$, in the positive or negative direction of the eigenvector corresponding to the negative (stable) eigenvalue λ_-. This eigenvector is given by $(U_-, V_-)^T$, where

$$\begin{pmatrix} 0 & 1 \\ -f'(0) & c \end{pmatrix}\begin{pmatrix} U_- \\ V_- \end{pmatrix} = \begin{pmatrix} V_- \\ -f'(0)U_- + cV_- \end{pmatrix} = \lambda_- \begin{pmatrix} U_- \\ V_- \end{pmatrix}$$

i.e. $(U_-, V_-) = \alpha(1, \lambda_-)$ for any constant α. Since λ_- is negative, this is in the second quadrant around $(1,0)$ for α negative, and there is thus a trajectory T entering $(1,0)$ as shown in Fig. 4.66.

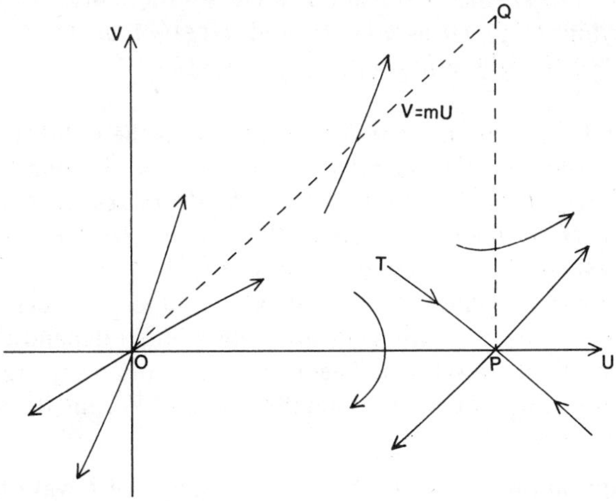

Fig. 4.66. The phase plane for travelling wave solutions of the generalised Fisher equation, i.e. for (4.65) where f satisfies (4.23).

Consider the triangle OPQ in Fig. 4.66. If it can be shown that, for a given value of c, no trajectory may enter this triangle, then the marked trajectory T must tend to O as $\xi \to -\infty$, and thus there must be a travelling wave solution for this value of c. But $U' > 0$ on PQ since $U' = V$, and $V' < 0$ on OP since $V' = -f(U)$ there, so it remains only to prove that no trajectories cross OQ into the triangle OPQ for some value of the gradient m. This is so if $V' - mU' > 0$ on OQ. But

$$V' - mU' = -(m-c)mU - f(U) \geqslant -(m-c)mU - kU$$

where

$$k = \sup_{U \in (0,1)} f(U)/U \geqslant f'(0)$$

Hence $V' - mU' > 0$ if

$$m^2 - cm + k < 0$$

or if the quadratic has two real roots and m lies between them. Hence m can be chosen to satisfy the required conditions if $c^2 > 4k$. Note that $c^2 > 4f'(0)$ (since $k \geqslant f'(0)$) in this case, so that our original hypothesis that the origin is a node is satisfied. Thus there exists a trajectory from $(0, 0)$ to $(1, 0)$ for every positive c satisfying $c^2 > 4k$, i.e. for every c satisfying $c > 2\sqrt{k}$. We may extend this to

THEOREM 4.67: The generalised Fisher equation $u_t = f(u) + u_{xx}$, where $f(u)$ satisfies (4.23), has a travelling wave solution $u(x, t) = U(x + ct) = U(\xi)$ satisfying $U(\xi) \to 0$ as $\xi \to -\infty$ and $U(\xi) \to 1$ as $\xi \to \infty$ for every c satisfying $c \geqslant 2\sqrt{k}$, where $k = \sup_{U \in (0,1)} f(U)/U$.

Proof: Let $T(2\sqrt{k})$ be the trajectory which enters the saddle point at $(1, 0)$ for $c = 2\sqrt{k}$, and consider its behaviour as $\xi = -\infty$. As long as it remains in the half-strip $H = \{0 < U < 1,\ V > 0\}$, U decreases as ξ decreases. It cannot leave H through $V = 0$, since $V' < 0$ there, so that it either leaves through $U = 0$, at $V = V^*$, say, or it tends to $(0, 0)$ as $\xi \to -\infty$. But we have shown that there are values of c arbitrarily close to $2\sqrt{k}$ such that $T(c) \to (0, 0)$ as $\xi \to -\infty$. Since solutions of the system (4.65) depend continuously on c and the initial conditions (Theorem 2.2), a continuity argument rules out the possibility that $T(2\sqrt{k})$ crosses the V-axis at V^*, and hence $T(2\sqrt{k}) \to (0, 0)$ as $\xi \to -\infty$.

This condition on c is sufficient for the existence of a wave front from 0 to 1, but is not in general necessary. In fact we may prove the following theorem.

THEOREM 4.68: For the generalised Fisher equation there is a speed $c^* \in [2\sqrt{f'(0)}, 2\sqrt{k}]$ such that a wave front from 0 to 1 exists if and only if $c \geq c^*$.

Proof: Define $c^* = \inf\{c \mid \text{there exists a wave front from 0 to 1 with speed } c\}$. c^* exists and $c^* \in [2\sqrt{f'(0)}, 2\sqrt{k}]$ since there are wave fronts from 0 to 1 for any $c \geq 2\sqrt{k}$ but not for $c < 2\sqrt{f'(0)}$ (since the origin is then no longer an unstable node). To show that c^* satisfies the conditions of the theorem, we must prove that (a) there is a travelling wave for $c = c^*$, and (b) there is a travelling wave for any $c > c^*$. The proof of (a) follows from a continuity argument, as in the proof of Theorem 4.67. For the proof of (b), assume that $c > c^*$. Consider Fig. 4.69, which represents the phase plane for (4.65)

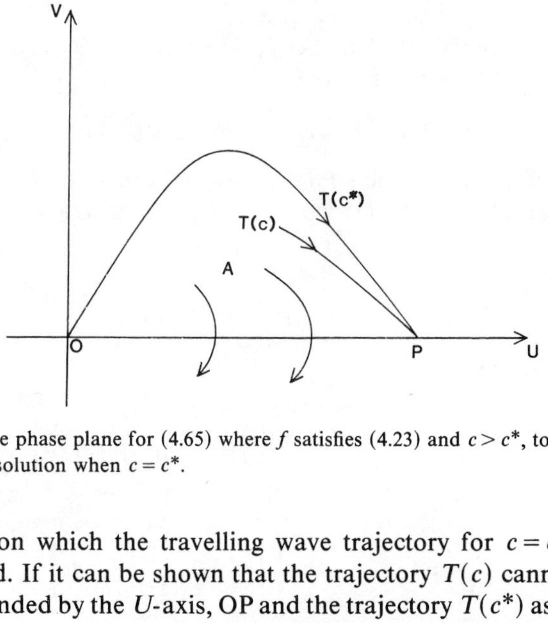

Fig. 4.69. The phase plane for (4.65) where f satisfies (4.23) and $c > c^*$, together with the travelling wave solution when $c = c^*$.

with $c > c^*$, on which the travelling wave trajectory for $c = c^*$ has been superimposed. If it can be shown that the trajectory $T(c)$ cannot leave the region A bounded by the U-axis, OP and the trajectory $T(c^*)$ as ξ *decreases*, then it follows that the trajectory $T(c)$, which tends to P in $V > 0$ as $\xi \to \infty$, must tend to 0 as $\xi \to -\infty$, and thus represents a travelling wave front. It is immediate that no trajectory can leave A as ξ decreases by crossing the U-axis between O and P. Note that we cannot immediately conclude from phase plane considerations that $T(c)$ and $T(c^*)$ cannot cross, since they are trajectories belonging to two separate phase planes. However, the proof may be accomplished as follows.

Consider the trajectory $T(c)$ for U in $(0, 1)$. It is monotone since $V > 0$, and hence the gradient of U is well-defined there as a function of U; that is, there exists a function $P(U)$ such that $P(U)=dU/d\xi$, or $P(U(\xi))=V(\xi)$. Then

$$P'=\frac{dP}{dU}=\frac{d^2U}{d\xi^2}\Big/\frac{dU}{d\xi}=\frac{V'}{U'}=\frac{cP-f}{P}$$

or

$$P+\frac{f}{P}=c \tag{4.70}$$

P also satisfies the boundary condition

$$P(1)=0 \tag{4.71}$$

We may now prove the following lemma:

LEMMA 4.72 (Kanel', 1962): Let $f(U)\geqslant 0$ for $1-U$ small and positive. Let $P_i(U)$, $i=1, 2$, be solutions of

$$P_i'+f/P_i=c_i \qquad \text{for} \quad U\in(0,1)$$

with $P_i(1)=0$. If $c_1=c_2$, then $P_1(U)=P_2(U)$ whenever $P_1(U)>0$, $U<1$. If $c_1<c_2$, then $P_1(U)>P_2(U)$ whenever $P_1(U)>0$, $U<1$.

Proof:

$$P_1'-P_2'-f(P_1-P_2)/(P_1P_2)=(c_1-c_2)$$

Defining $G(U)$ by

$$G(U)=(P_1-P_2)\exp\left\{\int_{1/2}^{U}-\frac{f(t)}{P_1(t)P_2(t)}\,dt\right\}$$

we have

$$\frac{dG}{dU}=(c_1-c_2)\exp\left\{\int_{1/2}^{U}-\frac{f(t)}{P_1(t)P_2(t)}\,dt\right\}$$

If $c_1=c_2$, then $G'(U)=0$ and $\lim_{U\to 1}G(U)=0$ (it is here that we need $f(U)\geqslant 0$ for $1-U$ small and positive), so that $G\equiv 0$ and $P_1\equiv P_2$. If $c_1<c_2$, then $G'(U)<0$ and $\lim_{U\to 1}G(U)=0$, so that $G>0$ for $U<1$ and $P_1>P_2$, as required.

It follows from this lemma that since $c^*<c$, $T(c^*)$ is always above $T(c)$ for $U\in(0,1)$. But this is what we needed to show to complete the proof, so we conclude that there is a travelling wave solution with speed c for any $c>c^*$.

Note the special case in which $k=f'(0)$, i.e. $\sup_{U\in(0,1)} f(U)/U=f'(0)$. Here c^* is known, since $c^*=2\sqrt{k}=2\sqrt{f'(0)}$, and thus there are travelling waves if and only if $c\geqslant 2\sqrt{f'(0)}$. This case includes the original Fisher equation, where $f(u)=u(1-u)$, since here $f'(0)=1$ and $k=\sup_{U\in(0,1)}(1-U)=1$, so that $c^*=2$.

(ii) *Degenerate Critical Points*: The behaviour is very similar even if the derivative $f'(0)=0$, as long as $f(U)$ is still positive in $(0,1)$. Let f satisfy (4.28), that is $f(0)=f(1)=0$, $f(U)>0$ for $U\in(0,1)$ and $f'(0)=0$, $f(1)<0$. Here the origin is a degenerate critical point, and the phase plane for the system (4.65) is shown in Fig. 4.73.

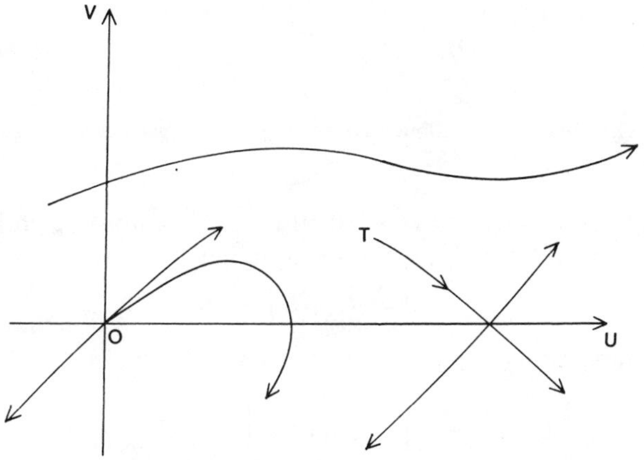

Fig. 4.73. The phase plane for travelling wave solutions of (4.64) in the degenerate case, i.e. for (4.65) where f satisfies (4.28).

THEOREM 4.74: If f satisfies (4.28), then there is a speed $c^*\in(0,2\sqrt{k}]$ such that (4.65) has a travelling wave solution from 0 to 1 if and only if $c\geqslant c^*$, where $k=\sup_{U\in(0,1)} f(U)/U$.

The proof is similar to that of Theorem 4.68. This result will be used in proving the existence of solitary travelling wave solutions of certain reaction-diffusion systems.

(iii) *Saddle–Saddle Orbits*: Let us now consider the generalised Nagumo equation or bistable equation, so that f satisfies (4.30), that is, it has an interior zero, $f(a)=0$, for some $a\in(0,1)$, $f(U)<0$ for $U\in(0,a)$, $f(U)>0$ for $U\in(a,1)$ and $F(1)=\int_0^1 f(s)\,ds>0$. We also have $f'(0)<0$, $f'(1)<0$ and both $(0,0)$ and $(1,0)$ are saddle points. The phase plane in this case is shown in Fig. 4.75.

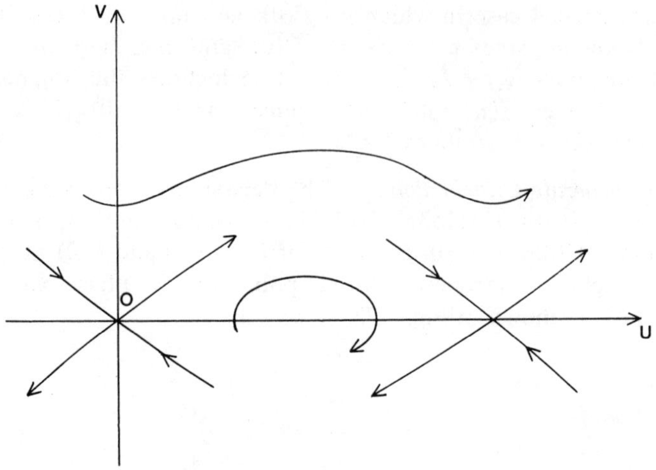

Fig. 4.75. The phase for travelling wave solutions of the generalised Nagumo equation, i.e. for (4.65) where f satisfies (4.30).

Note that for a trajectory from $(0,0)$ to $(1,0)$, c must again be positive. This is easily seen by multiplying the equation

$$U'' - cU' + f(U) = 0$$

by U' and integrating from $\xi = -\infty$ to $\xi = \infty$. We obtain

$$\left[\frac{1}{2}U'^2\right]_{-\infty}^{\infty} - c\int_{-\infty}^{\infty} U'^2\, d\xi + \int_{-\infty}^{\infty} f(s)\frac{ds}{d\xi}\, d\xi = 0$$

or since $U' = 0$ at $\xi = \pm\infty$,

$$c = \int_0^1 f(s)\, ds \Big/ \int_{-\infty}^{\infty} U'^2\, d\xi$$

Since $\int_0^1 f(s)\, ds$ and $\int_{-\infty}^{\infty} U'^2\, d\xi$ are both positive, the wave speed c is positive as required.

THEOREM 4.76: *For the generalised Nagumo equation, where f satisfies (4.30), there is a* unique $c > 0$ *such that there exists a wave front from 0 to 1.*

Proof: To show that there is at most one c, assume the contrary—that there are trajectories $T(c_1)$ and $T(c_2)$ for speeds c_1 and c_2, respectively, from $(0,0)$ to $(1,0)$. Then by Lemma 4.72 one of these trajectories is strictly below the other for $U \in (0,1)$. But consideration of the eigenvectors at $(0,0)$ and $(1,0)$ shows that if $T(c_1)$ is below $T(c_2)$ at $(0,0)$, it must be above it at (1, 0), and *vice versa.* Hence we have arrived at a contradiction, and there is at most one value of c. To prove the existence of such a c we use a continuity argument. We show that if c is sufficiently large, then the trajectory leaves the half strip $H=\{0<U<1,\ V>0\}$ through $\{U=1,\ V>0\}$, and if c is sufficiently small, then it leaves H through $\{0<U<1,\ V=0\}$. Since the solution is continuously dependent on c, it follows that there is some intermediate value of c such that it tends to (1, 0) as $\xi \to 0$, and is therefore a travelling wave front from (0, 0) to (1, 0).

Consider c to be large. We wish to compare the solution of

$$U'=V, \qquad V'=cV-f(U) \tag{4.77}$$

which leaves the origin in the positive quadrant with a solution of

$$U'=V, \qquad V'=aV-bU \tag{4.78}$$

where a and b are to be determined. The relevant trajectories, T_1 and T_2, respectively, are shown in Fig. 4.79. We wish to choose a and b in such a way that T_2 is a straight line with positive slope and that if c is sufficiently large, then T_1 leaves the origin above T_2 and cannot subsequently cross to below it. It must then leave the half strip H through $\{U=1,\ V>0\}$. Let T_2 be the trajectory corresponding to the eigenvalue $\{a+\sqrt{a^2-4b}\}/2$. Then

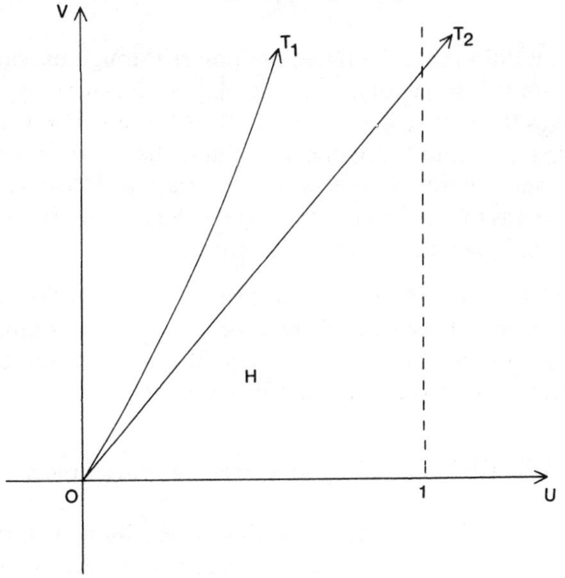

Fig. 4.79. Trajectories for the systems (4.77) and (4.78).

this is real and positive, so that T_2 is a straight line with positive slope if

$$a^2 > 4b \tag{4.80}$$

T_1 leaves the origin above T_2 if the positive eigenvalue of (4.77) is greater than the eigenvalue of T_2, that is, if

$$\{c+\sqrt{c^2-4f'(0)}\}/2 > \{a+\sqrt{a^2-4b}\}/2 \tag{4.81}$$

T_1 cannot subsequently cross to below T_2 if its gradient at any point is greater than that of T_2, that is, if

$$(cV-f(U))/V > (aV-bU)/V \tag{4.82}$$

But (4.82) is satisfied in H if $b = k = \sup_{U\in(0,1)} f(U)/U$ and $c > a$, (4.80) is satisfied if $a^2 > 4k$ and (4.81) is then satisfied for c sufficiently large, as required.

Consider now $c = 0$. The system of equations becomes

$$U' = V, \qquad V' = -f(U)$$

which has a first integral for the trajectory leaving the origin in the positive quadrant given by

$$\frac{U'^2}{2} + \int_0^U f(s)\,ds = 0$$

on using the condition that the trajectory passes through the origin. Hence $\int_0^U f(s)\,ds < 0$ on this trajectory, and since $\int_0^1 f(s)\,ds > 0$ by hypothesis, the trajectory leaves the half-strip $\{0 < U < 1, V > 0\}$ where $V = 0$ and $U < 1$. It follows by using a continuity argument similar to that for the Fisher equation that there is some intermediate c where the trajectory leaving the saddle point at $(0, 0)$ tends to $(1, 0)$ as $\xi \to \infty$. There is thus a wave front from $U = 0$ to $U = 1$ for a unique value of the wave speed c.

For a review of results in the bistable case see Fife (1979b). In the above discussion we have not considered the case in which f has more than one intermediate zero. Fife and McLeod (1977) derive conditions for the existence of travelling wave fronts in this case.

4.7 Global Stability of Travelling Wave Fronts

The question of stability of the travelling wave solutions in the previous section has been the subject of much research. Kolmogorov *et al.* (1937) proved that the unique bounded solution of the Cauchy problem for the

Fisher-type equation with $k=\sup_{u\in(0,1)} f(u)/u = f'(0)$ and initial conditions a Heaviside function tends to a travelling wave front $U(x+ct)$ in the following sense:

THEOREM 4.83: There exists a function $\psi(t)$ such that $\lim_{t\to\infty}\psi'(t)=0$ and

$$|u(x,t)-U(x+c^*t-\psi(t))|\to 0$$

as $t\to\infty$; $c^*=2\sqrt{f'(0)}$ is the wave speed of the previous section.

Since $\psi(t)\not\to 0$ as $t\to\infty$, the convergence to a wave front is not uniform; this type of convergence is known as convergence "in wave form". Larson (1978), following McKean (1975), Kanel' (1962) and Hoppensteadt (1975), showed that for other initial conditions convergence to travelling waves with speeds c satisfying $c>c^*$ was possible, and that the particular wave which resulted depended crucially on the behaviour of the initial conditions when these were close to $u=0$. For other stability results on such equations see Stokes (1976), Kametaka (1976), Sattinger (1976), who considers functions f depending on u_x as well as u, Bramson (1983) and Lau (1985).

In the Nagumo-type equation, uniform convergence to the unique (up to translation) wave front can be proved for a large class of initial data [see Fife and McLeod (1977)].

THEOREM 4.84: Let $f(0)=f(1)=0$, $f'(0)<0$, $f'(1)<0$, and let f have one interior zero at a. Then the bounded solution of the Cauchy problem

$$u_t=f(u)+u_{xx}$$

with initial data

$$u(x,0)=u_0(x), \qquad 0\leq u_0\leq 1,$$

satisfying

$$\limsup_{x\to-\infty} u_0(x)<a, \qquad \liminf_{x\to+\infty} u_0(x)>a$$

tends uniformly in x and exponentially in t to a travelling wave solution, i.e. there exist constants c, ξ_0, β and K such that

$$|u(x,t)-U(x+ct-\xi_0)|<Ke^{-\beta t}$$

where U is the travelling wave solution of the last section.

The proof of stability of U is by a comparison argument. The estimates of the rate of convergence are proved by introducing a Lyapunov function. [For details, see the original paper of Fife and McLeod (1977).] Jones (1983a, b) proves similar results in the case of more than one spatial dimension.

5. Analytic Techniques for Systems of Parabolic Partial Differential Equations

5.1 Introduction

In the last chapter the maximum principle and the comparison theorem for a single nonlinear diffusion equation were discussed. In the case of the initial-boundary-value problems these results led to existence and uniqueness theorems by supplying *a priori* bounds on the solution of the equation, and also to stability results as in the spruce budworm system and the discussion on travelling waves. The method is capable of extension to certain systems of parabolic partial differential equations, but has some limitations and in general leads to weaker results than in the case of the single equation.

Early work on such methods for systems of equations was carried out by McNabb (1961). More recently contributions have been made by Alikakos (1981), Amann (1978a,b), Capasso and Maddalena (1981, 1982a, 1983), Cerrai (1983), Cosner and Lazer (1984), Chandra and Davis (1979), Chueh *et al.* (1977), Conway and Smoller (1977a,b), Deimling and Lakshmikantham (1980), Fife (1979a), Fife and Tang (1981), Gardner (1981), Kuiper (1977, 1980), Lakshmikantham (1979, 1984), Pao (1980, 1981, 1984) and Weinberger (1975).

DEFINITION 5.1: If $\mathbf{u}, \mathbf{v} \in \mathbb{R}^m$, then $\mathbf{u} \geqslant \mathbf{v}$ if $u_i \geqslant v_i$ for each $i = 1, 2, \ldots, m$. The other order relations and the operators max, min, sup and inf, when referred to vectors, are also defined componentwise. Note that we shall always use m to denote the dimension of phase space and n to denote the dimension of the physical space in which the equations hold.

DEFINITION 5.2: The various mathematical problems are defined as for the scalar case (Definition 4.3). Solutions are defined as for the scalar case (Definition 4.4), as are regular and non-regular sub- and supersolutions (Definition 4.8), the inequalities holding componentwise as in Definition 5.1. Stability properties are defined as for the scalar case (Definitions 4.6).

DEFINITION 5.3: The boundary operator B is defined by

$$(B\mathbf{u})_i \equiv c_i(\mathbf{x}, t)u_i(\mathbf{x}, t) + d_i(\mathbf{x}, t)\,\partial u_i/\partial \nu_i(\mathbf{x}, t)$$

where $(B\mathbf{u})_i$ is the ith component of the vector $B\mathbf{u}$; $c_i \geqslant 0$, $d_i \geqslant 0$, $c_i^2 + d_i^2 > 0$ and $\partial/\partial\nu_i$ denotes any outward derivative. Note that B is a diagonal operator, i.e. $(B\mathbf{u})_i$ depends only on the ith component of $\mathbf{u}$.

DEFINITION 5.4: An *invariant set* $\Sigma \subset \mathbb{R}^m$ for the various mathematical problems is defined as for the scalar case (Definition 4.17). The condition on the boundary values is that

$$B\mathbf{u}(\mathbf{x}, t) \in \mathbf{c}(\mathbf{x}, t)\Sigma \text{ for each point } (\mathbf{x}, t) \in \Gamma$$

where the set $\mathbf{c}\Sigma = \mathbf{c}(\mathbf{x}, t)\Sigma$ is defined by

$$\mathbf{c}\Sigma = \{(c_1 v_1, c_2 v_2, \ldots, c_m v_m)^T \,|\, \mathbf{v} \in \Sigma\} \text{ for each point } (\mathbf{x}, t) \in \Gamma$$

DEFINITION 5.5: The function $\mathbf{f}(\mathbf{u})$ is *quasi-monotone non-decreasing* if each component function $f_i(\mathbf{u})$ is non-decreasing in u_j for each $j \neq i$. The function $\mathbf{f}(\mathbf{u})$ has the *mixed quasi-monotone property* if each component function $f_i(\mathbf{u})$ is monotonic in u_j for each $j \neq i$. Define $\mathbf{U}^i$ to be the vector of all elements of $\mathbf{u}$ except the ith, so that we may write $\mathbf{f}(\mathbf{u}) = \mathbf{f}(u_i, \mathbf{U}^i)$.

DEFINITION 5.6: For any scalar ε, we define the vector $\boldsymbol{\varepsilon}$ to be the vector $(\varepsilon, \varepsilon, \ldots, \varepsilon)^{\mathrm{T}}$ with all components equal to ε.

In the next three sections we shall consider the application of comparison theorems to systems of reaction–diffusion equations. We shall concentrate our attention on the initial-boundary-value problems for reaction–diffusion systems, although the techniques may be applied to other problems. Thus our equations will be defined in a spatio-temporal domain $Q_T \cup S_T$, where $Q_T = \Omega \times (0, T)$, $\Omega \subset \mathbb{R}^n$ is bounded, and we define $S_0 = \Omega \times \{0\}$, $S_T = \Omega \times \{T\}$ and $\Gamma = \partial\Omega \times (0, T)$. We shall consider systems of the form

$$N\mathbf{u} \equiv L\mathbf{u} - \mathbf{f}(\mathbf{u}) \equiv \mathbf{u}_t - D\,\nabla^2\mathbf{u} - \mathbf{f}(\mathbf{u}) = \mathbf{0} \quad \text{in} \quad Q_T \cup S_T \tag{5.7}$$

with initial conditions

$$\mathbf{u}(\mathbf{x}, 0) = \mathbf{u}_0(\mathbf{x}) \quad \text{on} \quad S_0 \tag{5.8}$$

and boundary conditions

$$B\mathbf{u} = \mathbf{b}(\mathbf{x}, t) \quad \text{on} \quad \Gamma \tag{5.9}$$

Here D is a diagonal diffusion matrix, whose diagonal elements need not all be non-zero, and B is a diagonal boundary operator. As in the scalar case, the results may easily be extended to functions $\mathbf{f}$ continuously dependent on $\mathbf{x}$ and t as well as $\mathbf{u}$ and to more general diagonal elliptic operators E. Each f_i may also depend on ∇u_i. Similar theorems may also be proved for non-diagonal elliptic operators and boundary operators and for more general dependence of $\mathbf{f}$ on $\nabla\mathbf{u}$ [see, for example, Chueh *et al.* (1977)], so that cross-diffusion may be included, but the results here are much less developed and in general difficult to apply.

An alternative technique for the analysis of the Cauchy problem, the method of fundamental solutions, is discussed in Section 5.5. In Section 5.6 we consider the stability of stationary solutions *via* spectral theory, and other methods and in Section 5.7 the asymptotic behaviour of solutions as $t \to \infty$ using the energy method.

5.2 Comparison Theorems

We may now state the comparison theorem for systems of reaction–diffusion equations. Throughout this section the problem (P) is defined to be the differential equations (5.7) with initial and boundary conditions (5.8) and (5.9).

THEOREM 5.10 (Comparison Theorem): If (i) $\underset{\sim}{\mathbf{u}}$ and $\tilde{\mathbf{u}}$ are sub- and supersolutions of (P), (ii) $\mathbf{f}$ is uniformly Lipschitz continuous in $\mathbf{u}$, (iii) Q_T satisfies the interior sphere property (unless we are concerned with the Dirichlet problem) and (iv) $\mathbf{f}$ is quasi-monotone non-decreasing, then

$$\underset{\sim}{\mathbf{u}} \leqslant \tilde{\mathbf{u}} \quad \text{in} \quad \bar{Q}_T$$

This is the weak comparison theorem. Moreover, the strong comparison theorem states that for each i either

(a) $\underset{\sim}{u}_i < \tilde{u}_i$ in $Q_T \cup S_T$, or
(b) $\underset{\sim}{u}_i \equiv \tilde{u}_i$ in $Q_{t^*} \cup S_{t^*}$ for some $t^* \leqslant T$.

A boundary point lemma, as in Theorem 4.10, also holds for each i.

Proof: We shall prove both parts of the theorem for the case of regular sub- and supersolutions. The extension to non-regular sub- and supersolutions is straightforward.

The idea of the proof is to apply the scalar maximum principle to $w_i = \tilde{u}_i - \underset{\sim}{u}_i + y$ to show that $w_i > 0$, where $y \to 0$ as some parameter $\varepsilon \to 0$. Let $\mathbf{y} = (y, y, \ldots, y)^T$, and let $y = y(t)$ satisfy the initial value problem

$$\dot{y} = K\|\mathbf{y}\|, \qquad y(0) = \varepsilon$$

where K is the Lipschitz constant for f (e.g. $\dot{y} = K\sqrt{m}\, y$, $y = \varepsilon e^{K\sqrt{m}}$, if the Euclidean norm is being used). We must first show that

$$\mathbf{v} = \underset{\sim}{\mathbf{u}} - \mathbf{y}$$

is a subsolution for the differential equation $L\mathbf{v}=\mathbf{f}(\mathbf{v})$. But for each i,

$$\begin{aligned} L_i v_i &\equiv \partial v_i/\partial t - D_i \nabla^2 v_i = L_i \underset{\sim}{u}_i - \dot{y} \\ &\leqslant f_i(\underset{\sim}{\mathbf{u}}) - K\|\mathbf{y}\| = f_i(\mathbf{v}+\mathbf{y}) - K\|\mathbf{y}\| \\ &\leqslant f_i(\mathbf{v}) + K\|\mathbf{y}\| - K\|\mathbf{y}\| = f_i(\mathbf{v}) \end{aligned}$$

as required. Now let $\mathbf{w}=\tilde{\mathbf{u}}-\underset{\sim}{\mathbf{u}}+\mathbf{y}$. Then $\mathbf{w}(\mathbf{x},0)=\tilde{\mathbf{u}}(\mathbf{x},0)-\underset{\sim}{\mathbf{u}}(\mathbf{x},0)+\mathbf{y}(0)\geqslant \mathbf{y}(0)=\boldsymbol{\varepsilon}>0$, and for any boundary operator B, $B\mathbf{w}=B\tilde{\mathbf{u}}-B\underset{\sim}{\mathbf{u}}+B\mathbf{y}\geqslant B\mathbf{y}=\mathbf{c}y\geqslant \mathbf{0}$, since $\mathbf{y}$ is spatially uniform and positive and $\mathbf{c}\geqslant\mathbf{0}$. Since $\mathbf{w}>\mathbf{0}$ initially, we may define $\tau\in(0,T]$ such that $\mathbf{w}(\mathbf{x},t)\geqslant\mathbf{0}$ for $t\in[0,\tau]$. Then, in $(0,\tau]$,

$$\begin{aligned} L_i w_i &= L_i\tilde{u}_i - L_i\underset{\sim}{u}_i + L_i y = L_i\tilde{u}_i - L_i v_i \\ &\geqslant f_i(\tilde{\mathbf{u}}) - f_i(\mathbf{v}) = f_i(\tilde{u}_i, \tilde{\mathbf{U}}^i) - f_i(v_i, \mathbf{V}^i) \\ &\geqslant f_i(\tilde{u}_i, \mathbf{V}^i) - f_i(v_i, \mathbf{V}^i) \end{aligned}$$

since $\mathbf{f}$ is quasi-monotone non-decreasing, and $\mathbf{w}\geqslant\mathbf{0}$, and so

$$L_i w_i \geqslant -K(\tilde{u}_i - v_i) = -Kw_i$$

since $\mathbf{f}$ is Lipschitz continuous. It follows from the scalar maximum principle (Theorem 4.9) that $w_i>0$ (since $w_i>0$ initially), so that $\mathbf{w}>0$. Hence if $\tau\in(0,T]$ and $\mathbf{w}\geqslant\mathbf{0}$ in $[0,\tau]$, then $\mathbf{w}>\mathbf{0}$ in $[0,\tau]$. It follows that $\tau=T$, and thus $\mathbf{w}>\mathbf{0}$, or $\tilde{\mathbf{u}}>\mathbf{v}=\underset{\sim}{\mathbf{u}}-\mathbf{y}$, for $t\in[0,T]$. Since $\mathbf{y}\to 0$ as $\varepsilon\to 0$, it follows that $\tilde{\mathbf{u}}\geqslant\underset{\sim}{\mathbf{u}}$ in $\bar{Q}_T$. This is the weak comparison theorem.

For the strong comparison theorem, let $N_i\mathbf{u}=L_iu_i-f_i(\mathbf{u})$, and define $\mathbf{v}=\tilde{\mathbf{u}}-\underset{\sim}{\mathbf{u}}$. Then

$$\begin{aligned} 0 &\leqslant N_i\tilde{\mathbf{u}} - N_i\underset{\sim}{\mathbf{u}} = L_i\tilde{u}_i - f_i(\tilde{\mathbf{u}}) - L_i\underset{\sim}{u}_i + f_i(\underset{\sim}{\mathbf{u}}) \\ &= L_iv_i - \{f_i(\tilde{u}_i, \tilde{\mathbf{U}}^i) - f_i(\underset{\sim}{u}_i, \underset{\sim}{\mathbf{U}}^i)\} \leqslant L_iv_i - \{f_i(\tilde{u}_i, \underset{\sim}{\mathbf{U}}^i) - f_i(\underset{\sim}{u}_i, \underset{\sim}{\mathbf{U}}^i)\} \end{aligned}$$

since $\mathbf{f}$ is quasi-monotone non-decreasing and $\tilde{\mathbf{u}}\geqslant\underset{\sim}{\mathbf{u}}$,

$$\begin{aligned} &= L_iv_i - \frac{f_i(\tilde{u}_i, \underset{\sim}{\mathbf{U}}^i) - f_i(\underset{\sim}{u}_i, \underset{\sim}{\mathbf{U}}^i)}{\tilde{u}_i - \underset{\sim}{u}_i} v_i \\ &= L_iv_i - h_iv_i \end{aligned}$$

where h_i is a bounded function since $\mathbf{f}$ is uniformly Lipschitz continuous. It follows from the scalar strong maximum principle (Theorem 4.9) that either

(a) $\underset{\sim}{u}_i < \tilde{u}_i$ in $Q_T \cup S_T$, or
(b) $\underset{\sim}{u}_i \equiv \tilde{u}_i$ in $Q_{t^*} \cup S_{t^*}$, for some $t^* \leqslant T$.

The boundary point lemma also follows immediately. But these are the contentions of the theorem.

As in the scalar case, uniqueness follows immediately from the comparison theorem, and *a priori* bounds may be proved which result in existence of the solution for all time.

COROLLARY 5.11 (Uniqueness): If hypotheses (ii)–(iv) of Theorem 5.10 hold, then the initial-boundary-value problem (P) has at most one solution.

Proof: Assume it has two, $\mathbf{v}$ and $\mathbf{w}$. Then $\mathbf{v} \leqslant \mathbf{w}$ and $\mathbf{w} \leqslant \mathbf{v}$, so $\mathbf{v} = \mathbf{w}$.

COROLLARY 5.12 (Invariant Sets): If hypotheses (ii)–(iv) of Theorem 5.10 hold, and $-\infty \leqslant \mathbf{a} \leqslant \mathbf{b} \leqslant \infty$ are constant vectors such that $\mathbf{f}(\mathbf{a}) \geqslant \mathbf{0}$, $\mathbf{f}(\mathbf{b}) \leqslant \mathbf{0}$, then $\Sigma = \{\mathbf{u} \mid \mathbf{a} \leqslant \mathbf{u} \leqslant \mathbf{b}\}$ is an invariant set for the initial-boundary-value problem (P).

Proof: Immediate.

THEOREM 5.13 (Global Existence): If hypotheses (ii)–(iv) of Theorem 5.10 hold, and there exists an invariant set $\Sigma = \{\mathbf{u} \mid \mathbf{a} \leqslant \mathbf{u} \leqslant \mathbf{b}\}$ for the initial-boundary-value problem (P), with $\mathbf{a}$ and $\mathbf{b}$ finite, then this problem has a solution for all time.

REMARK 5.13a: We emphasise that our proof is constructive, and may be used as a basis for the numerical calculation of a solution in any given case.

Proof: We define the sequences $\{\mathbf{v}^n\}$ and $\{\mathbf{w}^n\}$ by

$$\mathbf{v}^0 = \mathbf{a}, \qquad (L_i + K)v_i^n = f_i(\mathbf{v}^{n-1}) + Kv_i^{n-1}$$
$$\mathbf{w}^0 = \mathbf{b}, \qquad (L_i + K)w_i^n = f_i(\mathbf{w}^{n-1}) + Kw_i^{n-1}$$

with boundary and initial conditions (5.8) and (5.9). The finiteness of $\mathbf{a}$ and $\mathbf{b}$ is required here. Then we make the inductive hypothesis that $\mathbf{v}^{n+1} - \mathbf{v}^n \geqslant \mathbf{0}$. Since

$$\begin{aligned}(L_i + K)(v_i^1 - v_i^0) &= f_i(\mathbf{v}^0) + Kv_i^0 - L_i v_i^0 - Kv_i^0 \\ &= -(L_i v_i^0 - f_i(\mathbf{v}^0)) = f_i(\mathbf{a}) \geqslant 0\end{aligned}$$

this is true by the scalar maximum principle for $n = 0$. Assume it is true up to $n-1$. Then

$$\begin{aligned}(L_i + K)(v_i^{n+1} - v_i^n) &= f_i(\mathbf{v}^n) + Kv_i^n - f_i(\mathbf{v}^{n-1}) - Kv_i^{n-1} \\ &\geqslant -K(v_i^n - v_i^{n-1}) + K(v_i^n - v_i^{n-1}) = 0\end{aligned}$$

using the monotonicity and Lipschitz properties of $\mathbf{f}$. Thus $\mathbf{v}^{n+1} - \mathbf{v}^n \geqslant 0$, as required. Similarly $\mathbf{w}^{n+1}-\mathbf{w}^n \leqslant \mathbf{0}$. It follows that $\{\mathbf{v}^n\}$ and $\{\mathbf{w}^n\}$ are monotonic increasing and decreasing, respectively, and hence that $\mathbf{v}=\lim_{n\to\infty}\mathbf{v}^n$ and $\mathbf{w}=\lim_{n\to\infty}\mathbf{w}^n$ exist. As in the scalar case (Chapter 4), $\mathbf{v}=\mathbf{w}$ is the unique solution of the problem.

Quasi-monotonicity is an essential requirement for the comparison theorem. For consider the following problem:

$$u_t = u + u_{xx}, \qquad v_t = -u + v + v_{xx}$$

in $(x_1, x_2)\times(0,\infty)$, with constant initial conditions $u=u_0$, $v=v_0$, and boundary conditions $u_x = v_x = 0$. The solution is given by

$$u = u_0 e^t, \qquad v = -u_0 t e^t + v_0 e^t$$

Thus an increase in the initial conditions for u and v leads ultimately to a decrease in v, so that a comparison theorem cannot hold. However, the uniqueness result does not depend on quasi-monotonicity.

THEOREM 5.14 (Uniqueness): The initial-boundary-value problem (P), where $\mathbf{f}$ is uniformly Lipschitz continuous and Q_T satisfies the interior sphere property (unless we are concerned with the Dirichlet problem), has at most one solution.

Proof: Let $\mathbf{v}$ and $\mathbf{w}$ be two solutions of (P), so that

$$L\mathbf{v}=\mathbf{f}(\mathbf{v}), \qquad L\mathbf{w}=\mathbf{f}(\mathbf{w}) \qquad \text{in} \quad Q_T \cup S_T$$

Define y to be the solution of the initial value problem

$$\dot{y} = K\|\mathbf{y}\|, \qquad y(0)=\varepsilon>0$$

where $\mathbf{y}=(y, y, \ldots, y)^T$ and K is the Lipschitz constant for $\mathbf{f}$, and let $p_i = v_i - w_i - y$, $q_i = w_i - v_i - y$. Then $p_i = q_i = -\varepsilon < 0$ initially for all i, so that there is a $\tau^* \in (0, T)$ such that $p_i < 0$ and $q_i < 0$, i.e. $|w_i - v_i| < y$, for all $t \in (0, \tau^*)$. Moreover, $B_i p_i \leqslant 0$ and $B_i q_i \leqslant 0$ on Γ, and in $Q_{\tau^*}\cup S_{\tau^*}$

$$\begin{aligned} L_i p_i &= L_i v_i - L_i w_i - L_i y = f_i(\mathbf{v}) - f_i(\mathbf{w}) - \dot{y} \\ &\leqslant K\|\mathbf{v}-\mathbf{w}\| - K\|\mathbf{y}\| \leqslant K\|\mathbf{y}\| - K\|\mathbf{y}\| = 0 \end{aligned}$$

and

$$L_i q_i \leqslant 0$$

It follows from the scalar maximum principle (Theorem 4.9) that $p_i < 0$, $q_i < 0$ in $Q_{\tau^*}\cup S_{\tau^*}$, (since they are strictly negative initially), so that there can be no time at which p_i or q_i first become equal to zero, and $p_i < 0$, $q_i < 0$ for all $t \leqslant T$. Thus $|v_i - w_i| < y$ for all $t \leqslant T$. But $y = \varepsilon e^{K\sqrt{m}\,t}$, (using the Euclidean norm), and ε is arbitrarily small, so $|v_i - w_i| = 0$, and $\mathbf{v}=\mathbf{w}$ as required.

Invariant sets Σ may also be proved to exist without the requirement of quasi-monotonicity, although the behaviour of $\mathbf{f}$ at the boundaries of Σ must then be more tightly controlled. This will be considered in the next section. In the presence of such sets a constructive proof of existence of solutions similar to Theorem 5.13 may be given in certain cases if the vector field $\mathbf{f}$ possesses the mixed quasi-monotone property (Definition 5.5). Bounds for the solutions may also be found. For details and applications see Deimling and Lakshmikantham (1980), Ladde *et al.* (1985), Grindrod and Sleeman (1984), Leung (1984), Pao (1982, 1985), and Terman (1983).

5.3 Invariant Sets

The requirement that a system be quasi-monotone non-decreasing is a very restrictive one, and we should therefore like to extend the use of the comparison theorem beyond such systems. To this end we state the following corollary of Theorem 5.10.

COROLLARY 5.15: Let $\underset{\sim}{\mathbf{f}}:\mathbb{R}^m \to \mathbb{R}^m$ be a quasi-monotone non-decreasing uniformly Lipschitz continuous function (to be constructed) satisfying $\underset{\sim}{\mathbf{f}}(\mathbf{v}) \leqslant \mathbf{f}(\mathbf{v})$ for all $\mathbf{v} \in \mathbb{R}^m$. Recalling that the problem (P) is (5.7) with (5.8) and (5.9), define ($\underset{\sim}{\mathrm{P}}$) to be (P) with $\mathbf{f}$ replaced by $\underset{\sim}{\mathbf{f}}$. Let $\underset{\sim}{\mathbf{u}}$ be a subsolution of ($\underset{\sim}{\mathrm{P}}$), and let $\mathbf{u}$ be a supersolution of (P). Then $\underset{\sim}{\mathbf{u}} \leqslant \mathbf{u}$ in $\bar{Q}_T$. Moreover, the corresponding strong comparison theorem holds. A similar statement holds for a supersolution of the problem ($\tilde{\mathrm{P}}$) with a super-reaction function $\tilde{\mathbf{f}}$.

Proof: Since $\mathbf{u}$ is a supersolution of (P),

$$L\mathbf{u} \geqslant \mathbf{f}(\mathbf{u}) \geqslant \underset{\sim}{\mathbf{f}}(\mathbf{u})$$

so that it is a supersolution of ($\underset{\sim}{\mathrm{P}}$). Since $\underset{\sim}{\mathbf{f}}$ has the required monotonicity properties, the maximum principle applies to $\underset{\sim}{\mathbf{u}}$ and $\mathbf{u}$, so that $\underset{\sim}{\mathbf{u}} \leqslant \mathbf{u}$ as claimed.

This corollary gives no hint as to how to construct the quasi-monotone bounding function $\underset{\sim}{\mathbf{f}}$. It is often profitable to define it as the infimum of $\mathbf{f}$ over an appropriate set, making use of the following lemma.

LEMMA 5.16: Let $\underset{\sim}{\mathbf{f}}$ be defined by

$$\underset{\sim}{f}_i(\mathbf{u}) = \inf_{\{\mathbf{v} | \mathbf{u} \leqslant \mathbf{v} \leqslant \tilde{\mathbf{u}}, v_i = u_i\}} f_i(\mathbf{v})$$

for some supersolution $\tilde{\mathbf{u}}$. Then $\mathbf{f}(\mathbf{u}) \geqslant \underset{\sim}{\mathbf{f}}(\mathbf{u})$ for all $\mathbf{u}$, $\underset{\sim}{\mathbf{f}}$ is quasi-monotone non-decreasing and is uniformly Lipschitz continuous if $\mathbf{f}$ is, with the same Lipschitz constant.

Proof: The proof is straightforward and is omitted.

Similarly, $\tilde{\mathbf{f}}$ may be defined by

$$\tilde{f}_i(\mathbf{u}) = \sup_{\{\mathbf{v}|\underset{\sim}{\mathbf{u}} \leq \mathbf{v} \leq \mathbf{u}, v_i = u_i\}} f_i(\mathbf{v})$$

for some subsolution $\underset{\sim}{\mathbf{u}}$.

We may now introduce an important application of the maximum principle in reaction–diffusion systems, which is in proving the existence of invariant sets. In the case of invariant rectangles, the comparison functions are simply constant vectors, and we may prove the following theorem.

THEOREM 5.17 (Invariant Rectangles): Let $\mathbf{f}$ be Lipschitz continuous and let Q_T satisfy the interior sphere property (unless we are considering the Dirichlet problem), and let $-\infty < \mathbf{a} < \mathbf{b} < \infty$ be two constant vectors in $\mathbb{R}^m$ such that

$$f_i(\mathbf{v}) \geq 0 \quad \text{for} \quad v_i = a_i, \qquad \mathbf{a} \leq \mathbf{v} \leq \mathbf{b}$$

$$f_i(\mathbf{v}) \leq 0 \quad \text{for} \quad v_i = b_i, \qquad \mathbf{a} \leq \mathbf{v} \leq \mathbf{b}$$

Then $\Sigma = \{\mathbf{v} | \mathbf{a} \leq \mathbf{v} \leq \mathbf{b}\} \subset \mathbb{R}^m$ is an invariant set for the initial-boundary-value problem (P), that is, (5.7) with (5.8) and (5.9).

Proof: For $\mathbf{u} \geq \mathbf{a}$, define $\tilde{\mathbf{f}}$ by

$$\tilde{f}_i(\mathbf{u}) = \sup_{\{\mathbf{v}|\mathbf{a} \leq \mathbf{v} \leq \mathbf{u}, v_i = u_i\}} f_i(\mathbf{v})$$

For $\mathbf{u} \leq \mathbf{b}$, define $\underset{\sim}{\mathbf{f}}$ by

$$\underset{\sim}{f}_i(\mathbf{u}) = \inf_{\{\mathbf{v}|\mathbf{u} \leq \mathbf{v} \leq \mathbf{b}, v_i = u_i\}} f_i(\mathbf{v})$$

Clearly $\tilde{\mathbf{f}}$ and $\underset{\sim}{\mathbf{f}}$ are quasi-monotone non-decreasing in their domains of definition. Moreover, by the hypotheses of the theorem,

$$\underset{\sim}{f}_i(\mathbf{a}) \geq 0, \quad \tilde{f}_i(\mathbf{b}) \leq 0$$

for each i (so that $\mathbf{a}$ is a subsolution of ($\underset{\sim}{\mathrm{P}}$), the problem (P) with $\mathbf{f}$ replaced by $\underset{\sim}{\mathbf{f}}$, and $\mathbf{b}$ is a supersolution of ($\tilde{\mathrm{P}}$)). Let $\Sigma' = \{\mathbf{v} | \mathbf{a}' \leq \mathbf{v} \leq \mathbf{b}'\}$ be a bounded rectangle in $\mathbb{R}^m$, with $\mathbf{a}' < \mathbf{a}$, $\mathbf{b}' > \mathbf{b}$. Since $\mathbf{f}$ is Lipschitz continuous it is uniformly Lipschitz in Σ'; let the Lipschitz constant be K. Let $\mathbf{y} = (y, y, \ldots, y)^T$, and let y satisfy the initial value problem

$$\dot{y} = K\|\mathbf{y}\| + 2Ky, \quad y(0) = \varepsilon > 0$$

Choose ε so small that $\{\mathbf{v} | \mathbf{a} - 2\mathbf{y} \leq \mathbf{v} \leq \mathbf{b} + 2\mathbf{y}\} \subset \Sigma'$ for all $t \in (0, T]$. Define

$\boldsymbol{\phi} = \mathbf{u} - \mathbf{a} + \mathbf{y}$, and $\boldsymbol{\psi} = \mathbf{u} - \mathbf{b} - \mathbf{y}$. It is easy to show that if the initial and boundary data for $\mathbf{u}$ satisfy $\mathbf{u} \in \Sigma$ initially, $B\mathbf{u} \in c\Sigma$ on the boundary Γ, then $\boldsymbol{\phi} > 0$ and $\boldsymbol{\psi} < 0$ at $t = 0$, and $B\boldsymbol{\phi} \geq 0$ and $B\boldsymbol{\psi} \leq 0$ on Γ. Let τ^* be such that $\boldsymbol{\phi} \geq 0$, $\boldsymbol{\psi} \leq 0$ in $[0, \tau^*]$. Then, in $Q_{\tau^*} \cup S_{\tau^*}$,

$$
\begin{aligned}
L_i\phi_i &= L_i u_i + \dot{y} \\
&= f_i(\mathbf{u}) + K\|\mathbf{y}\| + 2Ky \\
&\geq f_i(\mathbf{u} - \mathbf{y}) - K\|\mathbf{y}\| + K\|\mathbf{y}\| + 2Ky \\
&\geq \underset{\sim}{f}_i(\mathbf{u} - \mathbf{y}) + 2Ky \\
&\geq \underset{\sim}{f}_i(\mathbf{u} - \mathbf{y}) - \underset{\sim}{f}_i(\mathbf{a}) + 2Ky \\
&\geq -K|u_i - y - a_i| + K|2y| \\
&\geq -K|u_i + y - a_i| = -K\phi_i,
\end{aligned}
$$

so that by the scalar maximum principle $\phi_i > 0$ in $[0, \tau^*]$ (since it is strictly positive initially). Thus $\boldsymbol{\phi} > \mathbf{0}$, and similarly $\boldsymbol{\psi} < \mathbf{0}$, in $[0, \tau^*]$. It follows that $\tau^* = T$ and since y is as small as we like $\mathbf{u} - \mathbf{a} \geq \mathbf{0}$, $\mathbf{u} - \mathbf{b} \leq \mathbf{0}$ in $[0, T]$, and Σ is an invariant set for (P), as required.

REMARK 5.17a: Note that it is sufficient to define $\underset{\sim}{\mathbf{f}}$ and $\tilde{\mathbf{f}}$ in the putative invariant set, and then show that this set is indeed invariant. Similarly, if the intention is to bound a solution simultaneously between putative non-constant sub- and supersolutions $\underset{\sim}{\mathbf{u}}$ and $\tilde{\mathbf{u}}$ then it is sufficient to define $\tilde{\mathbf{f}}$ and $\underset{\sim}{\mathbf{f}}$ by

$$
\tilde{f}_i(\underset{\sim}{\mathbf{u}}) = \sup_{\{\mathbf{v} | \underset{\sim}{\mathbf{u}} \leq \mathbf{v} \leq \mathbf{u}, v_i = u_i\}} f_i(\mathbf{v})
$$

$$
\underset{\sim}{f}_i(\underset{\sim}{\mathbf{u}}) = \inf_{\{\mathbf{v} | \mathbf{u} \leq \mathbf{v} \leq \tilde{\mathbf{u}}, v_i = u_i\}} f_i(\mathbf{v}),
$$

and then to show that indeed $\underset{\sim}{\mathbf{u}} \leq \mathbf{u} \leq \tilde{\mathbf{u}}$.

Note that this proof fails if any of the components of $\mathbf{a}$ or $\mathbf{b}$ are infinite, for in this case the supremum or the infimum are taken over infinite sets and therefore may not exist. An example of this is discussed in Section 5.4. Global existence of solutions again follows if such an invariant set exists.

THEOREM 5.18 (Global Existence): Let the conditions of Theorem 5.17 hold. Then the initial-boundary-value problem (P) with initial conditions $\mathbf{u}_0 \in \varepsilon\Sigma$ and boundary conditions $B\mathbf{u} \in c\Sigma$ has a (unique, by Theorem 5.14) global solution.

Proof: See Amann (1978b).

As an example of the use of Theorem 5.17, consider the simple ecological model for competing species (Volterra, 1926) with self-regulation (logistic-type terms) and diffusion

$$\begin{aligned}\partial u_1/\partial t &= u_1(1-u_1-\alpha u_2)+D_1\nabla^2 u_1 = f_1(u_1, u_2)+D_1\nabla^2 u_1\\ \partial u_2/\partial t &= \theta u_2(1-\beta u_1-u_2)+D_2\nabla^2 u_2 = f_2(u_1, u_2)+D_2\nabla^2 u_2\end{aligned} \tag{5.19}$$

in $Q_T \cup S_T$, where θ, α, $\beta > 0$, with zero-flux boundary conditions

$$\partial u_1/\partial n(\mathbf{x}, t) = \partial u_2/\partial n(\mathbf{x}, t) = 0 \qquad \text{on} \quad \Gamma \tag{5.20}$$

and initial data

$$u_1(\mathbf{x}, 0) = u_{1,0}(\mathbf{x}), \qquad u_2(\mathbf{x}, 0) = u_{2,0}(\mathbf{x}) \qquad \text{on} \quad S_0 \tag{5.21}$$

Then we have

THEOREM 5.22: If the initial data are non-negative and bounded, that is, if

$$\mathbf{0} \leqslant (u_{1,0}, u_{2,0}) \leqslant \mathbf{M} = (M_1, M_2) \tag{5.23}$$

then the logistic diffusional Lotka–Volterra system (5.19) with zero-flux boundary conditions (5.20) has solutions satisfying

$$\mathbf{0} \leqslant (u_1, u_2) \leqslant \mathbf{b} = (b_1, b_2) \tag{5.24}$$

where $b_1 = \max(M_1, 1)$, $b_2 = \max(M_2, 1)$.

Proof: The conditions for Theorem 5.17 hold with $\mathbf{a} = \mathbf{0}$ and $\mathbf{b}$ as given. For when $u_1 = 0$, $f_1(u_1, u_2) = f_1(0, u_2) = 0$ which is non-negative, and when $u_2 = 0$, $f_2(u_1, u_2) = f_2(u_1, 0) = 0$. When $u_1 = b_1$, $f_1(u_1, u_2) = f_1(b_1, u_2) = b_1(1-b_1-\alpha u_2) \leqslant 0$, since $b_1 \geqslant 1$, and when $u_2 = b_2$, $f_2(u_1, u_2) = f_2(u_1, b_2) = \theta b_2(1-\beta u_1 - b_2) \leqslant 0$, since $b_2 \geqslant 1$. The requirements on the boundary conditions are satisfied, since

$$\partial \mathbf{a}/\partial n = \partial \mathbf{u}/\partial n = \partial \mathbf{b}/\partial n = \mathbf{0}$$

and those on the initial conditions, since

$$\mathbf{a} = \mathbf{0} \leqslant (u_{1,0}, u_{2,0}) \leqslant \mathbf{M} \leqslant \mathbf{b}$$

Invariant sets other than rectangles are hard to find, since invariance in this case does not follow so simply from the spatially non-uniform problem. The following theorem (Chueh *et al.*, 1977) sheds some light on the reasons for this.

THEOREM 5.25: Let $G_i: \mathbb{R}^m \to \mathbb{R}$, $i = 1, 2, \ldots, p$, be smooth functions, and let $\Sigma = \{\mathbf{u} \mid G_i(\mathbf{u}) \leq 0 \text{ for all } i = 1, 2, \ldots, p\} \subset \mathbb{R}^m$ be an invariant set for the system

$$\mathbf{u}_t = \mathbf{f}(\mathbf{u}) + D\,\nabla^2\mathbf{u} \qquad \text{in} \quad Q_T \cup S_T$$

with boundary conditions

$$B\mathbf{u}(\mathbf{x}, t) = \mathbf{b}(\mathbf{x}, t) \qquad \text{on} \quad \Gamma$$

and initial conditions

$$\mathbf{u}(\mathbf{x}, 0) = \mathbf{u}_0(\mathbf{x}) \qquad \text{on} \quad S_0$$

Then the following conditions are satisfied, where $\mathbf{u}^* \in \partial\Sigma$ is such that $G_i(\mathbf{u}^*) = 0$, and ∇ represents the gradient with respect to the $\mathbf{u}$ variables:

(i) $\nabla G_i(\mathbf{u}^*).\ \mathbf{f}(\mathbf{u}^*) \leq 0$.
(ii) Σ is convex.
(iii) $\nabla G_i(\mathbf{u}^*)$ is a left eigenvector of D.

REMARK 5.26: If Σ is the rectangle $\mathbf{a} \leq \mathbf{u} \leq \mathbf{b}$, then there are $2m$ functions $G_i(\mathbf{u})$ given by

$$\begin{aligned} G_{2i-1}(\mathbf{u}) &= a_i - u_i, \qquad i = 1, 2, \ldots, m \\ G_{2i}(\mathbf{u}) &= u_i - b_i, \qquad i = 1, 2, \ldots, m \end{aligned} \tag{5.27}$$

since it is easy to see that $\Sigma = \{\mathbf{u} \mid G_i(\mathbf{u}) \leq 0 \text{ for all } i = 1, 2, \ldots, 2m\}$.

REMARK 5.28: The first conclusion of the theorem states that the vector field of $\mathbf{f}$ does not point out of the set Σ, and it is therefore necessary to ensure that spatially uniform solutions remain in Σ. This is easily seen in the case of the rectangle defined by the functions G_i in (5.27), since here

$$\begin{aligned} \nabla G_{2i-1}(\mathbf{u}^*) &= -\mathbf{e}_i = \mathbf{n} \qquad \text{if} \quad G_{2i-1}(\mathbf{u}^*) = 0, \qquad \text{i.e.} \quad u_i^* = a_i, \\ \nabla G_{2i}(\mathbf{u}^*) &= \mathbf{e}_i = \mathbf{n} \qquad \text{if} \quad G_{2i}(\mathbf{u}^*) = 0, \qquad \text{i.e.} \quad u_i^* = b_i, \end{aligned} \tag{5.29}$$

where $\mathbf{e}_i$ is the ith unit vector and $\mathbf{n}$ is the unit outward normal to the set Σ. Thus condition (i) becomes $\mathbf{n} \cdot \mathbf{f} \leq 0$ (cf. Lemma 2.19).

REMARK 5.30: The second conclusion of the theorem is necessary because of the nature of diffusion as an averaging operator.

Proof of (iii): The third conclusion of the theorem presents the most problems to satisfy. To see why it must be satisfied, note that at any boundary point $\mathbf{u}^*$, where $G_i(\mathbf{u}^*) = 0$, any solution $\mathbf{u}$ with data in Σ must be changing in such a way as to decrease G_i (since Σ is an invariant set), that is,

$\nabla G_i \cdot \mathbf{u}_t \le 0$. But if $\nabla G_i(\mathbf{u}^*)$ is not a left eigenvector of D, we can find a vector $\boldsymbol{\phi}$ such that $\nabla G_i(\mathbf{u}^*) \cdot \boldsymbol{\phi} < 0$, but $\nabla G_i(\mathbf{u}^*) \cdot D\boldsymbol{\phi} > 0$. Consider the solution of the initial-boundary-value problem in a small domain containing the origin, with initial data

$$\mathbf{u}(\mathbf{x}, 0) = \mathbf{u}^* + \lambda|\mathbf{x}|^2\boldsymbol{\phi}/2 = \mathbf{u}_0(\mathbf{x})$$

where $\lambda > 0$ is to be determined. Then

$$G_i(\mathbf{u}_0(\mathbf{x})) = G_i(\mathbf{u}^*) + \lambda|\mathbf{x}|^2 \nabla G_i(\mathbf{u}^*) \cdot \boldsymbol{\phi}/2 + o(|\mathbf{x}|^2) \le 0$$

for $|\mathbf{x}|$ sufficiently small, which proves that the initial data $\mathbf{u}_0 \in \Sigma$. However, at $(\mathbf{x}, t) = (\mathbf{0}, 0)$,

$$\nabla G_i(\mathbf{u}_0) \cdot \mathbf{u}_t = \lambda \nabla G_i(\mathbf{u}^*) \cdot D\boldsymbol{\phi} + \nabla G_i(\mathbf{u}^*) \cdot \mathbf{f} > 0$$

if λ is sufficiently large, so that $\mathbf{u}$ immediately moves out of Σ. Clearly we may choose a domain so small that both inequalities hold, so that this contradicts the assumption that Σ is an invariant set.

REMARK 5.31: If Σ is the rectangle $\mathbf{a} \le \mathbf{u} \le \mathbf{b}$, then the functions $G_i(\mathbf{u})$ are given by (5.27), and $\nabla G_{2i-1} = -\mathbf{e}_i$, $\nabla G_{2i} = \mathbf{e}_i$ from (5.29). If D is a diagonal matrix (which we shall assume henceforth), ∇G_i is therefore a left eigenvector of D for each i. If Σ is not a rectangle, this is not in general true.

REMARK 5.32: Σ is an invariant set if there is a comparison theorem for each G_i, because this implies that if $G_i \le 0$ initially and on the boundaries, then $G_i \le 0$ everywhere. If D is diagonal, then such theorems may be proved for each u_i, since the diffusion term in the equation for u_i is $D_i \nabla^2 u_i$ and is an averaging operator, and hence may be proved for each G_i in (5.27). However, diffusion may not be an averaging operator for more general functions G_i, so that comparison theorems cannot be proved.

There is an exceptional case when D is a scalar matrix so that all vectors are its eigenvectors, but this is unrealistic if, for example, D is a matrix of diffusion coefficients of various chemical substances, since *exactly* equal diffusion coefficients do not occur in practice.

Further results on invariant sets may be found in Redheffer and Walter (1978), who consider mixed boundary conditions, Chueh *et al.* (1977) and Smoller (1983). When invariant sets cannot be found, global existence theorems may be proved using other *a priori* bounds [see e.g. Rothe (1982, 1984)]. It is of interest to note that invariant sets do exist for many systems of reaction–diffusion equations modelling biological phenomena, for example the Hodgkin–Huxley (1952) equation for nerve conduction, the Fitzhugh–Nagumo system of Chapter 4, the spruce budworm equation of

Ludwig *et al.* (1978), realistic predator–prey models and the Field–Noyes model for the Belousov–Zhabotinskii reaction (the Oregonator, Chapter 2). See Smoller (1983) for some proofs.

5.4 Nested Rectangles and Time-dependent Comparison Functions

So far we have considered constant bounds on $\mathbf{u}$. Systems with reaction terms independent of $\mathbf{x}$ and zero-flux boundary conditions have solutions which are time-dependent but spatially independent, and these may often be used as comparison functions for such systems.

THEOREM 5.33: Let $\mathbf{u}$ be a solution of the system

$$\mathbf{u}_t = \mathbf{f}(\mathbf{u}) + D\,\nabla^2\mathbf{u} \qquad \text{in} \quad Q_T \cup S_T$$

where D is a diagonal diffusion matrix, $\mathbf{f}$ is uniformly Lipschitz continuous and quasi-monotone non-decreasing, and Q_T satisfies the interior sphere property, with zero-flux boundary conditions

$$\partial\mathbf{u}/\partial n = 0 \qquad \text{on} \quad \Gamma$$

and initial conditions

$$\mathbf{u}(\mathbf{x}, 0) = \mathbf{u}_0(\mathbf{x}) \qquad \text{on} \quad S_0$$

where $\mathbf{v}_0 \leqslant \mathbf{u}_0(\mathbf{x}) \leqslant \mathbf{w}_0$. Define $\mathbf{v}(t)$ to be the solution of the kinetic system $\mathbf{v}_t = \mathbf{f}(\mathbf{v})$ satisfying $\mathbf{v}(0) = \mathbf{v}_0$, and $\mathbf{w}(t)$ that satisfying $\mathbf{w}(0) = \mathbf{w}_0$. Then

$$\mathbf{v}(t) \leqslant \mathbf{u}(\mathbf{x}, t) \leqslant \mathbf{w}(t) \qquad \text{in} \quad \bar{Q}_T$$

Moreover, the corresponding strong comparison theorem holds.

Proof: The theorem follows directly from two applications of the comparison theorem, in the first taking $\mathbf{v}$ as a subsolution and $\mathbf{u}$ as a supersolution, and in the second taking $\mathbf{u}$ as a subsolution and $\mathbf{w}$ as a supersolution. It is stated as a theorem in its own right because of its importance in applications.

As an example of its use, consider the ecological competition model (5.19)

$$\partial u_1/\partial t = u_1(1 - u_1 - \alpha u_2) + D_1\,\nabla^2 u_i = f_1(u_1, u_2) + D_1\,\nabla^2 u_1$$

$$\partial u_2/\partial t = \theta u_2(1 - \beta u_1 - u_2) + D_2\,\nabla^2 u_2 = f_2(u_1, u_2) + D_2\,\nabla^2 u_2$$

with zero-flux boundary conditions. We have already shown (Theorem 5.22) that $\mathbf{u} \geqslant \mathbf{0}$ is an invariant set for this system. Therefore, the system with non-negative initial conditions is quasi-monotone, since $\partial f_1/\partial u_2 = -\alpha u_1 \leqslant 0$, and $\partial f_2/\partial u_1 = -\theta\beta u_2 \leqslant 0$, but is quasi-monotone non-increasing rather than non-decreasing, and the theorem therefore cannot be applied directly. However the change of variables $u_1^* = -u_1$, $u_2^* = u_2$ leads to

$$\partial u_1^*/\partial t = u_1^*(1+u_1^* - \alpha u_2^*) + D_1 \nabla^2 u_1^* = f_1^*(u_1^*, u_2^*) + D_1 \nabla^2 u_1^*$$

$$\partial u_2^*/\partial t = \theta u_2^*(1+\beta u_1^* - u_2^*) + D_2 \nabla^2 u_2^* = f_2^*(u_1^*, u_2^*) + D_2 \nabla^2 u_2^*$$

which has an invariant set $\{\mathbf{u}^* \mid -\infty < u_1^* \leq 0, 0 \leq u_2^* < \infty\}$. The system with initial conditions is now quasi-monotone non-decreasing, since $\partial f_1^*/\partial u_2^* = -\alpha u_1^* \geq 0$, and $\partial f_2^*/\partial u_1^* = \theta\beta u_2^* \geq 0$. Such a change of variables was used by Pao (1981).

The comparison theorem may now be applied to the starred system, and the results carried across to the original one. It may be used, for example, to prove the following theorem:

THEOREM 5.34 [see also Zhou and Pao (1982)]: Consider the system (5.19) with zero-flux boundary conditions and non-negative initial conditions. Let $\alpha < 1$ and $\beta > 1$. Then the non-trivial homogeneous steady state is asymptotically stable.

Proof: The phase plane for the kinetic equations is shown in Fig. 5.35. A straightforward analysis shows that the rectangle $\Sigma = P_1Q_1P_2Q_2$ is an invariant set for the kinetic system. To attempt to use the comparison theorem on the system as it stands is to attempt to bound the solutions of

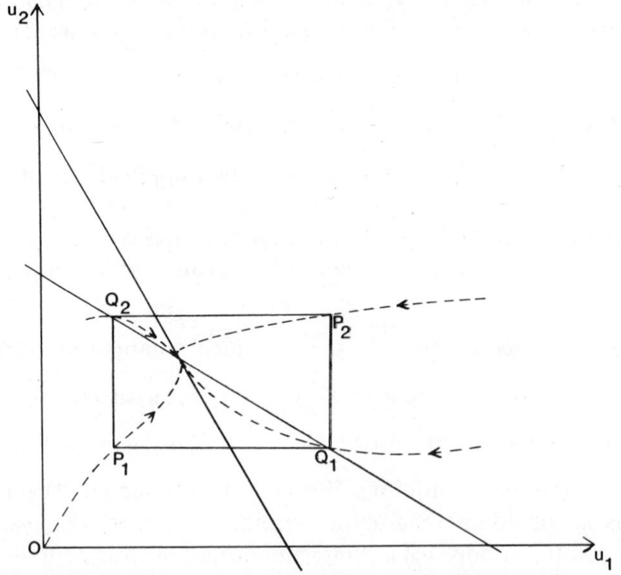

Fig. 5.35. The phase plane for spatially independent solutions of the system (5.19).

(5.19) with initial conditions in Σ by the spatially uniform solutions with initial conditions at P_1 and P_2, respectively. This cannot work if the trajectories are as shown since the value of u_1 on the trajectory from P_2 becomes smaller than its value on the trajectory from P_1. This corresponds mathematically to the inapplicability of the theorem due to the incorrect monotonicity properties of $\mathbf{f}$. The change of variables $u_1^* = -u_1$, $u_2^* = u_2$ corresponds to a reflection of the phase plane in the u_2 axis. The bounding functions for the comparison theorem are now the trajectories passing through Q_1 and Q_2, or to be more accurate, their reflections, which tend monotonically to the steady state. These are valid bounding functions (since the system now has the correct monotonicity properties) and hence the solution of the full reaction–diffusion system also tends to the steady state, as required.

Note that the proof depends in an essential way on the fact that the stable steady state in the kinetic system is a node. If it were a focus, then it would not be possible to find monotonic functions corresponding to those starting at Q_1 and Q_2 to bound the solution of the reaction–diffusion system. In this case, then, it is possible for diffusion to de-stabilise the steady state. This phenomenon of diffusive instability is discussed further in Chapter 6.

Comparison techniques are therefore valuable in proving asymptotic stability of constant solutions of reaction–diffusion systems. They may also be used in considering the stability of non-constant solutions. Klaasen and Troy (1981) use them together with Schauder estimates to prove the stability of travelling wave solutions of a certain class of reaction–diffusion equations. The proof is similar to that of Fife and McLeod (1977) in the scalar case.

Theorem 5.33 may be extended in the same way as Theorem 5.10 to systems which are not quasi-monotone non-decreasing, by constructing sub- and super-reaction functions $\underset{\sim}{\mathbf{f}}$ and $\tilde{\mathbf{f}}$. It can then be used to construct invariant sets in quite general classes of prey–predator and other ecological systems (Conway and Smoller, (1977a,b)). It may also be used to give bounds on the rate of combustion in a prototype combustion problem (Chandra and Davis, 1979).

A related method for proving stability of homogeneous steady state solution is that of nested rectangles. Consider the classic epidemic model of Kermack and McKendrick (1927) (discussed in Chapter 3), which is given by

$$\partial S/\partial t = -\beta IS, \qquad \partial I/\partial t = \beta IS - \gamma I, \qquad \partial R/\partial t = \gamma I$$

Here S is the number of susceptibles in a population, I the number of infectives and R the number removed from the population through immunity, isolation or death. Since the R equation is uncoupled from the system, we shall not consider R further. These quantities are all assumed normalised with respect to some reference population. When the spatial distribution of the disease is considered and diffusion is added, the equations become a reaction–diffusion system for the population densities u_1 of susceptibles and u_2 of infectives, given by

$$\begin{aligned}\partial u_1/\partial t &= -\beta u_1 u_2 + D_1 \nabla^2 u_1 = f_1(u_1, u_2) + D_1 \nabla^2 u_1 \\ \partial u_2/\partial t &= \beta u_1 u_2 - \gamma u_2 + D_2 \nabla^2 u_2 = f_2(u_1, u_2) + D_2 \nabla^2 u_2\end{aligned} \tag{5.36}$$

We shall consider zero-flux boundary conditions and initial conditions

$$u_1(\mathbf{x}, 0) = u_{1,0}(\mathbf{x}), \qquad u_2(\mathbf{x}, 0) = u_{2,0}(\mathbf{x}) \tag{5.37}$$

where

$$0 \leqslant u_{1,0}(\mathbf{x}) \leqslant M_1, \qquad 0 \leqslant u_{2,0}(\mathbf{x}) \leqslant M_2 \tag{5.38}$$

We look for an invariant rectangle of the form $\Sigma = \{\mathbf{u} \mid \mathbf{0} \leqslant \mathbf{u} \leqslant \mathbf{b}\}$ with $\mathbf{u}(\mathbf{x}, 0) \in \Sigma$. The requirements for such a set to be invariant are satisfied at $u_1 = 0$, since $f_1(0, u_2) = 0$, at $u_2 = 0$, since $f_2(u_1, 0) = 0$ and at $u_1 = b_1$, since $f_1(b_1, u_2) = -\beta b_1 u_2 \leqslant 0$. Hence it remains to satisfy

$$f_2(u_1, b_2) = \beta u_1 b_2 - \gamma b_2 = (\beta u_1 - \gamma) b_2 \leqslant 0 \tag{5.39}$$

for $0 \leqslant u_1 \leqslant b_1$. The existence of a bounded invariant rectangle Σ with $\mathbf{u}(\mathbf{x}, 0) \in \Sigma$ thus depends on the size of M_1 in (5.38). If $M_1 \leqslant \gamma/\beta$, then we may choose b_1 and b_2 such that $M_1 \leqslant b_1 \leqslant \gamma/\beta$, $M_2 \leqslant b_2$, and it follows that $\mathbf{0} \leqslant \mathbf{u}(\mathbf{x}, t) \leqslant \mathbf{b}$. In particular, we may take $\mathbf{b} = \mathbf{M}$. We have proved

THEOREM 5.40: Let $\mathbf{u}$ be a solution of the diffusional Kermack and McKendrick epidemic model (5.36) in $Q_T \cup S_T$ with zero-flux boundary conditions. If the initial conditions satisfy

$$\mathbf{0} \leqslant \mathbf{u}(\mathbf{x}, 0) \leqslant \mathbf{M} \qquad \text{on} \quad S_0$$

where $M_1 \leqslant \gamma/\beta$, then

$$\mathbf{0} \leqslant \mathbf{u}(\mathbf{x}, t) \leqslant \mathbf{M} \qquad \text{in} \quad \bar{Q}_T \tag{5.41}$$

The case $M_1 > \gamma/\beta$ will be considered later. The Kermack-McKendrick system thus has arbitrarily small invariant rectangles. The following theorem holds.

THEOREM 5.42 (Nested Invariant Rectangles): Let $\Sigma = \{\mathbf{v} \mid \mathbf{0} \leqslant \mathbf{v} \leqslant \mathbf{M}\} \subset \mathbb{R}^m$ and let $\theta\Sigma$ be an invariant rectangle for the initial boundary value problem for the system $\mathbf{u}_t = \mathbf{f}(\mathbf{u}) + D \nabla^2 \mathbf{u}$ for each $\theta \in [0, 1]$. Then zero is stable as a solution of the system with homogeneous boundary conditions and non-negative initial conditions.

Proof: This is immediate from the definition of invariant rectangles. However it is of interest for the sequel to give a proof using Lyapunov's direct method. For any given t let us define a functional V on the bounded non-negative functions $\mathbf{v}(\cdot, t)\colon \Omega \to \mathbb{R}^m$ by

$$V(\mathbf{v}(\cdot, t)) = \max_{1 \leq i \leq m} \sup_{\mathbf{x} \in \Omega} v_i(\mathbf{x}, t)/M_i.$$

We claim that V is a Lyapunov functional for the system $\mathbf{u}_t = \mathbf{f}(\mathbf{u}) + D\,\nabla^2\mathbf{u}$. Clearly $V \geq 0$ and $V = 0$ if and only if $\mathbf{v}(\cdot, t) \equiv 0$. We must show that V is non-increasing with t. The upper Dini derivative $\bar{D}V(\mathbf{v}(\cdot, t))$ is defined by

$$\bar{D}V(\mathbf{v}(\cdot, t)) = \limsup_{h \to 0} \{V(\mathbf{v}(\cdot, t+h)) - V(\mathbf{v}(\cdot, t))\}/h.$$

Clearly V is non-increasing if $\bar{D}V \leq 0$. Now for some solution $\mathbf{u}(\mathbf{x}, t)$ of the system $\mathbf{u}_t = \mathbf{f}(\mathbf{u}) + D\,\nabla^2\mathbf{u}$ with homogeneous boundary conditions and some time t_0 let $V(\mathbf{u}(\cdot, t_0)) = \theta$. Then $\mathbf{u}(\cdot, t_0) \leq \theta\mathbf{M}$ by the definition of V, and by the invariance of the set $\theta\Sigma$ we have $\mathbf{u}(\cdot, t_0 + h) \leq \theta\mathbf{M}$ for all $h > 0$. Thus

$$\bar{D}V(\mathbf{v}(\cdot, t_0)) \leq 0$$

as required.

Note that we have not proved asymptotic stability. In fact zero is not asymptotically stable for (5.36) with zero flux boundary conditions: $\mathbf{u} \equiv (M_1, 0)$ is a solution for any M_1, representing a uniformly distributed population with no infectives. Asymptotic stability may be proved under stronger assumptions on $\mathbf{f}$ and Σ.

DEFINITION 5.43: A bounded rectangle $\Sigma \subset \mathbb{R}^m$ is *contracting*, or an *attractor*, for the vector field $\mathbf{f}$ if for every point $\mathbf{v}^* \in \partial\Sigma$ we have $\mathbf{f}(\mathbf{v}^*) \cdot \mathbf{n}(\mathbf{v}^*) < 0$, where $\mathbf{n}(\mathbf{v}^*)$ is the outward pointing normal at $\mathbf{v}^*$.

THEOREM 5.44 (Nested Contracting Rectangles): Let Σ be a rectangle with $\mathbf{0} \in \Sigma \backslash \partial\Sigma$, and let $\theta\Sigma$ be contracting for the vector field $\mathbf{f}$ for each $\theta \in (0, 1]$. Then zero is asymptotically stable as a solution of $\mathbf{u}_t = \mathbf{f}(\mathbf{u}) + D\,\nabla^2\mathbf{u}$ with homogeneous boundary conditions $B\mathbf{u} = \mathbf{0}$.

Proof: The proof follows by Lyapunov's direct method: the upper Dini derivative of the Lyapunov function is now negative definite. See Smoller (1983) for the details.

Such methods have been used to prove asymptotic stability of the steady state solution for the Fitzhugh–Nagumo equations (Rauch and Smoller, 1978), for an epidemic model (Capasso and Maddalena, 1982b) and for ecological systems and prey-predator chains (Brown, 1980, 1983, 1985).

We shall now return to (5.36).

We proved in Theorem 5.40 that the set $\Sigma = \{\mathbf{v} \mid \mathbf{0} \leq \mathbf{v} \leq \mathbf{M}\}$ is invariant for these equations on the assumption that $M_1 \leq \gamma/\beta$. If $M_1 > \gamma/\beta$, then $f_2(u_1, b_2)$ given by (5.39) cannot be always non-positive, and there is no bounded invariant rectangle Σ, although it is intuitively clear that Σ is an (unbounded) invariant set of the *kinetic*, or diffusionless, system if $b_1 \geq M_1$, $b_2 = \infty$. It would be interesting to know whether this property of invariance carries over to the full reaction–diffusion system. Note that we do not have an existence proof for this system. We shall assume existence of a solution in $Q_T \cup S_T$, and attempt to justify this later. By definition the solution is continuous, and hence bounded, and so $\mathbf{f}$ is uniformly Lipschitz continuous in the required domain. We attempt the usual definitions for sub- and super-reaction functions $\underset{\sim}{\mathbf{f}}$ and $\tilde{\mathbf{f}}$.

Since $f_2(u_1, u_2)$ is non-decreasing in u_1, it follows that $\tilde{f}_2 = \underset{\sim}{f}_2 = f_2$. We may define $\tilde{f}_1$ by

$$\tilde{f}_1(\mathbf{u}) = \sup_{0 \leq U_2 \leq u_2} f_1(u_1, U_2) = \sup_{0 \leq U_2 \leq u_2} -\beta u_1 U_2 = 0$$

but an attempt to define $\underset{\sim}{f}_1$ fails, since

$$\underset{\sim}{f}_1(\mathbf{u}) = \inf_{u_2 \leq U_2 < \infty} f_1(u_1, U_2) = \inf_{u_2 \leq U_2 < \infty} -\beta u_1 U_2 = -\infty$$

We therefore have no lower bound for u_1 and no hope of justifying our existence assumption. The failure is occasioned by the fact that it is f_2 which is quasi-monotone non-decreasing, so that extreme values of f_1 have to be found over the whole range of u_2, which is infinite. If f_1 had the correct monotonicity properties, then extreme values of f_2 would have to be found over the range of u_1, which is finite. But such monotonicity properties may be achieved by the transformation

$$u_1^* = u_1, \qquad u_2^* = -u_2$$

The invariant set becomes $\Sigma^* = \{\mathbf{v}^* \mid 0 \leq v_1^* \leq M_1, -\infty < v_2^* \leq 0\}$, and the equations become

$$\partial u_1^*/\partial t = \beta u_1^* u_2^* + D_1 \nabla^2 u_1^* = f_1^*(u_1^*, u_2^*) + D_1 \nabla^2 u_1^*$$
$$\partial u_2^*/\partial t = \beta u_1^* u_2^* - \gamma u_2^* + D_2 \nabla^2 u_2^* = f_2^*(u_1^*, u_2^*) + D_2 \nabla^2 u_2^*$$

Thus $\partial f_1^*/\partial u_2^* = \beta u_1^* \geq 0$, so that f_1^* is quasi-monotone non-decreasing, and $\tilde{f}_1^* = \underset{\sim}{f}_1^* = f_1^*$. We may define $\tilde{f}_2^*$ and $\underset{\sim}{f}_2^*$ by

$$\begin{aligned}\tilde{f}_2^*(\mathbf{u}^*) &= \sup_{0 \leq U_1^* \leq u_1^*} f_2^*(U_1^*, u_2^*) \\ &= \sup_{0 \leq U_1^* \leq u_1^*} (\beta U_1^* u_2^* - \gamma u_2^*) = -\gamma u_2^*\end{aligned}$$

$$\underset{\sim}{f}{}_2^*(u^*) = \inf_{u_1^* \leq U_1^* \leq M_1} f_2^*(U_1^*, u_2^*)$$
$$= \inf_{u_1^* \leq U_1^* \leq M_1} (\beta U_1^* u_2^* - \gamma u_2^*) = (\beta M_1 - \gamma) u_2^*$$

recalling in both cases that $u_2^* \leq 0$.

A lower bound $\mathbf{v}^*$ and an upper bound $\mathbf{w}^*$ for $\mathbf{u}^*$ may now be found, where $\mathbf{v}^*$ is the solution of

$$\partial \mathbf{v}^*/\partial t = \underset{\sim}{\mathbf{f}}^*(\mathbf{v}^*), \qquad \mathbf{v}^*(0) = (0, -M_2)$$

and $\mathbf{w}^*$ is the solution of

$$\partial \mathbf{w}^*/\partial t = \tilde{\mathbf{f}}^*(\mathbf{w}^*), \qquad \mathbf{w}^*(0) = (M_1, 0)$$

Solving these, we have

$$\mathbf{v}^* = (0, -M_2 \exp\{(\beta M_1 - \gamma)t\}), \qquad \mathbf{w}^* = (M_1, 0)$$

so that

$$\mathbf{v}^*(t) \leq \mathbf{u}^*(\mathbf{x}, t) \leq \mathbf{w}^*(t)$$

or converting to the original variables

$$0 \leq u_1(\mathbf{x}, t) \leq M_1, \qquad 0 \leq u_2(\mathbf{x}, t) \leq M_2 e^{(\beta M_1 - \gamma)t}$$

These are the putative bounds for $\mathbf{u}$. But an argument similar to that of Theorem 5.17 (see Remark 5.17a) shows that they are indeed bounds for $\mathbf{u}$ and that the assumption of existence of $\mathbf{u}$ is justified. We have proved

THEOREM 5.45: Let $\mathbf{u}$ be a solution of the diffusional Kermack and McKendrick epidemic model (5.36) in $Q_T \cup S_T$ with zero-flux boundary conditions. If the initial conditions satisfy

$$\mathbf{0} \leq \mathbf{u}(\mathbf{x}, t) \leq \mathbf{M} \qquad \text{on} \quad S_0$$

then

$$\mathbf{0} \leq \mathbf{u}(\mathbf{x}, t) \leq (M_1, M_2 e^{(\beta M_1 - \gamma)t}) \qquad \text{in} \quad \bar{Q}_T \tag{5.46}$$

Note that even when Theorem 5.40 applies, i.e. when $M_1 \leq \gamma/\beta$, (5.46) represents an improvement in the bounds which have been proved. In epidemiology, $\sigma = \beta/\gamma$ is the infectious contact parameter. We may conclude that if $\sigma M_1 < 1$ (i.e. the initial density of susceptibles is everywhere less than $1/\sigma$), then the epidemic will die out, whereas if $\sigma M_1 > 1$, there is the possibility of exponential growth of the epidemic. This parallels results obtained in the spatially independent case (Chapter 3), where there is either an outbreak of the disease or it dies out depending on whether $\sigma u_1(0)$ is greater or less than unity.

REMARK 5.47: For a general reaction-diffusion system $\mathbf{u}_t = \mathbf{f}(\mathbf{u}) + D\nabla^2\mathbf{u}$ with the dimension of u equal to m, there are 2^m possible transformations of the form:

$$\mathbf{u}^* = (\pm u, \pm u_2, \ldots, \pm u_m)^T$$

which may lead to better comparison functions than the original variables. Making such transformations is equivalent to the method of Gardner (1981).

5.5 The Cauchy Problem and Fundamental Solutions

A comparison theorem and its corollaries may be proved for the Cauchy problem as well as for the initial-boundary-value problem, as long as all solutions and sub- and supersolutions are assumed to be bounded. Existence for all time and uniqueness of solutions of the Cauchy problem for certain systems of reaction–diffusion equations may also be proved by the method of fundamental solutions. Consider the diagonal system

$$(\partial_t - D_i \nabla^2) u_i \equiv L_i u_i = f_i(\mathbf{u}) \tag{5.48}$$

for $(\mathbf{x}, t) \in \mathbb{R}^n \times (0, \infty)$, $i = 1, 2, \ldots, m$, where each $D_i > 0$. Some *a priori* bounds are required on $\mathbf{u}$; we shall assume that there exists some bounded invariant set $\mathbf{a} \leq \mathbf{u} \leq \mathbf{b}$ for the system. The initial conditions are given by

$$\mathbf{u}(\mathbf{x}, 0) = \boldsymbol{\phi}(\mathbf{x}) \tag{5.49}$$

for $x \in \mathbb{R}^n$, where $\mathbf{a} \leq \boldsymbol{\phi} \leq \mathbf{b}$. Without loss of generality we shall take $\mathbf{f}(\mathbf{0}) = \mathbf{0}$. We may now state

THEOREM 5.50: Consider the Cauchy problem (5.48) with (5.49). Let $\Sigma = \{\mathbf{u} \mid \mathbf{a} \leq \mathbf{u} \leq \mathbf{b}\}$ be an invariant set for (5.48), and let $\mathbf{u}(\mathbf{x}, 0) \in \Sigma$. Let $\mathbf{f}$ satisfy a Lipschitz condition. Then there exists a unique bounded solution of the Cauchy problem (5.48) with (5.49).

Proof: Essentially the proof involves replacing (5.48) and (5.49) by an integral equation $\mathbf{u} = \mathbf{G} * \boldsymbol{\phi} + \mathbf{G} ** \mathbf{f}(\mathbf{u})$, where $\mathbf{G} = \mathbf{G}(\mathbf{x}, t)$ is the Green's function or fundamental solution $*$ and $**$ are spatial and spatio-temporal convolutions to be defined, and then proving that the operator S defined by $S\mathbf{u} = \mathbf{G} * \boldsymbol{\phi} + \mathbf{G} ** \mathbf{f}(\mathbf{u})$ has a unique fixed point.

We have $|\mathbf{f}(\mathbf{u}^1) - \mathbf{f}(\mathbf{u}^2)| \leq K|\mathbf{u}^1 - \mathbf{u}^2|$ for some constant K and any two vectors $\mathbf{a} \leq \mathbf{u}^1, \mathbf{u}^2 \leq \mathbf{b}$. The fundamental solution of the operator L_i in (5.48) is

$$G_i(\mathbf{x}, t) = (4\pi D_i t)^{-n/2} \exp(-|\mathbf{x}|^2/4D_i t)$$

so that

$$\int_{\mathbb{R}^n} G_i(\mathbf{x}, t)\, d\mathbf{x} = 1 \tag{5.51}$$

Let u_i^0, for $i=1,2,\ldots,m$, be the solution of the linear problem

$$L_i u_i^0 = 0, \qquad u_i^0(\mathbf{x},0)=\phi_i(\mathbf{x})$$

then, defining the spatial convolution of G_i with a function of $\mathbf{x}$ by

$$(G_i * g)(\mathbf{x},t)=\int_{\mathbb{R}^n} G_i(\mathbf{x}-\boldsymbol{\xi},t)g(\boldsymbol{\xi})\,d\boldsymbol{\xi}$$

we have

$$u_i^0(\mathbf{x},t)=(G_i * \phi_i)(\mathbf{x},t) \tag{5.52}$$

Let us also define the spatio-temporal convolution of G_i with a function of $\mathbf{x}$ and t by

$$(G_i ** h)(\mathbf{x},t)=\int_0^t\int_{\mathbb{R}^n} G_i(\mathbf{x}-\boldsymbol{\xi},t-\tau)h(\boldsymbol{\xi},\tau)\,d\boldsymbol{\xi}\,d\tau$$

Define the sequence of functions $\{\mathbf{u}^p(\mathbf{x},t)\}$, $p=1,2,\ldots$, by

$$u_i^{p+1}(\mathbf{x},t)=u_i^0(\mathbf{x},t)+(G_i ** f_i(\mathbf{u}^0))(\mathbf{x},t)$$

which, on applying the operator L_i, gives

$$L_i u_i^{p+1}=L_i u_i^0+f_i(\mathbf{u}^p)=f_i(\mathbf{u}^p)$$

We also have

$$u_i^{p+1}(\mathbf{x},0)=u_i^0(\mathbf{x},0)=\phi_i(\mathbf{x})$$

Introducing

$$M^{p+1}(t)=\sup_{\mathbf{x}\in\mathbb{R}^n,\sigma\leq t}|\mathbf{u}^{p+1}(\mathbf{x},\sigma)-\mathbf{u}^p(\mathbf{x},\sigma)|$$

we have

$$\begin{aligned}
M^{p+1}(t) &= \sup_{\mathbf{x}\in\mathbb{R}^n,\sigma\leq t}\sum_{i=1}^m |G_i ** \{f_i(\mathbf{u}^p)-f_i(\mathbf{u}^{p-1})\}| \\
&\leq \sup_{\mathbf{x}\in\mathbb{R}^n,\sigma\leq t}\sum_{i=1}^m G_i ** |f_i(\mathbf{u}^p)-f_i(\mathbf{u}^{p-1})| \\
&\leq \sup_{\mathbf{x}\in\mathbb{R}^n,\sigma\leq t} K\sum_{i=1}^m G_i ** |\mathbf{u}^p-\mathbf{u}^{p-1}| \\
&\leq \sup_{\mathbf{x}\in\mathbb{R}^n,\sigma\leq t} K\sum_{i=1}^m G_i ** M^p \\
&\leq \sup_{\sigma\leq t} Km\int_0^\sigma M^p(\tau)\,d\tau \\
&= Km\int_0^t M^p(\tau)\,d\tau
\end{aligned}$$

Also,

$$M^1(t) = \sup_{\mathbf{x}\in\mathbb{R}^n, \sigma\leq t} \sum_{i=1}^{m} |G_i ** f_i(\mathbf{u}^0)|$$

$$\leq \sup_{\mathbf{x}\in\mathbb{R}^n, \sigma\leq t} K \sum_{i=1}^{m} G_i ** |\mathbf{u}^0|$$

$$\leq \sup_{\mathbf{x}\in\mathbb{R}^n, \sigma\leq t} K \sum_{i=1}^{m} G_i ** \left(\left| \sum_{j=1}^{m} G_j * \phi_j \right| \right)$$

$$\leq Km \sup_{\mathbf{x}\in\mathbb{R}^n} |\boldsymbol{\phi}(\mathbf{x})| \int_0^t d\tau$$

$$= Kmkt$$

where

$$k = \sup_{\mathbf{x}\in\mathbb{R}^n} |\boldsymbol{\phi}(\mathbf{x})|$$

Inductively, it follows that

$$M^p(t) \leq Kkm^p t^p/p! \to 0 \qquad \text{as} \quad p\to\infty$$

Thus the sequence $\{\mathbf{u}^p(\mathbf{x}, t)\}$ converges as $p\to\infty$ to a limit $\mathbf{u}(\mathbf{x}, t)$ which satisfies

$$u_i(\mathbf{x}, t) = u_i^0(\mathbf{x}, t) + (G_i ** f_i(\mathbf{u}))(\mathbf{x}, t) \tag{5.53}$$

with $\mathbf{u}^0$ given by (5.52), and

$$\mathbf{u}(\mathbf{x}, 0) = \boldsymbol{\phi}(\mathbf{x})$$

Applying the operator L_i to (5.48), we obtain

$$L_i u_i = L_i u_i^0 + f_i(\mathbf{u}) = f_i(\mathbf{u})$$

It follows that $\mathbf{u}(\mathbf{x}, t)$ is a solution of the Cauchy problem.

For uniqueness, suppose that $\mathbf{u}^1$ and $\mathbf{u}^2$ are two bounded solutions of the Cauchy problem. Then it is a standard result that they both satisfy (5.53), so that

$$u_i^1 - u_i^2 = G_i ** \{f_i(\mathbf{u}^1) - f_i(\mathbf{u}^2)\}$$

Defining

$$N(t) = \sup_{\mathbf{x}\in\mathbb{R}^n, \sigma\leq t} |\mathbf{u}^1(\mathbf{x}, \sigma) - \mathbf{u}^2(\mathbf{x}, \sigma)|$$

we have

$$N(t) \leq \sup_{\mathbf{x}\in\mathbb{R}^n, \sigma\leq t} K \sum_{i=1}^{m} (G_i ** |\mathbf{u}^1 - \mathbf{u}^2|)(\mathbf{x}, \sigma)$$

$$\leq Km \int_0^t N(\tau)\, d\tau$$

Since $N(0)=0$, it follows that $N(t)\equiv 0$, so that $\mathbf{u}^1(\mathbf{x}, t) \equiv \mathbf{u}^2(\mathbf{x}, t)$, and the solution of the Cauchy problem is unique.

It also follows from the technique of fundamental solutions that the solution of the Cauchy problem (5.48) depends continuously on the initial conditions. From (5.52) and (5.53) $\mathbf{u}$ satisfies the integral equation

$$u_i = G_i * \phi_i + G_i ** f_i(\mathbf{u})$$

or

$$u_i(\mathbf{x}, t) = \int_{\mathbb{R}^n} G_i(\mathbf{x}-\boldsymbol{\xi}, t)\phi_i(\boldsymbol{\xi})\, d\boldsymbol{\xi} + \int_0^t \int_{\mathbb{R}^n} G_i(\mathbf{x}-\boldsymbol{\xi}, s) f_i(\mathbf{u}(\boldsymbol{\xi}, s))\, d\boldsymbol{\xi}\, ds$$

so that two solutions $\mathbf{u}$ and $\mathbf{v}$ with initial conditions $\mathbf{u}(\mathbf{x}, 0) = \boldsymbol{\phi}(\mathbf{x})$ and $\mathbf{v}(\mathbf{x}, 0) = \boldsymbol{\psi}(\mathbf{x})$ satisfy

$$\begin{aligned} u_i(\mathbf{x}, t) - v_i(\mathbf{x}, t) &= \int_{\mathbb{R}^n} G_i(\mathbf{x}-\boldsymbol{\xi}, t)(\phi_i(\boldsymbol{\xi}) - \psi_i(\boldsymbol{\xi}))\, d\boldsymbol{\xi} \\ &+ \int_0^t \int_{\mathbb{R}^n} G_i(\mathbf{x}-\boldsymbol{\xi}, s)(f_i(\mathbf{u}(\boldsymbol{\xi}, s)) - f_i(\mathbf{v}(\boldsymbol{\xi}, s)))\, d\boldsymbol{\xi}\, ds \end{aligned}$$

Defining a norm by $\|\mathbf{u}(\cdot, t)\| = \max_{\mathbf{x}\in\mathbb{R}^n} |\mathbf{u}(\mathbf{x}, t)|$, we have

$$\|\mathbf{u}(\cdot, t) - \mathbf{v}(\cdot, t)\| \leq \|\boldsymbol{\phi}(\cdot) - \boldsymbol{\psi}(\cdot)\| + K \int_0^t \|\mathbf{u}(\cdot, s) - \mathbf{v}(\cdot, s)\|\, ds$$

so that Gronwall's inequality leads to

$$\|\mathbf{u}(\cdot, t) - \mathbf{v}(\cdot, t)\| \leq e^{Kt} \|\boldsymbol{\phi}(\cdot) - \boldsymbol{\psi}(\cdot)\|$$

and we have continuous dependence of the solution on the initial conditions.

In some applications it happens that one or more of the diffusion coefficients D_i is zero. Rauch and Smoller (1978) shows that such a system may be treated in exactly the same way as a purely parabolic system, with $G_i(\mathbf{x}, t) = \delta(\mathbf{x})$ if $D_i = 0$, since the solution of (5.48) is continuously dependent on the diffusion coefficients.

5.6 Stationary Solutions and Their Stability

The stability of stationary solutions of reaction–diffusion equations may be analysed by considering the eigenvalues of the linearised system. We shall be concerned with systems of the form

$$\mathbf{u}_t = \mathbf{f}(\mathbf{u}) + D\,\nabla^2\mathbf{u} \tag{5.54}$$

in $\Omega\times(0,\infty)$, where Ω is a bounded domain and D is a diagonal matrix, with zero-flux boundary conditions

$$\partial\mathbf{u}/\partial n(\mathbf{x}, t) = 0 \qquad \text{on} \quad \partial\Omega\times(0,\infty) \tag{5.55}$$

and initial conditions

$$\mathbf{u}(\mathbf{x}, 0) = \mathbf{u}_0(\mathbf{x}) \qquad \text{on} \quad \Omega\times\{0\} \tag{5.56}$$

Note that we are assuming that the solution $\mathbf{u}$ is a global solution, i.e. it is defined for all $t \geq 0$. Homogeneous Dirichlet conditions may also be considered. Neumann boundary conditions were considered by Casten and Holland (1977), and we shall follow their approach. Problems may also be considered where $\mathbf{f}(\mathbf{u})$ depends on values of $\mathbf{u}$ not only at $\mathbf{x}$ but also at other points of the domain Ω [see, for example, Capasso and Fortunato (1980)].

It is first necessary to define stability of a stationary solution of the system (5.54) with (5.55). It is defined here in terms of the $L^\infty(\Omega)$ norm, $\|\cdot\|_{L^\infty(\Omega)}$, given by

$$\|\mathbf{h}(\cdot)\|_{L^\infty(\Omega)} = \sup_{\mathbf{x}\in\Omega} |\mathbf{h}(\mathbf{x})|$$

where $|\cdot|$ is a vector norm, e.g. the usual Euclidean norm, of the vector $\mathbf{h}$. Henceforth in this section we shall denote this norm simply by $\|\cdot\|$.

Definition 5.57: A stationary solution $\mathbf{v}(\mathbf{x})$ of (5.54) with (5.55) is *stable* if $\mathbf{u}(\mathbf{x}, t) = \mathbf{v}(\mathbf{x})$ is stable as a solution of the corresponding initial boundary value problem with initial conditions $\mathbf{u}(\mathbf{x}, 0) = \mathbf{u}_0(\mathbf{x}) = \mathbf{v}(\mathbf{x})$. See definitions 4.6, with the obvious extension to vector-valued functions (Global) asymptotic stability and instability are similarly defined.

We shall assume that $\mathbf{u} = \mathbf{0}$ is a steady state of the equations and linearise about zero. We decompose $\mathbf{f}(\mathbf{u})$ into a linear and a nonlinear part by

$$\mathbf{f}(\mathbf{u}) = A\mathbf{u} + \mathbf{N}(\mathbf{u}) \tag{5.58}$$

where A is a matrix and $\mathbf{N}(\mathbf{u}) = o(\mathbf{u})$ as $\mathbf{u}\to\mathbf{0}$, i.e. $|\mathbf{N}(\mathbf{u})|/|\mathbf{u}| \to 0$ as $\mathbf{u}\to\mathbf{0}$. The linearised system is defined to be

$$\mathbf{u}_t = A\mathbf{u} + D\,\nabla^2\mathbf{u} \tag{5.59}$$

This system with boundary and initial conditions (5.55) and (5.56) may be solved by the method of eigenfunction expansions. Denote the eigenvalues of $-\nabla^2$ in Ω with homogeneous Neumann boundary conditions by $0 = \lambda_0 < \lambda_1 \leqslant \lambda_2 \leqslant \cdots$ and the corresponding normalised eigenfunctions by ϕ_0, ϕ_1, $\phi_2, \ldots$. Define the vector coefficient $\mathbf{u}_{0n}$ of the eigenfunction ϕ_n in the expansion of the initial condition $\mathbf{u}_0(\mathbf{x})$ by

$$\mathbf{u}_{0n} = \int_\Omega \mathbf{u}_0(\mathbf{x})\phi_n(\mathbf{x})\, d\mathbf{x}$$

and let $\exp(A_n t)$ be the matrix solution of the differential equation

$$\frac{d}{dt}\{\exp(A_n t)\} = A_n \exp(A_n t)$$

where $A_n = A - \lambda_n D$, with initial conditions $\exp(A_n 0) = I$. Then the solution of the linear system (5.59) with zero-flux boundary conditions may be written in the form

$$\mathbf{u}(\mathbf{x}, t) = \sum_{n=0}^{\infty} \phi_n(\mathbf{x}) \exp(A_n t)\mathbf{u}_{0n}$$

If follows after some analysis that the stability of the zero solution depends on the eigenvalues of the matrices A_n, and we state the theorem:

THEOREM 5.60 (Casten and Holland, 1977): (i) The zero solution of the linear system (5.59) with zero-flux boundary conditions is globally asymptotically stable if for each non-negative integer n the eigenvalues of $A_n = A - \lambda_n D$ have negative real parts. Furthermore there exist positive constants K and ω such that for any $t > 0$

$$\|\mathbf{u}(\cdot, t)\| \leqslant K e^{-\omega t} \|\mathbf{u}_0(\cdot)\|$$

(ii) The zero solution is stable if for each non-negative integer n the eigenvalues of A_n have non-positive real parts and those with zero real parts have simple elementary divisors.

(iii) The zero solution is unstable if for some n there exists an eigenvalue of A_n with either positive real part or zero real part with a non-simple elementary divisor.

It is easy to see that if D is a scalar matrix, $D = dI$, then diffusion cannot destabilise the zero solution, since if μ is an eigenvalue of A, then $\mu_n = \mu - \lambda_n d$ is the corresponding eigenvalue of A_n, and each λ_n is non-negative. However, if D is not scalar, this statement does not hold, as we shall see in Chapter 6.

Turning now to the nonlinear system (5.54), we can show that, just as in the spatially uniform case (Theorem 2.14), asymptotic stability of the zero solution of the nonlinear follows from that of the linear system:

THEOREM 5.61 [Linearised Stability (Casten and Holland, 1977)]: The zero solution of (5.54) with (5.55) is asymptotically stable if the zero solution of (5.59) with (5.55) is.

Proof: Define a semi-group $T(t)$ by

$$T(t)\mathbf{v}(\mathbf{x}) = \sum_{n=0}^{\infty} \phi_n(\mathbf{x}) \exp(A_n t)\mathbf{v}_n$$

where

$$\mathbf{v}_n = \int_\Omega \mathbf{v}(\mathbf{x})\phi_n(\mathbf{x})\,d\mathbf{x}$$

Then the solution of the linear system is given by

$$\mathbf{u}(\mathbf{x}, t) = T(t)\mathbf{u}_0(\mathbf{x})$$

and the solution of the nonlinear system satisfies

$$\mathbf{u}(\mathbf{x}, t) = T(t)\mathbf{u}_0(\mathbf{x}) + \int_0^t T(t-s)\mathbf{N}(\mathbf{u}(\mathbf{x}, s))\,ds$$

Since $\mathbf{N}(\mathbf{u}) = o(\mathbf{u})$ as $\mathbf{u} \to \mathbf{0}$, there exists a positive constant γ such that $|\mathbf{N}(\mathbf{u})| < (\omega/2K)|\mathbf{u}|$ when $|\mathbf{u}| < 2\gamma$, where ω and K are as in Theorem 5.50. For $\|\mathbf{u}_0(\cdot)\| < \gamma$, there is a time T such that $\|\mathbf{u}(\cdot, t)\| < 2\gamma$ for $0 \leqslant t \leqslant T$. Then on $[0, T]$ we have

$$\|\mathbf{u}(\cdot, t)\| \leqslant Ke^{-\omega t}\|\mathbf{u}_0(\cdot)\| + \int_0^t \frac{\omega e^{-\omega(t-s)}\|\mathbf{u}(\cdot, s)\|}{2}\,ds$$

Defining

$$R(t) = \|\mathbf{u}(\cdot, t)\|e^{\omega t}$$

Gronwall's inequality yields

$$R(t) \leqslant R(0)K \exp(\omega t/2)$$

or in terms of $\mathbf{u}$,

$$\|\mathbf{u}(\cdot, t)\| \leqslant \gamma \exp(-\omega t/2)$$

This holds as long as $\|\mathbf{u}(\cdot, t)\| < 2\gamma$; but the right-hand side is always less than 2γ, so that the inequality is valid for all $t > 0$. It follows that the zero solution is asymptotically stable, as required.

Another method of proving asymptotic stability of an equilibrium state is to find a Lyapunov function for the system (c.f. Section 3.3). De Mottoni and Rothe (1979) consider the predator-prey system

$$u_t = h(u)(f(u) - a(v)) + D_1 \nabla^2 u$$
$$v_t = k(v)(-g(v) + b(u)) + D_2 \nabla^2 v$$

with zero-flux boundary conditions, where $h, k > 0$ in $(0, \infty)$, f is strictly monotone decreasing, and a, b and g are strictly monotone increasing. They also assume that there exists an invariant set Σ in the positive quadrant, that there exists a unique steady state (u_s, v_s) and that $(a(v) - a(v_s))/k(v)$ and $(b(u) - b(u_s))/h(u)$ are monotone increasing in Σ. Then

$$\tilde{V}(u, v) = \int \frac{b(u) - b(u_s)}{h(u)}\, du + \int \frac{a(v) - a(v_s)}{k(v)}\, dv$$

is a Lyapunov function (Definition 2.13) for the spatially uniform system, and

$$V(u, v) = \int_\Omega \tilde{V}(u, v)\, d\mathbf{x}$$

is a Lyapunov functional with dV/dt negative definite for the system with diffusion. Thus $V(u, v) \to 0$ as $t \to \infty$, so $u \to u_s$, $v \to v_s$, and the steady state is asymptotically stable. Note that the theorem does not apply to the Lotka–Volterra system (3.6) since there $f(u) = 1$ and $g(v) = 1$, neither of which is *strictly* monotone. For this reason we cannot prove asymptotic stability of the steady state of the Lotka–Volterra system, only that it tends to a spatially uniform state as $t \to \infty$ (Section 3.3).

There are few theorems proving the existence of non-trivial stationary solutions of the system (5.54). An exception is that of Hadeler *et al.* (1979), who proved that (5.54) with boundary conditions

$$\mathbf{u} = \boldsymbol{\psi} \qquad \text{or} \qquad \partial \mathbf{u}/\partial n + c(\mathbf{x})\mathbf{u} = c(\mathbf{x})\boldsymbol{\psi}$$

has a non-constant stationary solution as long as there exists a convex compact set containing the boundary data and invariant for $\dot{\mathbf{v}} = D^{-1}\mathbf{f}(\mathbf{v})$. This theorem does not require the hypothesis that $\mathbf{f}(\mathbf{0}) = \mathbf{0}$.

A system which admits a reasonably comprehensive analysis of stationary solutions is that of Maginu (1975) given by

$$u_t = f(u) - v + k\nabla^2 u, \qquad v_t = u - v + \nabla^2 v \tag{5.62}$$

where f is such that the phase plane is as shown in Fig. 5.63. It is easy to show that the steady state is always stable to homogeneous perturbations

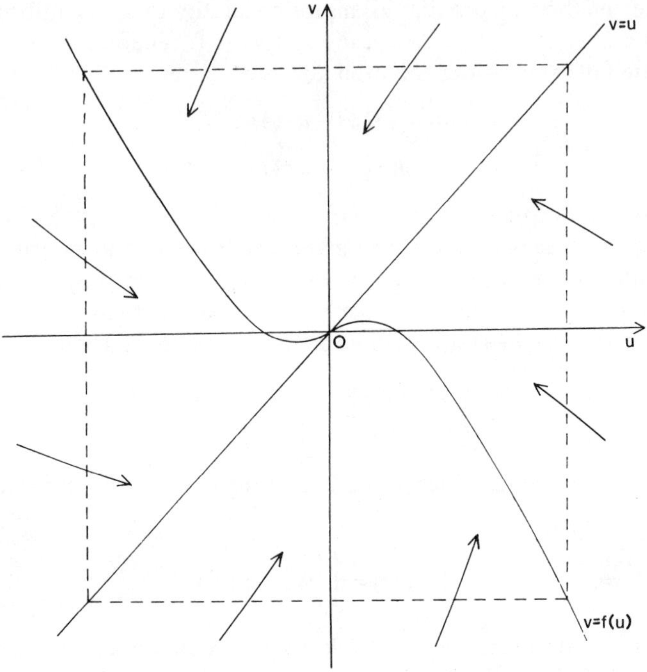

Fig. 5.63. The phase plane for (5.62).

but is unstable to spatially inhomogeneous perturbations if k is sufficiently small, and that there is an invariant rectangle Σ so that all solutions with initial and boundary conditions inside Σ are bounded *a priori.* Rothe and de Mottoni (1979) then use a Lyapunov functional to prove convergence to stationary solutions. Rothe (1981) considers such stationary solutions and noting that $u = v - \nabla^2 v$ with zero Dirichlet boundary conditions defines a linear operator $v = Bu$ (other homogeneous boundary conditions may also be considered), considers the problem

$$k\,\nabla^2 u - Bu + f(u) = k\,\nabla^2 u - Bu + \lambda u - uh(u) = 0$$

splitting f up into its linear and nonlinear parts. He shows that if $\{\lambda_m\}_{m=1}^{\infty}$ are the eigenvalues of $-k\,\nabla^2 + B$ and h is even and obeys a polynomial growth restriction, then this has at least m pairs of solutions $\{\pm u_j\}_{j=1}^{m}$ when $\lambda_m < \lambda \leqslant \lambda_{m+1}$. Lazer and McKenna (1982) have improved this result for $m = 1$, relaxing the restrictions on h and using degree theory to prove that there are *exactly* three solutions in this case, and showing the stability of the non-constant solutions by using a Lyapunov functional. Some other papers on similar lines are those of Blat and Brown (1984), Brown (1983), Conway *et al.* (1982), de Mottoni (1984) and Leung (1983).

5.7 The Energy Method

We have seen that the comparison theorem, spectral theory and other techniques may be used in certain circumstances to prove the convergence of solutions of systems of reaction–diffusion equations to time-independent functions, and in particular to the steady state. If the reaction terms have no explicit dependence on **x**, and homogeneous Neumann boundary conditions are applied, another possibility for the asymptotic behaviour of solutions as $t\to\infty$ is convergence to a *spatially* homogeneous solution. This might be expected to happen when the diffusion coefficients are large, since diffusion tends to damp out spatial variations. This is indeed the case, as has been shown by Othmer (1977) and Conway *et al.* (1978). The treatment here follows the latter paper.

The basic technique is the energy method [e.g. Courant and Hilbert (1962)]. It involves consideration of the so-called energy function $\Phi(t)$ defined by

$$\Phi(t)=\frac{1}{2}\int_\Omega \langle\nabla\mathbf{u},\nabla\mathbf{u}\rangle\,d\mathbf{x}=\frac{1}{2}\int_\Omega \|\nabla\mathbf{u}\|^2\,d\mathbf{x}=\frac{1}{2}\|\nabla\mathbf{u}\|^2_{L^2(\Omega)}$$

where integration is over the spatial domain Ω of the problem. The inner product $\langle\cdot,\cdot\rangle$ is defined by

$$\langle\nabla\mathbf{u},\nabla\mathbf{v}\rangle=\sum_{i=1}^m \nabla u_i\cdot\nabla v_i$$

and in this section the norm $\|\cdot\|$ is the usual matrix norm defined by

$$\|\nabla\mathbf{u}\|^2=\langle\nabla\mathbf{u},\nabla\mathbf{u}\rangle$$

We shall apply the technique to the reaction–diffusion system

$$\mathbf{u}_t=\mathbf{f}(\mathbf{u})+D\,\nabla^2\mathbf{u}\qquad\text{in}\quad\Omega\times(0,\infty)\tag{5.64}$$

where $D>0$ is a diagonal matrix, with zero-flux boundary conditions

$$\partial\mathbf{u}/\partial n(\mathbf{x},t)=0\qquad\text{on}\quad\partial\Omega\times(0,\infty)\tag{5.65}$$

and initial conditions

$$\mathbf{u}(\mathbf{x},0)=\mathbf{u}_0(\mathbf{x})\qquad\text{for}\quad\mathbf{x}\in\Omega\tag{5.66}$$

Extensions may be made to non-diagonal diffusion matrices, systems involving convection and mixed boundary conditions (Conway *et al.*, 1978), and also to functions $\mathbf{f}$ depending on $\mathbf{x}$ and t as well as $\mathbf{u}$ (Yu, 1985). We shall assume that the system has a bounded invariant set Σ. Define the quantity σ by

$$\sigma = d\lambda - M \tag{5.67}$$

where d is the smallest diffusion coefficient of the system, λ is the smallest positive eigenvalue of $-\nabla^2$ on Ω with homogeneous Neumann boundary conditions and

$$M = \sup_{\mathbf{u}\in\Sigma}\{\|\mathbf{f}_{\mathbf{u}}(\mathbf{u}\,|\,\mathbf{u})\|\}$$

where

$$\|\mathbf{f}_{\mathbf{u}}(\mathbf{u}\,|\,\mathbf{u})\|^2 = \sum_{i=1}^{m}\sum_{j=1}^{m}\left|\frac{\partial f_i}{\partial u_j}\right|^2$$

Then we have the following theorem:

THEOREM 5.68: Let σ be positive, let Σ be an invariant set of (5.64) with (5.65) and let $\mathbf{u}$ be any solution of (5.64) with (5.65) and initial values $\mathbf{u}_0(\mathbf{x})\in\Sigma$. Then there are constants c_1, c_2 and c_3 such that

$$\|\nabla\mathbf{u}(\cdot,t)\|_{L^2(\Omega)} \le c_1 e^{-\sigma t}, \qquad \|\mathbf{u}(\cdot,t)-\bar{\mathbf{u}}(t)\|_{L^2(\Omega)} \le c_2 e^{-\sigma t}$$

where $\bar{\mathbf{u}}$, the average of $\mathbf{u}$ over Ω, satisfies the system of ordinary differential equations

$$d\bar{\mathbf{u}}/dt = \mathbf{f}(\bar{\mathbf{u}}) + \mathbf{g}(t)$$

with initial conditions

$$\bar{\mathbf{u}}(0) = \frac{1}{|\Omega|}\int_{\Omega}\mathbf{u}_0(\mathbf{x})\,d\mathbf{x}.$$

Where $\mathbf{g}(t)$ satisfies

$$\mathbf{g}(t) \le c_3 e^{-\sigma t}$$

and $|\Omega|$ denotes the measure of Ω. The constants c_1, c_2 and c_3 are given by

$$c_1 = \|\nabla\mathbf{u}_0(\cdot)\|_{L^2(\Omega)}$$
$$c_2 = \lambda^{-1/2}\|\nabla\mathbf{u}_0(\cdot)\|_{L^2(\Omega)}$$
$$c_3 = M\lambda^{-1/2}|\Omega|^{-1/2}\|\nabla\mathbf{u}_0(\cdot)\|_{L^2(\Omega)}$$

Proof: We shall need the following technical result:

$$\lambda^{-1}\int_\Omega |\nabla^2\mathbf{w}|^2\,d\mathbf{x} \geq \int_\Omega \|\nabla\mathbf{w}\|^2\,d\mathbf{x} \geq \lambda\int_\Omega |\mathbf{w}-\bar{\mathbf{w}}|^2\,d\mathbf{x}$$

for any $\mathbf{w}\in C^2(\Omega)$ satisfying zero-flux boundary conditions on $\partial\Omega$, where

$$\bar{\mathbf{w}} = |\Omega|^{-1}\int_\Omega \mathbf{w}\,d\mathbf{x}$$

To prove the first inequality let $\{\phi_k\}$, $k=0,1,2,\ldots$, be a complete set of orthonormal eigenfunctions of $-\nabla^2$ on Ω with homogeneous Neumann boundary conditions; let the corresponding eigenvalues be $\{\lambda_k\}$. Then $\lambda_0=0$, $\phi_0=|\Omega|^{-1/2}$, and $0<\lambda=\lambda_1\leq\lambda_2\leq\cdots$. Moreover, $\nabla^2\mathbf{w}$ is continuous and may be expanded as a series in the ϕ_k,

$$\nabla^2\mathbf{w} = \sum_{k=0}^{\infty} \mathbf{a}_k\phi_k$$

where

$$\mathbf{a}_k = \int_\Omega \phi_k\,\nabla^2\mathbf{w}\,d\mathbf{x}$$

Thus

$$\mathbf{a}_0 = \int_\Omega \phi_0\,\nabla^2\mathbf{w}\,d\mathbf{x} = \phi_0\int_\Omega \frac{\partial\mathbf{w}}{\partial n}\,d\mathbf{x} = \mathbf{0}$$

Integrating the series for $\nabla^2\mathbf{w}$, using the fact that $\nabla^2\phi_k+\lambda_k\phi_k=0$ and the zero-flux boundary conditions,

$$\mathbf{w} = \sum_{k=1}^{\infty} -\mathbf{a}_k\lambda_k^{-1}\phi_k + \mathbf{b}_0\phi_0$$

where $\mathbf{b}_0$ is an arbitrary constant vector, so that

$$\begin{aligned}\int_\Omega \|\nabla\mathbf{w}\|^2\,d\mathbf{x} &= \int_\Omega \langle\nabla\mathbf{w},\nabla\mathbf{w}\rangle\,d\mathbf{x}\\ &= \int_{\partial\Omega}\left(\mathbf{w},\frac{\partial\mathbf{w}}{\partial n}\right)d\mathbf{x} - \int_\Omega (\mathbf{w},\nabla^2\mathbf{w})\,d\mathbf{x}\\ &= 0+\sum_{k=1}^{\infty}\frac{|\mathbf{a}_k|^2}{\lambda_k} \leq \lambda_1^{-1}\sum_{k=1}^{\infty}|\mathbf{a}_k|^2 = \lambda^{-1}\int_\Omega |\nabla^2\mathbf{w}|^2\,d\mathbf{x}\end{aligned}$$

where $(\cdot,\cdot)$ denotes the usual scalar product of two vectors, as required.

To prove the second inequality, note that with **w** as given above

$$\bar{\mathbf{w}} = |\Omega|^{-1} \int_\Omega \mathbf{w}\, d\mathbf{x} = \left\{ \int_\Omega \phi_0 \mathbf{w}\, d\mathbf{x} \right\} \phi_0 = \mathbf{b}_0 \phi_0$$

Thus

$$\begin{aligned}
\int_\Omega \|\nabla \mathbf{w}\|^2\, d\mathbf{x} &= \int_\Omega \langle \nabla \mathbf{w}, \nabla \mathbf{w} \rangle\, d\mathbf{x} \\
&= \int_\Omega \left(\mathbf{w}, \frac{\partial \mathbf{w}}{\partial n} \right) d\mathbf{x} - \int_\Omega (\mathbf{w}, \nabla^2 \mathbf{w})\, d\mathbf{x} \\
&= 0 + \sum_{k=1}^\infty \frac{|\mathbf{a}_k|^2}{\lambda_k} \geqslant \lambda_1 \sum_{k=1}^\infty \frac{|\mathbf{a}_k|^2}{\lambda_k^2} \\
&= \lambda \int_\Omega |\mathbf{w} - \bar{\mathbf{w}}|^2\, d\mathbf{x} = \lambda \|\mathbf{w}(\cdot, t) - \bar{\mathbf{w}}(t)\|_{L^2(\Omega)}
\end{aligned}$$

as required.

We now define the energy function, $\Phi(t)$, by

$$\Phi(t) = \frac{1}{2} \int_\Omega \langle \nabla \mathbf{u}, \nabla \mathbf{u} \rangle\, d\mathbf{x}$$

where **u** is a solution of

$$\mathbf{u}_t = \mathbf{f}(\mathbf{u}) + D\, \nabla^2 \mathbf{u} \qquad \text{in} \quad \Omega \times (0, \infty)$$

with zero-flux boundary conditions. Then

$$\begin{aligned}
\dot{\Phi}(t) &= \int_\Omega \langle \nabla \mathbf{u}, \nabla \mathbf{u}_t \rangle\, d\mathbf{x} \\
&= \int_\Omega \langle \nabla \mathbf{u}, \nabla (\mathbf{f}(\mathbf{u}) + D\, \nabla^2 \mathbf{u}) \rangle\, d\mathbf{x} \\
&= \int_\Omega \langle \nabla \mathbf{u}, \mathbf{f}_\mathbf{u}(\mathbf{u}|\mathbf{u})\, \nabla \mathbf{u} \rangle\, d\mathbf{x} \\
&\quad + \int_{\partial\Omega} \left(\frac{\partial \mathbf{u}}{\partial n}, D\, \nabla^2 \mathbf{u} \right) d\mathbf{x} - \int_\Omega (\nabla^2 \mathbf{u}, D\, \nabla^2 \mathbf{u})\, d\mathbf{x} \\
&\leqslant M \int_\Omega \langle \nabla \mathbf{u}, \nabla \mathbf{u} \rangle\, d\mathbf{x} - d \int_\Omega (\nabla^2 \mathbf{u}, \nabla^2 \mathbf{u})\, d\mathbf{x} \\
&\leqslant (M - \lambda d) \int_\Omega \|\nabla \mathbf{u}\|^2\, d\mathbf{x} = -2\sigma \Phi(t)
\end{aligned}$$

It follows that

$$\Phi(t) \leqslant \Phi(0) e^{-2\sigma t}$$

or, taking the square root,

$$\|\nabla \mathbf{u}(\cdot, t)\|_{L^2(\Omega)} \leqslant c_1 e^{-\sigma t} = \|\nabla \mathbf{u}_0(\cdot)\|_{L^2(\Omega)} e^{-\sigma t}$$

as required. Also

$$\|\mathbf{u}(\cdot, t) - \bar{\mathbf{u}}(t)\|^2_{L^2(\Omega)} \leqslant \lambda^{-1} \|\nabla \mathbf{u}(\cdot, t)\|_{L^2(\Omega)} \leqslant c_1^2 \lambda^{-1} e^{-2\sigma t}$$

so that

$$\|\mathbf{u}(\cdot, t) - \bar{\mathbf{u}}(t)\|_{L^2(\Omega)} \leqslant c_2 e^{-\sigma t} = \lambda^{-1/2} \|\nabla \mathbf{u}_0(\cdot)\|_{L^2(\Omega)} e^{-\sigma t}$$

It remains to find the differential equation satisfied by $\bar{\mathbf{u}}$. We have

$$\bar{\mathbf{u}}(t) = |\Omega|^{-1} \int_\Omega \mathbf{u}(\mathbf{x}, t)\, d\mathbf{x}$$

so that

$$\begin{aligned}
\dot{\bar{\mathbf{u}}}(t) &= |\Omega|^{-1} \int_\Omega \frac{\partial \mathbf{u}}{\partial t}(\mathbf{x}, t)\, d\mathbf{x} \\
&= |\Omega|^{-1} \int_\Omega (\mathbf{f}(\mathbf{u}) + D\, \nabla^2 \mathbf{u})\, d\mathbf{x} \\
&= |\Omega|^{-1} \int_\Omega (\mathbf{f}(\bar{\mathbf{u}}) + \mathbf{f}(\mathbf{u}) - \mathbf{f}(\bar{\mathbf{u}}))\, d\mathbf{x} \\
&= \mathbf{f}(\bar{\mathbf{u}}) + |\Omega|^{-1} \int_\Omega (\mathbf{f}(\mathbf{u}) - \mathbf{f}(\bar{\mathbf{u}}))\, d\mathbf{x}
\end{aligned}$$

But

$$\begin{aligned}
&\left| \frac{1}{|\Omega|} \int_\Omega (\mathbf{f}(\mathbf{u}) - \mathbf{f}(\bar{\mathbf{u}}))\, d\mathbf{x} \right| \\
&\qquad \leqslant \frac{M}{|\Omega|} \int_\Omega |\mathbf{u} - \bar{\mathbf{u}}|\, d\mathbf{x} \\
&\qquad \leqslant \frac{M}{|\Omega|^{1/2}} \left\{ \int_\Omega |\mathbf{u} - \bar{\mathbf{u}}|^2\, d\mathbf{x} \right\}^{1/2} \qquad \text{by the Cauchy–Schwarz inequality} \\
&\qquad \leqslant c_3 e^{-\sigma t} = \frac{M \|\nabla \mathbf{u}_0(\cdot)\|_{L^2(\Omega)}}{\lambda^{1/2} |\Omega|^{1/2}} e^{-\sigma t}
\end{aligned}$$

so that **u** satisfies the differential equation

$$\dot{\bar{\mathbf{u}}} = \mathbf{f}(\bar{\mathbf{u}}) + \mathbf{g}$$

where

$$\mathbf{g}(t) = |\Omega|^{-1} \int_\Omega (\mathbf{f}(\mathbf{u}) - \mathbf{f}(\bar{\mathbf{u}}))\, d\mathbf{x}$$

satisfies

$$\mathbf{g}(t) \leqslant c_3 e^{-\sigma t}$$

This completes the proof of the theorem.

We have thus shown that $\mathbf{u}(\mathbf{x}, t)$ decays exponentially to its spatial average in the sense of the $L^2(\Omega)$ norm. It can be shown that this is also true in the sense of the $L^\infty(\Omega)$ norm, and we have

THEOREM 5.69 (Conway *et al.*, 1978): Let Σ be an invariant set of (5.64) and (5.65), and let **u** be any solution of (5.64) and (5.65) with initial values $\mathbf{u}_0(\mathbf{x}) \in \Sigma$. Then

$$\|\mathbf{u}(\cdot, t) - \bar{\mathbf{u}}(t)\|_{L^\infty(\Omega)} \leqslant c\|\nabla \mathbf{u}_0(\cdot)\|_{L^\infty(\Omega)} e^{-\sigma' t}$$

where

$$\bar{\mathbf{u}}(t) = |\Omega|^{-1} \int_\Omega \mathbf{u}(\mathbf{x}, t)\, d\mathbf{x}$$

c is a constant, and $\sigma' > 0$ whenever $\sigma = d\lambda - M > 0$.

We refer the reader to the original paper for a proof of this result. It may be used to prove the stability of certain spatially independent solutions of systems of reaction-diffusion equations, where it is more natural to use the $L^\infty(\Omega)$ norm. To do this we first need some definitions relevant to the kinetic system

$$\dot{\mathbf{v}} = \mathbf{f}(\mathbf{v}) \tag{5.70}$$

DEFINITION 5.71: Given a set U, the ω-limit set of U is the intersection over $t \geqslant 0$ of the closures of all orbits of (5.68) starting in U at time $t = 0$.

DEFINITION 5.72: A bounded invariant set $\Gamma \subset \mathbb{R}^m$ is called an attractor for (5.70) if there is an open set $U \supset \Gamma$ such that the ω-limit set of U is Γ.

We also need to define stability of a spatially uniform solution of a reaction-diffusion system.

DEFINITION 5.73: A bounded solution $\boldsymbol{\gamma}(t)$ of (5.70) is said to be a stable spatially uniform solution of

$$\mathbf{u}_t = \mathbf{f}(\mathbf{u}) + D\,\nabla^2\mathbf{u} \qquad \text{in} \quad \Omega \times (0, \infty)$$

with zero-flux boundary conditions if for any $\varepsilon > 0$ there exist constants $\delta > 0$, $T > 0$ and t_0 such that

$$\|\mathbf{u}(\cdot, t) - \boldsymbol{\gamma}(t - t_0)\|_{L^\infty(\Omega)} \leq \varepsilon$$

for all $t \geq T$ whenever

$$\operatorname{dist}(\mathbf{u}_0, \Gamma) + \|\boldsymbol{\nabla}\mathbf{u}_0(\cdot)\|_{L^\infty(\Omega)} < \delta$$

where $\Gamma = \{\boldsymbol{\gamma}(t)\}_{t \in \mathbb{R}}$ and

$$\operatorname{dist}(\mathbf{u}_0, r) = \inf_{\mathbf{x} \in \Omega, t \in \mathbb{R}} |\mathbf{u}_0(\mathbf{x}) - \boldsymbol{\gamma}(t)|.$$

Asymptotic stability is defined in the obvious way.

THEOREM 5.74: Let $\Gamma = \{\boldsymbol{\gamma}(t)\}$ be an attractor for the system (5.70). Then if $\sigma > 0$, $\boldsymbol{\gamma}(t)$ is a stable solution of (5.64) with zero-flux boundary conditions.

Proof: It can be shown (Churchill, 1972; Conley and Easton, 1971; Montgomery, 1973) that there is a compact neighbourhood B of Γ such that Γ is the maximal positively and negatively invariant set for (5.68) in B, (i.e. invariant as $t \to \infty$ and also as $t \to -\infty$), and $\mathbf{f}(\mathbf{p}) \cdot \mathbf{n}(\mathbf{p}) < -\delta_1 < 0$ for $\mathbf{p} \in \partial B$, where $\mathbf{n}$ is the outward normal to B. We define $\delta_2 = \operatorname{dist}(\partial B, \Gamma)$, and let

$$\operatorname{dist}(\mathbf{u}_0, \Gamma) + \|\boldsymbol{\nabla}\mathbf{u}_0(\cdot)\|_{L^\infty(\Omega)} < \delta$$

where

$$\delta = \min\{\lambda^{1/2}\delta_1/M, \delta_2\}$$

Let $\mathbf{g}$ be as in Theorem 5.66, so that the spatial average $\bar{\mathbf{u}}(t)$ satisfies

$$\bar{\mathbf{u}}_t = \mathbf{f}(\bar{\mathbf{u}}) + \mathbf{g}$$

where

$$|\mathbf{g}| \leq c_3 e^{-\sigma t} \leq c_3 = M\lambda^{-1/2}\|\boldsymbol{\nabla}\mathbf{u}_0(\cdot)\|_{L^\infty(\Omega)} < M\lambda^{-1/2}\delta \leq \delta_1$$

for $t \geq 0$. Then for $\mathbf{p} \in \partial B$

$$(\mathbf{f}(\mathbf{p}) + \mathbf{g}) \cdot \mathbf{n}(\mathbf{p}) < 0$$

and B is an invariant set for the spatial average $\bar{\mathbf{u}}(t)$. Since

$$\operatorname{dist}(\mathbf{u}_0, \Gamma) < \delta \leq \delta_2 = \operatorname{dist}(\partial B, \Gamma)$$

then $\mathbf{u}_0 \in B$, and so $\bar{\mathbf{u}}(t) \in B$ for all t. It follows (Markus, 1956) that Γ is the

ω-limit set of $\bar{\mathbf{u}}$ in B, so that

$$\|\bar{\mathbf{u}}(t)-\boldsymbol{\gamma}(t-t_0)\|_{L^\infty(\Omega)}\to 0 \qquad \text{as} \quad t\to\infty$$

From Theorem 5.67

$$\|\mathbf{u}(\cdot,t)-\bar{\mathbf{u}}(t)\|_{L^\infty(\Omega)}\to 0 \qquad \text{as} \quad t\to\infty$$

so that

$$\|\mathbf{u}(\cdot,t)-\boldsymbol{\gamma}(t-t_0)\|_{L^\infty(\Omega)}\to 0 \qquad \text{as} \quad t\to\infty$$

and $\boldsymbol{\gamma}(t)$ is an asymptotically stable solution of the reaction-diffusion system, as required.

REMARK 5.75: We have thus shown that if diffusion is large enough, the behaviour of the reaction-diffusion system depends to a large extent on the reaction mechanism alone. For example, if the kinetic system admits a stable limit cycle solution, then this is also stable as a solution of the reaction-diffusion system.

It is of interest to know when diffusion is large enough, or equivalently, when the domain in which the reaction is taking place is small enough, for this result to hold in physical situations. It is of course not possible to put a precise value on this, since it depends on the particular reaction concerned. However, if typical reaction rates are of the order of $10\ \mathrm{s}^{-1}$ and diffusion constants of $10^5\ \mathrm{cm}^2\ \mathrm{s}^{-1}$ then the critical domain width is about 30 μm. Thus we might expect spatially uniform conditions in cells (unless some mechanism other than reaction and diffusion is important) but not necessarily in a chemical experiment in a test-tube.

Results similar to those obtained in this section may be proved using centre manifold theory (Carr and Muncaster, 1983a,b).

6. Bifurcation Theory

6.1 Introduction

Bifurcation theory is the study of equilibrium solutions of evolution equations of the form

$$\mathbf{u}_t = \mathbf{F}(\lambda, \mathbf{u}) \tag{6.1}$$

where $\mathbf{u}$ is in some Hilbert space H (or more generally some Banach space B), $\mathbf{F}(\lambda, \cdot)$ is some operator, and λ is a parameter which we shall take to be scalar for simplicity. In the context of reaction–diffusion systems, the equations may normally be written as

$$\mathbf{u}_t = \mathbf{f}(\lambda, \mathbf{u}) + D\,\nabla^2\mathbf{u} \tag{6.2}$$

where $\mathbf{f}(\lambda, \mathbf{u})$ represents the nonlinear reaction terms. We shall take D to be independent of λ, although this restriction is not necessary. We shall also assume that $\mathbf{f}$ is not explicitly dependent on $\mathbf{x}$ or t.

The Hilbert space H will normally be the completion under the norm derived from a suitable inner product $\langle \cdot, \cdot \rangle$ of the space of functions $\mathbf{u}(\mathbf{x}, t)$ defined on a domain $Q = \Omega \times (-\infty, \infty)$, where $\Omega \subset \mathbb{R}^n$, which are sufficiently smooth for the relevant derivatives to be continuous and such that the correct boundary conditions, or conditions at infinity, are satisfied.

Equilibrium solutions may be

(i) steady solutions
(ii) periodic solutions
(iii) sub-harmonic solutions
(iv) asymptotically quasi-periodic solutions

The last two of these are bifurcations of periodic solutions. We shall mainly be concerned with the first two categories. It will be necessary to consider both the existence and the stability of such solutions since an unstable equilibrium solution of the equations will not be observed in reality. Our treatment is based on the book of Iooss and Joseph (1980). For more rigorous treatment of bifurcation theory, see Crandall and Rabinowitz (1971, 1973, 1977), Sattinger (1971, 1973), or Chow and Hale (1982). For the particular case of bifurcations in reaction–diffusion systems, the work of

Mimura and co-workers (e.g. Mimura *et al.*, 1979) and the Brussels group (e.g. Nicolis and Prigogine, 1977) should be considered.

We shall first consider bifurcating solutions and their stability in one dimension, extend this to bifurcation of steady solutions from real simple eigenvalues in m dimensions and in infinite dimensions, and give some examples from biochemistry and developmental biology. We shall then consider the (periodic) Hopf bifurcation and the stability of the bifurcating solutions in two or more dimensions. The examples discussed will be drawn from physiology and chemistry.

6.2 Bifurcation in One Dimension

We shall first restrict our attention to spatially uniform solutions of (6.2) to obtain

$$\mathbf{u}_t = \mathbf{f}(\lambda, \mathbf{u}) \tag{6.3}$$

Such a problem is finite-dimensional and therefore represents a considerable simplification of the infinite-dimensional problem (6.2). However, many infinite-dimensional problems in bifurcation theory are essentially finite-dimensional in nature, so that (6.3) is worth studying both for its own sake and for the light it throws on (6.2). Moreover many finite-dimensional problems are essentially one- or two-dimensional, so that we consider first the one-dimensional problem

$$u_t = \dot{u} = f(\lambda, u) \tag{6.4}$$

The equilibrium solutions in this case can only be steady solutions and satisfy $f(\lambda, u) = 0$. The Banach space B is therefore just $\mathbb{R}^1$, and the bifurcation problem is equivalent to the study of the curves $f(\lambda, u) = 0$ in the (λ, u)-plane and their singular points. A bifurcation diagram shows the possible solutions u for any given value of the parameter λ. Let us assume that we have a solution $u = u^*$ for $\lambda = \lambda^*$, so that $f^* = f(\lambda^*, u^*) = 0$. We shall use asterisks to denote evaluation of f and its derivatives at $u = u^*$ and $\lambda = \lambda^*$. The main tool for the proof of the existence of solutions in bifurcation theory is the implicit function theorem, which may be stated for our present purposes as follows.

THEOREM 6.5 (Implicit Function Theorem in $\mathbb{R}^1$): Let $f(\lambda^*, u^*) = 0$ and let f be C^1 in some open neighbourhood of (λ^*, u^*) in the (λ, u)-plane. Then, if $f_u^* \neq 0$, there exist $\alpha, \beta > 0$ such that

(i) the equation $f(\lambda, u) = 0$ has a unique solution $u = u(\lambda)$ with $u^* - \beta < u < u^* + \beta$ whenever $\lambda^* - \alpha < \lambda < \lambda^* + \alpha$: moreover, this solution $u \in$

$C^1(\lambda^* - \alpha, \lambda^* + \alpha)$; and
(ii) $u_\lambda(\lambda) = -f_\lambda(\lambda, u(\lambda))/f_u(\lambda, u(\lambda))$.

Proof: See any elementary book on analysis.

REMARK 6.6: If $f_\lambda^* \neq 0$, we can solve for $\lambda = \lambda(u)$.

REMARK 6.7: If f is analytic, then so is $u(\lambda)$ or $\lambda(u)$.

REMARK 6.8: The theorem may be extended to vector-valued functions **u** and parameters **λ**.

We require the following definitions.

DEFINITIONS 6.9 (Classification of Points on Solution Curves): (i) A *regular point* (λ^*, u^*) of $f(\lambda, u) = 0$ is a point where the implicit function theorem may be applied, that is, either $f_u^* \neq 0$ or $f_\lambda^* \neq 0$. In this case there is a unique curve $u = u(\lambda)$ or $\lambda = \lambda(u)$ of solutions of $f(\lambda, u) = 0$ in the (λ, u)-plane passing through the point (λ^*, u^*), by the implicit function theorem. A *regular turning point* is a regular point at which $\lambda_u(u)$ changes sign. Here $f_u^* = 0$ but $f_\lambda^* \neq 0$ (see Fig. 6.10).

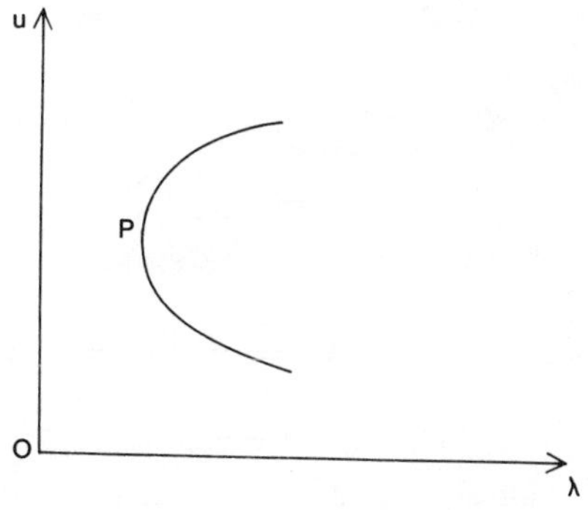

Fig. 6.10. A regular turning point.

(ii) A *singular point* is a point which is not regular, so that the implicit function theorem does not apply. Here $f_u^* = f_\lambda^* = 0$.
(iii) A *bifurcation point* is a singular point through which pass two or more branches of $f(\lambda, u) = 0$.
(iv) A *double point* is a singular point through which pass two and only two branches of $f(\lambda, u) = 0$ possessing distinct tangents, and where all second derivatives of f do not simultaneously vanish. A *turning double point*

is a double point at which λ_u changes sign on one branch (Fig. 6.11). The origin of the term bifurcation is obvious from this figure.

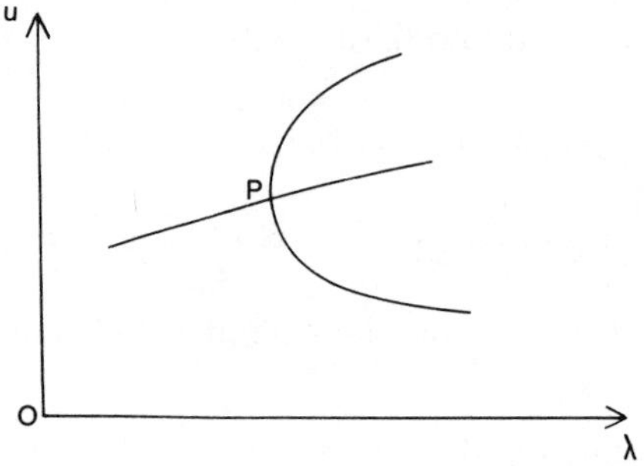

Fig. 6.11. A turning double point.

(v) A *cusp point* is a point of second-order contact between two branches of the curve (Fig. 6.12).

(vi) A *conjugate point* is an isolated singular point solution of $f=0$.

(vii) A *higher-order singular point* is a singular point at which all three second derivatives of f are zero.

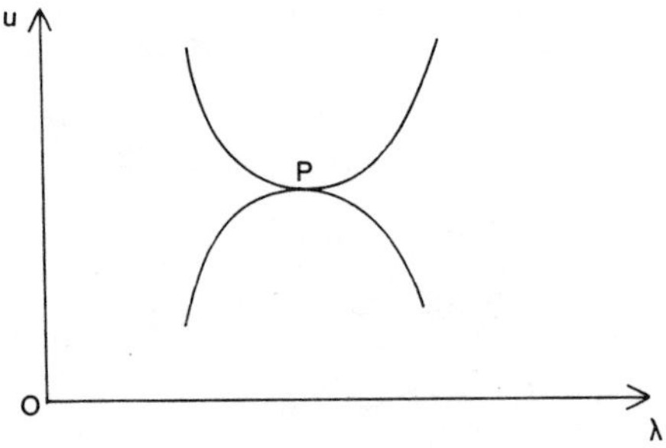

Fig. 6.12. A cusp point.

We shall consider first bifurcation at a double point. Our aim is to characterise those points at which a double point bifurcation occurs in terms of the function f. By definition we have $f_u^* = f_\lambda^* = 0$, but not all of the second derivatives of f are zero. Hence to leading order, the requirement that $f(\lambda, u) = 0$ becomes

$$f_{uu}^* \, du^2/2 + f_{\lambda u}^* \, du \, d\lambda + f_{\lambda\lambda}^* \, d\lambda^2/2 = 0 \qquad (6.13)$$

We require that this quadratic have two distinct real roots, corresponding to the two distinct tangents of the curve required at a double point. This implies that

$$\Delta = f_{\lambda u}^{*2} - f_{uu}^{*} f_{\lambda\lambda}^{*} > 0 \tag{6.14}$$

This is therefore a necessary condition for a double point bifurcation. In fact we may prove

THEOREM 6.15: The conditions $f^* = f_u^* = f_\lambda^* = 0$, $\Delta > 0$ are necessary and sufficient for a double point bifurcation if f is sufficiently smooth.

REMARK 6.16: In most applications in reaction–diffusion f is analytic, so that here and in the sequel we shall not consider exactly the smoothness conditions which are to be imposed on f, but shall assume that f is as smooth as is necessary.

Proof of sufficiency: We must show that if the conditions hold, then there exist two and only two branches of the curve $f(\lambda, u) = 0$ passing through the point (λ^*, u^*). Let us assume first that $f_{uu}^* \neq 0$. Then there are two solutions for the slope $du/d\lambda$ of the curve $f(\lambda, u) = 0$ given by solving

$$f_{uu}^*(du/d\lambda)^2 + 2f_{\lambda u}^*\, du/d\lambda + f_{\lambda\lambda}^* = 0 \tag{6.17}$$

We use this fact and the implicit function theorem to show the existence of the two branches of $f(\lambda, u) = 0$. Define $g(\lambda, v)$ by

$$\begin{aligned} g(\lambda, v) &= f(\lambda, u^* + (\lambda - \lambda^*)v)/(\lambda - \lambda^*)^2 \\ &= f_{uu}^* v^2/2 + f_{u\lambda}^* v + f_{\lambda\lambda}^*/2 + O(|\lambda - \lambda^*|) \end{aligned}$$

and let v^* be one or other of the solutions of

$$f_{uu}^* v^2/2 + f_{u\lambda}^* v + f_{\lambda\lambda}^*/2 = 0 \tag{6.18}$$

that is,

$$v^* = (-f_{u\lambda}^* \pm \sqrt{\Delta})/f_{uu}^*$$

Then

$$g^* = g(\lambda^*, v^*) = 0$$

and

$$g_v^* = f_{uu}^* v^* + f_{u\lambda}^* = \pm\sqrt{\Delta} \neq 0$$

Thus the implicit function theorem may be applied to g, and there exists a unique function $v = v(\lambda)$ in a neighbourhood of $\lambda = \lambda^*$ which satisfies $v(\lambda^*) = v^*$ and $g(\lambda, v) = 0$. It follows from the definition of $g(\lambda, v)$ that $u(\lambda) = u^* + (\lambda - \lambda^*)v(\lambda)$ satisfies $f(\lambda, u) = 0$ and from differentiating u that

$du/d\lambda = v^*$ at $\lambda = \lambda^*$, for either solution v^* of (6.18). But these are the two branches with distinct tangents required for a double point.

Let us now consider the other case and assume that $f^*_{uu} = 0$. Then $f^*_{u\lambda} \neq 0$, from (6.14), and on a branch of $f(\lambda, u) = 0$ either $d\lambda = 0$ or $du/d\lambda = -f^*_{\lambda\lambda}/2f^*_{u\lambda}$. A turning double point is of this form, since in this case $d\lambda/du$ changes sign on one branch, so that $d\lambda = 0$ on this branch at the singular point. The proof of the existence of the branch whose slope at the double point is finite goes through in exactly the same way as the proof for $f^*_{uu} \neq 0$. To prove the existence of the other branch, define

$$\begin{aligned} h(\mu, u) &= f(\lambda^* + \mu(u-u^*), u)/(u-u^*)^2 \\ &= f^*_{\lambda\lambda}\mu^2/2 + f^*_{u\lambda}\mu + O(|u-u^*|) \end{aligned}$$

Since $h^* = 0$ and $h^*_\mu = f^*_{u\lambda} \neq 0$, the implicit function theorem proves the existence of a unique function $\mu(u)$ such that $\mu(u^*) = 0$ which satisfies $h(\mu, u) = 0$, and thus the function $\lambda(u) = \lambda^* + (u-u^*)\mu(u)$ satisfies $f(\lambda, u) = 0$ and $d\lambda/du = 0$ at $u = u^*$. This proves the existence of the second branch and therefore that the singular point is a double point as required. Similar analyses may be carried out for bifurcation at other singular points.

6.3 Questions of Stability: The One-dimensional Case

The stability of the bifurcating solutions may often be determined by considering the linearised equation

$$v_t = f_u(\lambda, u)v$$

where u is the given solution; $\sigma = f_u(\lambda, u)$ is the eigenvalue of the linearised equation.

THEOREM 6.19 (Theorem 2.14): When $\sigma = f_u(\lambda, u) > 0$, the solution u is unstable, whereas when $\sigma < 0$, it is asymptotically stable. When $\sigma = 0$ stability is not determined by the linearised equation and higher-order terms must be considered.

REMARK 6.20: It follows immediately that the stability of solutions on the curve $f(u, \lambda) = 0$ cannot change except possibly at points where $f_u = 0$, assuming f is sufficiently smooth.

The following theorem, simple to state and prove in the present one-dimensional case, in conjunction with Theorem 6.19, allows us to make many deductions about the stability of bifurcating solutions. Note that it is sometimes convenient to solve $f(\lambda, u)=0$ for u as a function of λ, and sometimes for λ as a function of u.

THEOREM 6.21 (Factorisation Theorem in One Dimension): For every solution $u=u(\lambda)$ of $f(\lambda, u)=0$ we consider the eigenvalue σ as a function of λ also and define $\sigma(\lambda)=f_u(\lambda, u(\lambda))$. Then

$$u_\lambda \sigma = u_\lambda f_u = -f_\lambda \tag{6.22}$$

For every solution $\lambda=\lambda(u)$ of $f(\lambda, u)=0$ we consider σ as a function of u and define $\sigma(u)=f_u(\lambda(u), u)$. Here $f_u(\lambda, u)$ denotes the *partial* derivative with respect to the second variable. Then

$$\sigma = f_u = -\lambda_u f_\lambda \tag{6.23}$$

REMARK 6.24: Strictly speaking, we should use a different notation for $\sigma(\lambda)$ and $\sigma(u)$; however no confusion should arise.

Proof: The results follow from total differentiation of $f(u, \lambda)=0$ with respect to λ and u, respectively.

COROLLARY 6.25: $\sigma(u)$ must change sign as u is varied across a regular turning point.

Proof: At a regular turning point $f_u=0$ but $f_\lambda \neq 0$ and λ_u changes sign. Thus the result follows from (6.23). It follows from Theorem 6.19 that there is an interchange of stability at a regular turning point (see Fig. 4.38 for an example of this).

COROLLARY 6.26 (Transversality): (i) Let $u=u(\lambda)$, $\sigma=\sigma(\lambda)$ on one branch of the curve $f(\lambda, u)=0$ passing through (λ^*, u^*), let (λ^*, u^*) be a singular point and assume that $\sigma_\lambda^* \neq 0$. Then (λ^*, u^*) is a double point.

(ii) The same conclusion holds if $\lambda=\lambda(u)$, $\sigma=\sigma(u)$ on one branch of the curve and $\sigma_u^* \neq 0$.

Proof: By Theorem 6.15, we must show that $\Delta>0$. For (i) we have, differentiating $\sigma=f_u$ with respect to λ

$$\sigma_\lambda^* = f_{uu}^* u_\lambda^* + f_{u\lambda}^* \neq 0 \tag{6.27}$$

If $u_\lambda^*=0$, then from (6.17) $f_{\lambda\lambda}^*=0$, so that

$$\Delta = f_{u\lambda}^{*2} = \sigma_\lambda^{*2} > 0$$

as required. If $u_\lambda^* \neq 0$, then from (6.27) and (6.17)

$$0 \neq f_{uu}^* u_\lambda^{*2} + f_{\lambda u}^* u_\lambda^* = -f_{\lambda u}^* u_\lambda^* - f_{\lambda\lambda}^*$$

so that from (6.27)

$$\begin{aligned} 0 < \sigma_\lambda^{*2} &= f_{uu}^{*2} u_\lambda^{*2} + 2 f_{uu}^* f_{\lambda u}^* u_\lambda^* + f_{\lambda u}^{*2} \\ &= f_{uu}^* (-f_{\lambda u}^* u_\lambda^* - f_{\lambda\lambda}^*) + f_{uu}^* f_{\lambda u}^* u_\lambda^* + f_{\lambda u}^{*2} \\ &= f_{\lambda u}^{*2} - f_{uu}^* f_{\lambda\lambda}^* = \Delta \end{aligned}$$

so that $\Delta > 0$ and we again have a double point. The proof of (ii) is similar.

The conditions $\sigma_\lambda^* \neq 0$ or $\sigma_u^* \neq 0$ are "strict crossing" or "transversality" conditions, so called because they ensure that the eigenvalue σ crosses from negative to positive values, or *vice versa*. They were introduced by Hopf (1942). Such conditions enable us to determine the exchange of stability of branches at a bifurcation point.

At a double point the possibilities for exchange of stability are limited. For simplicity we shall restrict attention without loss of generality to problems of bifurcation from the trivial solution, that is to problems of the form

$$u_t = f(\lambda, u), \qquad f(\lambda, 0) = 0 \tag{6.28}$$

THEOREM 6.29 (Factorisation Theorem at a Double Point): At a double point of the problem (6.28), (i) the trivial solution changes stability (if it has eigenvalue σ, then $\sigma(\lambda) \sim f_{u\lambda}^*(\lambda - \lambda^*)$ as $\lambda \to \lambda^*$ and $f_{u\lambda}^* \neq 0$) and (ii) the non-trivial solution is stable for values of λ where the trivial solution is unstable, and unstable where it is stable (if it has eigenvalue $\tilde{\sigma}$, then $\tilde{\sigma}(u) \sim -n f_{u\lambda}^*(\lambda(u) - \lambda^*)$, or $\tilde{\sigma}(u) \sim -n\sigma(\lambda(u))$, as $u \to 0$ and $\lambda \to \lambda^*$, where n is the order of the first non-zero derivative of λ on the non-trivial branch with respect to u). If $\lambda \equiv \lambda^*$ on the non-trivial branch, then $\tilde{\sigma} = 0$. Note that we are comparing σ and $\tilde{\sigma}$ at the same value of λ, namely $\lambda(u)$.

REMARK 6.30: By $f(\lambda) \sim g(\lambda)$ as $\lambda \to \lambda^*$, we mean that $f(\lambda)/g(\lambda) \to 1$ as $\lambda \to \lambda^*$.

Proof of (*i*): From (6.28) $f_{\lambda\lambda}^* = 0$, so that the double point condition $\Delta > 0$ (Theorem 6.15) implies that

$$f_{u\lambda}^* \neq 0$$

On the trivial branch $\sigma = f_u(\lambda, 0)$, so that

$$\sigma = f_{u\lambda}^*(\lambda - \lambda^*)(1 + O(|\lambda - \lambda^*|)) \tag{6.31}$$

Thus the eigenvalue on the trivial branch changes sign as we pass through the double point, as required. Note that $\sigma_\lambda^* = f_{u\lambda}^* \neq 0$ so that we automatically have a transversality condition on this branch.

Proof of (ii): On the non-trivial branch λ and hence $\tilde{\sigma}$ may be found in terms of u (as shown in the proof of Theorem 6.15), so that we have, using the factorisation theorem (Theorem 6.21),

$$\begin{aligned}\tilde{\sigma} = f_u &= -f_\lambda \lambda_u = -f_{\lambda u}^* \lambda_u (u + O(|u|^2)) \\ &= -nf_{\lambda u}^*(\lambda - \lambda^*)(1 + O(|u|))\end{aligned} \tag{6.32}$$

if the nth derivative of λ at the bifurcation point is the first non-zero one, by comparing the Taylor expansions of λ_u and $\lambda - \lambda^*$ (on the non-trivial branch). A comparison of (6.31) with (6.32) immediately yields the result of the theorem. If $\lambda \equiv \lambda^*$ then $\lambda_u \equiv 0$, so that $\tilde{\sigma} = 0$.

EXAMPLE 6.33:

$$u_t = \lambda u - u^2 = u(\lambda - u) = f(\lambda, u) \tag{6.34}$$

Here the equilibrium solutions are given by $u = 0$ and $\lambda = u$. On $u = 0$

$$\sigma = f_u = \lambda - 2u = \lambda$$

On $\lambda = u$

$$\tilde{\sigma} = f_u = \lambda - 2u = -u = -\lambda$$

Here $n = 1$ since the first derivative of λ, λ_u, is non-zero. The transversality condition on the non-trivial branch is $\tilde{\sigma}_u^* \neq 0$, which holds. The bifurcation diagram is shown in Fig. 6.35.

EXAMPLE 6.36:

$$u_t = \lambda u - u^3 = u(\lambda - u^2) = f(\lambda, u) \tag{6.37}$$

Here $u = 0$ or $\lambda = u^2$. On $u = 0$,

$$\sigma = f_u = \lambda - 3u^2 = \lambda$$

On $\lambda = u^2$,

$$\tilde{\sigma} = f_u = \lambda - 3u^2 = -2u^2 = -2\lambda$$

Here $n = 2$ since on $\lambda = u^2$ we have $\lambda_u = 2u$ which is zero at the bifurcation point, but $\lambda_{uu} = 2 \neq 0$. The transversality condition on the non-trivial branch is $\tilde{\sigma}_u^* \neq 0$, which does not hold since $\tilde{\sigma}_u = -4u$, so that we do not know from transversality considerations whether an exchange of stability occurs on this branch. In fact it does not, by Theorem 6.29, as shown in the bifurcation diagram in Fig. 6.38.

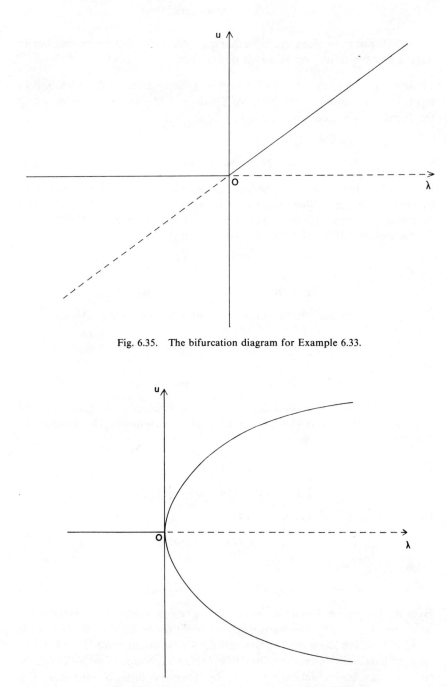

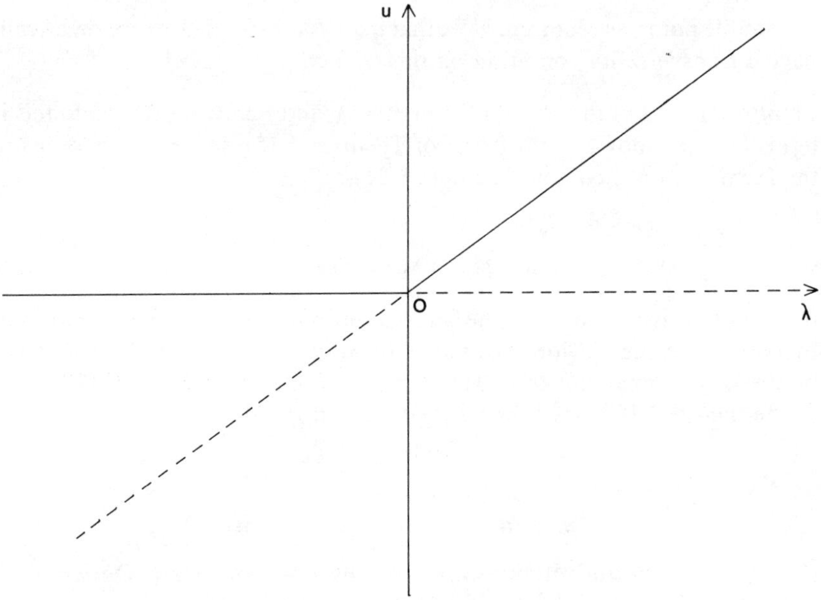

Fig. 6.35. The bifurcation diagram for Example 6.33.

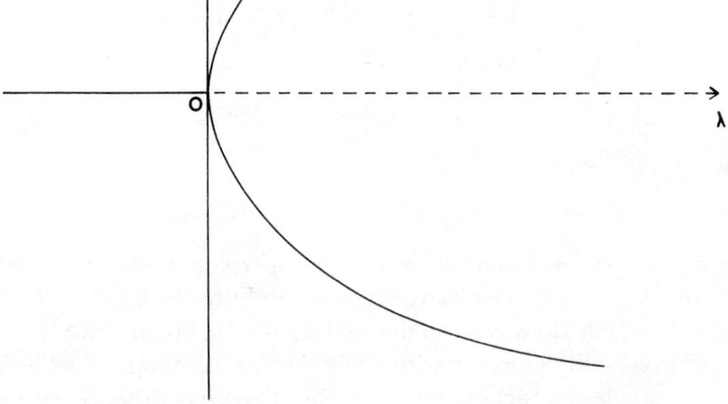

Fig. 6.38. The bifurcation diagram for Example 6.36.

EXAMPLE 6.39:

$$u_t = \lambda u - u^4 = u(\lambda - u^3) = f(\lambda, u) \tag{6.40}$$

On $u=0$, $\sigma=\lambda$. On $\lambda = u^3$, $\tilde{\sigma} = 3u^3 = -3\lambda$. Again $\tilde{\sigma}_u^* = 0$, and transversality considerations are insufficient to decide whether an exchange of stability takes place; Theorem 6.29 must again be used. The bifurcation diagram is shown in Fig. 6.41.

Exchange of stability at bifurcation points other than the double point may be dealt with similarly.

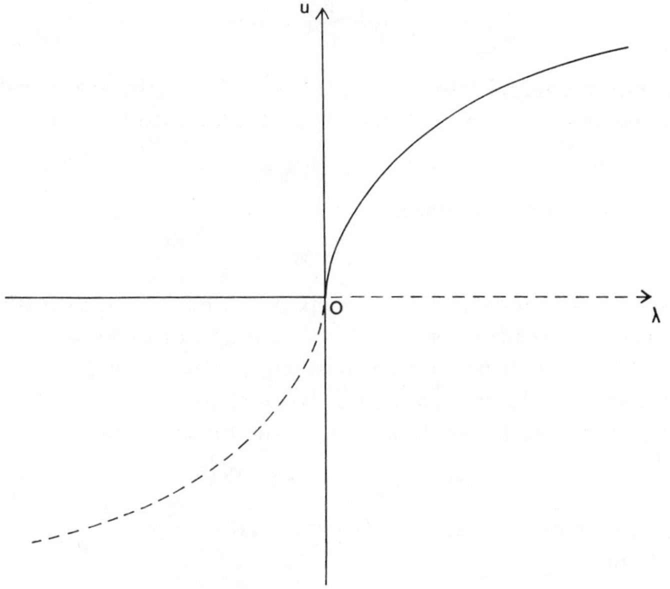

Fig. 6.41. The bifurcation diagram for Example 6.39.

6.4 Bifurcation in *m* Dimensions and in Infinite Dimensions

We now wish to consider the implications of our one-dimensional analysis to bifurcation in more than one dimension. We shall again consider bifurcation from the trivial solution, so that the problem may be written

$$\mathbf{u}_t = \mathbf{F}(\lambda, \mathbf{u}) = \mathbf{f}(\lambda, \mathbf{u}) + D\,\nabla^2 \mathbf{u}, \qquad \mathbf{f}(\lambda, \mathbf{0}) = \mathbf{0} \tag{6.42}$$

where $\mathbf{u} = (u_1, u_2, \ldots, u_m)^T$ and $\mathbf{u} \in B$, some Banach space. Stability of the trivial solution of the nonlinear system is often determined by the corresponding linear system

$$\mathbf{v}_t = \mathbf{F}_\mathbf{u}(\lambda, \mathbf{0}|\mathbf{v}) = L(\lambda)\mathbf{v}$$
$$= \mathbf{f}_\mathbf{u}(\lambda, \mathbf{0}|\mathbf{v}) + D\,\nabla^2\mathbf{v} = \mathbf{M}(\lambda)\mathbf{v} + D\,\nabla^2\mathbf{v} \tag{6.43}$$

where the linear operators $L(\lambda) = \mathbf{F}_\mathbf{u}(\lambda, \mathbf{0}|\cdot)$ and $M(\lambda) = \mathbf{f}_\mathbf{u}(\lambda, \mathbf{0}|\cdot)$ are defined by

$$L(\lambda)\mathbf{v} \equiv \frac{d}{dh}\mathbf{F}(\lambda, h\mathbf{v})|_{h=0} \tag{6.44}$$

$$M(\lambda)\mathbf{v} \equiv \frac{d}{dh}\mathbf{f}(\lambda, h\mathbf{v})|_{h=0} \tag{6.45}$$

In the m-dimensional case $D = 0$ and $L(\lambda) = M(\lambda)$, the $m \times m$ matrix whose (i, j)th element is given by $\partial f_i/\partial u_j(\lambda, \mathbf{0})$. Then the equation becomes

$$\mathbf{u}_t = \mathbf{f}(\lambda, \mathbf{u}) \tag{6.46}$$

and the linearised equation is

$$\mathbf{v}_t = M(\lambda)\mathbf{v}$$

Theorem 2.14 states that if the matrix $M(\lambda)$ has eigenvalues all of whose real parts are negative, then the trivial solution of (6.46) is asymptotically stable, whereas if it has one or more eigenvalues with positive real parts then the trivial solution is unstable. We wish to extend this theorem to the infinite-dimensional case. Here $L(\lambda)$ is the linear operator

$$L(\lambda) = M(\lambda) + D\,\nabla^2 \tag{6.48}$$

The eigenvalues $\sigma(\lambda)$ and eigenvectors $\boldsymbol{\phi}(\lambda)$ of this operator satisfy the spectral problem

$$L\boldsymbol{\phi} = \sigma\boldsymbol{\phi} \tag{6.49}$$

for $\boldsymbol{\phi} \in B$.

DEFINITION 6.50: The *resolvent set* of L is the set of complex numbers ρ for which $L - \rho I$ has a bounded inverse defined on the whole space.

DEFINITION 6.51: The *spectrum* $\Sigma(L)$ of L is the complement in the complex plane of the resolvent set.

It follows that if σ is an eigenvalue of L, then $\sigma \in \Sigma(L)$. For the simplest case of a system of reaction–diffusion equations defined on a bounded spatial domain, the spectrum consists entirely of eigenvalues, of which there are countably many.

We may now quote a theorem which corresponds to Theorem 2.14 for finite-dimensional systems, and gives criteria for asymptotic stability and instability of the trivial solution of (6.42).

THEOREM 6.52 (Principle of Linearised Stability, Kielhöfer, 1976): If $\Sigma(L)$ is contained in the left half plane and bounded away from the imaginary axis, then the trivial solution of (6.42) is asymptotically stable. If $\Sigma(L)$ contains a point in the right open half plane, the trivial solution is unstable.

REMARK 6.53: For systems on bounded spatial domains, we merely have to check that all eigenvalues are in the left half plane for asymptotic stability. See Theorem 5.61, which may easily be extended to deal with other boundary conditions than Neumann ones.

If it happens that at $\lambda = \lambda^*$ one or more eigenvalues cross from the left to the right half plane we will usually, but not always, get a bifurcation. The analysis of the bifurcating solutions depends on the nature of the eigenvalue, or eigenvalues, concerned. We require some definitions.

In the finite-dimensional case where $L = M$, a matrix, the eigenvalues of L (or M) satisfy

$$P(\sigma) = \det(M - \sigma I) = 0$$

DEFINITION 6.54: The *algebraic multiplicity* m_i of the root σ_i of $P(\sigma) = 0$ is the number of factors $\sigma - \sigma_i$ in the polynomial $P(\sigma) = 0$.

This quantity may also be defined in the more general context of infinite-dimensional linear operators.

DEFINITION 6.55: The *algebraic multplicity* m_i of the eigenvalue σ_i is defined to be the dimension of the union of the null spaces of $(L - \sigma_i I)^k$, where $k = 1, 2, \ldots$. It can be seen that this is equivalent to Definition 6.54 in the case in which L is the matrix M.

DEFINITION 6.56: σ_i is a *simple* eigenvalue of L if $m_i = 1$.

We shall usually be concerned with simple eigenvalues since the eigenvalues of reaction–diffusion systems in bounded domains are often simple.

6.5 Bifurcation of Steady Solutions from Real Simple Eigenvalues

Let $\sigma(\lambda)$ be an isolated real simple eigenvalue of $L(\lambda)$. For definiteness we shall take it to be the principal eigenvalue, or eigenvalue of greatest real part, of $L(\lambda)$. In this section we consider bifurcation from the trivial solution when this eigenvalue increases past zero. It is thus the m-dimensional or infinite-dimensional analogue of bifurcation in the one-dimensional case. We shall construct the non-trivial bifurcating solutions under a transversality condition.

DEFINITION 6.57.: The *adjoint operator* $L^*(\lambda)$ of $L(\lambda)$ is defined to be that operator which satisfies

$$\langle L\mathbf{u}, \mathbf{v}\rangle = \langle \mathbf{u}, L^*\mathbf{v}\rangle \tag{6.58}$$

for all $\mathbf{u}$ in the domain of L and $\mathbf{v}$ in the domain of L^*.

In the finite-dimensional case $L = M$, a real matrix. We shall again consider steady solutions so that $\mathbf{u} \in B = \mathbb{R}^m$. The appropriate Hilbert space H is $\mathbb{C}^m$ with inner product $\langle \cdot, \cdot \rangle$ given by $\langle \mathbf{u}, \mathbf{v}\rangle = \bar{\mathbf{u}}^{\mathrm{T}}\bar{\mathbf{v}}$, and $L^* = M^{\mathrm{T}}$ where the superscript T denotes the transpose. In the infinite-dimensional case $L = M + D\nabla^2$, where M and D are real matrices, and $\mathbf{u} \in B = C^2(\Omega, \mathbb{R}^m)$ together with appropriate boundary conditions. The appropriate Hilbert space H is $L^2[\Omega]$, with inner product given by $\langle \mathbf{u}, \mathbf{v}\rangle = \int_\Omega \mathbf{u}^{\mathrm{T}}\bar{\mathbf{v}}\, d\mathbf{x}$, and $L^* = M^{\mathrm{T}} + D^{\mathrm{T}}\nabla^2$. It is easy to show that σ defined above is also the principal eigenvalue of L^* and is a simple isolated eigenvalue of L^*. Let $\boldsymbol{\phi}(\lambda)$ be the eigenvector of L and $\boldsymbol{\phi}^*(\lambda)$ that of L^* associated with σ. Then

$$L\boldsymbol{\phi} = \sigma\boldsymbol{\phi}, \qquad L^*\boldsymbol{\phi}^* = \sigma\boldsymbol{\phi}^* \tag{6.59}$$

and $\boldsymbol{\phi}$ and $\boldsymbol{\phi}^*$ may be normalised so that

$$\langle \boldsymbol{\phi}, \boldsymbol{\phi}^*\rangle = 1 \tag{6.60}$$

We are looking for a bifurcation from the trivial solution at a critical value of the bifurcation parameter λ, which we take to be $\lambda = 0$ without loss of generality. We shall assume that $\sigma'(0) > 0$ so that the trivial solution loses stability as λ increases past zero; thus on this branch we have $\sigma = \sigma(\lambda)$ and

$$\sigma(0) = 0, \qquad \sigma'(0) > 0 \tag{6.61}$$

The case for which $\sigma'(0) < 0$ can be dealt with similarly. The condition $\sigma'(0) \neq 0$ is the appropriate transversality condition. From (6.59) and (6.60) it follows that $\langle L\boldsymbol{\phi}, \boldsymbol{\phi}^*\rangle = \sigma$, so that

$$\langle L(0)\boldsymbol{\phi}(0), \boldsymbol{\phi}^*(0)\rangle = \sigma(0) = 0$$

and, differentiating and noting that $(d/d\lambda)\langle \boldsymbol{\phi}, \boldsymbol{\phi}^*\rangle = 0$,

$$\langle L'(0)\boldsymbol{\phi}(0), \boldsymbol{\phi}^*(0)\rangle = \sigma'(0) > 0 \tag{6.62}$$

We now look for the bifurcating steady solution $(\lambda, \mathbf{u})$ of

$$\mathbf{u}_t = \mathbf{F}(\lambda, \mathbf{u}) = \mathbf{0} \tag{6.63}$$

Rather than look for $\mathbf{u}$ as a function of λ (just as in the one-dimensional case $\mathbf{u}$ may not be well-defined as such a function), we expand both $\mathbf{u}$ and λ in terms of a small parameter ε, so that

$$\mathbf{u} = \sum_{n=1}^{\infty} \varepsilon^n \mathbf{u}_n, \qquad \lambda = \sum_{n=1}^{\infty} \varepsilon^n \lambda_n \tag{6.64}$$

We are free to define ε, which from the first of (6.64) is a measure of the amplitude of the bifurcating solution, as convenient. To motivate our definition, consider the linearisation of the governing equation (6.63) at the bifurcation point, namely

$$\mathbf{F}_{\mathbf{u}}(0, \mathbf{0} \mid \mathbf{u}) = \mathbf{0} \qquad \text{or} \qquad L(0)\mathbf{u} = \mathbf{0}$$

This has solution $\mathbf{u} = \alpha\boldsymbol{\phi}(0)$, where α is any scalar. Since we wish $\mathbf{u}$ to be an $O(\varepsilon)$ quantity, let us take $\alpha = \varepsilon$, so that $\mathbf{u} = \varepsilon\boldsymbol{\phi}(0)$. Then, taking the inner product with $\boldsymbol{\phi}^*(0)$,

$$\varepsilon = \langle \mathbf{u}, \boldsymbol{\phi}^*(0) \rangle \tag{6.65}$$

This is the definition of ε which we use in the fully nonlinear problem. By substituting the power series (6.64) for $\mathbf{u}$ into this definition and equating powers of ε, we obtain

$$\langle \mathbf{u}_1, \boldsymbol{\phi}^*(0) \rangle = 1, \qquad \langle \mathbf{u}_n, \boldsymbol{\phi}^*(0) \rangle = 0, \qquad n > 1 \tag{6.66}$$

These may be thought of as normalisation conditions. Now, substituting (6.64) into (6.63) and equating powers of ε we have, assuming that $\mathbf{F}(\lambda, \mathbf{u})$ is analytic,

$$\mathbf{F}_{\mathbf{u}}(0, \mathbf{0} \mid \mathbf{u}_1) = \mathbf{0} \tag{6.67}$$

$$\mathbf{F}_{\mathbf{u}}(0, \mathbf{0} \mid \mathbf{u}_2) + \lambda_1 \mathbf{F}_{\mathbf{u}\lambda}(0, \mathbf{0} \mid \mathbf{u}_1) + \mathbf{F}_{\mathbf{uu}}(0, \mathbf{0} \mid \mathbf{u}_1 \mid \mathbf{u}_1)/2 = 0 \tag{6.68}$$

and in general

$$\mathbf{F}_{\mathbf{u}}(0, \mathbf{0} \mid \mathbf{u}_n) + \lambda_{n-1} \mathbf{F}_{\mathbf{u}\lambda}(0, \mathbf{0} \mid \mathbf{u}_1) + \mathbf{G}_n = \mathbf{0} \tag{6.69}$$

where $\mathbf{G}_n$ depends on lower-order terms. The analyticity assumption may be weakened, but since $\mathbf{F}$ is analytic in virtually all reaction–diffusion systems, we do not consider this extension. Equation (6.67) is just the linear problem considered above, and with the first equation of (6.66) gives

$$\mathbf{u}_1 = \boldsymbol{\phi}(0) \tag{6.70}$$

It is not immediately obvious that (6.68) has a solution $\mathbf{u}_2$, nor whether the solution, if it exists, is unique. The answers to these questions are given by the following fundamental theorem:

THEOREM 6.71 (Fredholm Alternative): If $L(0) = \mathbf{F_u}(0, \mathbf{0}|\cdot)$ has a simple isolated eigenvalue $\sigma(0) = 0$ with eigenvector $\boldsymbol{\phi}(0)$ and adjoint eigenvector $\boldsymbol{\phi}^*(0)$, then, given $\mathbf{g} \in H$, the equation

$$\mathbf{F_u}(0, \mathbf{0}|\mathbf{v}) = \mathbf{g} \tag{6.72}$$

is soluble for $\mathbf{v} \in H$ if and only if

$$\langle \mathbf{g}, \boldsymbol{\phi}^*(0) \rangle = 0 \tag{6.73}$$

This is known as a Fredholm orthogonality condition. If this condition holds, then the solution is unique if it is subject to a normalisation condition

$$\langle \mathbf{v}, \boldsymbol{\phi}^*(0) \rangle = k \tag{6.74}$$

for any given constant k.

REMARK 6.75: The necessity of condition (6.73) follows immediately from (6.72) by taking the inner product with $\boldsymbol{\phi}^*(0)$:

$$\begin{aligned}\langle \mathbf{g}, \boldsymbol{\phi}^*(0) \rangle &= \langle \mathbf{F_u}(0, \mathbf{0}|\mathbf{v}), \boldsymbol{\phi}^*(0) \rangle = \langle L(0)\mathbf{v}, \boldsymbol{\phi}^*(0) \rangle \\ &= \langle \mathbf{v}, L^*(0)\boldsymbol{\phi}^*(0) \rangle = \sigma(0)\langle \mathbf{v}, \boldsymbol{\phi}^*(0) \rangle = 0\end{aligned}$$

The sufficiency follows from the fact that $L(0)$ restricted to the complement of the eigenvector $\boldsymbol{\phi}(0)$ is invertible. Uniqueness follows since, given any solution $\mathbf{v}_0$ of (6.72), the general solution is $\mathbf{v}_0 + c\boldsymbol{\phi}(0)$ for an arbitrary constant c, and the normalisation condition (6.74) gives

$$\langle \mathbf{v}_0 + c\boldsymbol{\phi}(0), \boldsymbol{\phi}^*(0) \rangle = k \qquad \text{or} \qquad c = k - \langle \mathbf{v}_0, \boldsymbol{\phi}^*(0) \rangle$$

so that c is determined.

We may now continue with the analysis of Eqs. (6.67)–(6.69) with the normalisation (6.66). The Fredholm alternative states that (6.68) has a solution if and only if

$$\langle \lambda_1 \mathbf{F_{u\lambda}}(0, \mathbf{0}|\mathbf{u}_1) + \mathbf{F_{uu}}(0, \mathbf{0}|\mathbf{u}_1|\mathbf{u}_1)/2, \boldsymbol{\phi}^*(0) \rangle = 0$$

that is, if

$$\lambda_1 = -\frac{1}{2} \frac{\langle \mathbf{F_{uu}}(0, \mathbf{0}|\mathbf{u}_1|\mathbf{u}_1), \boldsymbol{\phi}^*(0) \rangle}{\langle \mathbf{F_{u\lambda}}(0, \mathbf{0}|\mathbf{u}_1), \boldsymbol{\phi}^*(0) \rangle} \tag{6.76}$$

as long as $\langle \mathbf{F_{u\lambda}}(0, \mathbf{0}|\mathbf{u}_1), \boldsymbol{\phi}^*(0) \rangle \neq 0$. But

$$\langle \mathbf{F_{u\lambda}}(0, \mathbf{0}|\mathbf{u}_1), \boldsymbol{\phi}^*(0) \rangle = \langle L'(0)\boldsymbol{\phi}(0), \boldsymbol{\phi}^*(0) \rangle = \sigma'(0) > 0$$

from the transversality condition (6.61), since $\mathbf{u}_1 = \boldsymbol{\phi}(0)$ (6.70). This determines λ_1. Then $\mathbf{u}_2$ may be obtained from (6.68) by using the normalisation condition (6.66). Similarly, λ_{n-1} may be obtained from (6.69) by taking the inner product with $\boldsymbol{\phi}^*(0)$, explicitly

$$\lambda_{n-1} = -\frac{\langle \mathbf{G}_n, \boldsymbol{\phi}^*(0)\rangle}{\langle \mathbf{F}_{\mathbf{u}\lambda}(0, \mathbf{0}\,|\,\mathbf{u}_1), \boldsymbol{\phi}^*(0)\rangle} = -\frac{\langle \mathbf{G}_n, \boldsymbol{\phi}^*(0)\rangle}{\sigma'(0)} \tag{6.77}$$

and $\mathbf{u}_n$ by solving (6.69) subject to the normalisation condition $\langle \mathbf{u}_n, \boldsymbol{\phi}^*(0)\rangle = 0$. It follows that the bifurcating solution $(\lambda, \mathbf{u})$ given by (6.63) may be constructed.

We have shown that bifurcation always takes place at an isolated real simple eigenvalue when a transversality condition holds. This is a special case of Krasnoselskii's (1964) theorem which states that under certain compactness conditions, bifurcation always takes place at an eigenvalue of odd multiplicity. Rabinowitz (1971) showed that if the same compactness conditions hold and λ ranges over the real line, then the bifurcating branch is either unbounded or meets $u = 0$ at a second bifurcation point $(\mu, 0)$, where $\mu \neq 0$. The proof in both cases is by degree theory.

6.6 Stability of the Bifurcating Solutions: The Factorisation Theorem

We are again interested in the stability of the bifurcating solutions, so we consider the linearisation of the system about the bifurcating solution $(\lambda(\varepsilon), \mathbf{u}(\varepsilon))$ which we have just shown how to construct. This gives

$$\mathbf{v}_t = \mathbf{F}_{\mathbf{u}}(\lambda(\varepsilon), \mathbf{u}(\varepsilon)\,|\,\mathbf{v}) = \tilde{L}(\varepsilon)\mathbf{v} \tag{6.78}$$

say. Note that $\tilde{L}$ is the linearisation about the bifurcating solution, whereas L was the linearisation about the trivial solution. At $\varepsilon = 0$, $\tilde{L}(0) = L(0)$, so that the eigenvalues of $\tilde{L}(0)$ are the same as those of $L(0)$. By assumption, $L(0)$ has a simple eigenvalue $\sigma(0) = 0$, and all its other eigenvalues are bounded away from the imaginary axis in the left half plane. It follows that for ε sufficiently small, $\tilde{L}(\varepsilon)$ has a simple real eigenvalue $\tilde{\sigma}(\varepsilon)$ (since complex eigenvalues occur in conjugate pairs), satisfying $\sigma(0) = 0$, and all its other eigenvalues are in the left half plane. Let $\tilde{\boldsymbol{\phi}}(\varepsilon)$ be the corresponding eigenvector, so that

$$\tilde{L}\tilde{\boldsymbol{\phi}} = \tilde{\sigma}\tilde{\boldsymbol{\phi}} \tag{6.79}$$

The adjoint operator $\tilde{L}^*$ also has eigenvalue $\tilde{\sigma}$ with eigenvector $\tilde{\boldsymbol{\phi}}^*$, say,

$$\tilde{L}^*\tilde{\boldsymbol{\phi}}^* = \tilde{\sigma}\tilde{\boldsymbol{\phi}}^* \tag{6.80}$$

We may now state the generalisation of the factorisation Theorem 6.21 (cf. Theorem 6.29).

THEOREM 6.81 (Factorisation Theorem in m Dimensions or in Infinite Dimensions): Consider the bifurcation problem

$$\mathbf{u}_t = \mathbf{F}(\lambda, \mathbf{u}) = \mathbf{0}, \qquad \mathbf{F}(\lambda, \mathbf{0}) = \mathbf{0} \tag{6.82}$$

where the linearisation of $\mathbf{F}(\lambda, \cdot\,)$ about the trivial solution has a simple real eigenvalue $\sigma(\lambda)$ of greatest real part satisfying

$$\sigma(0) = 0, \qquad \sigma'(0) > 0 \tag{6.83}$$

Let $\tilde{\sigma}(\varepsilon)$ be the eigenvalue associated with the non-trivial bifurcating solution $(\lambda(\varepsilon), \mathbf{u}(\varepsilon))$ as defined above. Then

$$\tilde{\sigma}(\varepsilon) = -\lambda_\varepsilon \frac{\langle \mathbf{F}_\lambda(\lambda, \mathbf{u}), \tilde{\boldsymbol{\phi}}^* \rangle}{\langle \mathbf{u}_\varepsilon, \tilde{\boldsymbol{\phi}}^* \rangle} \tag{6.84}$$

as long as $\langle \mathbf{u}_\varepsilon, \boldsymbol{\phi}^* \rangle \neq 0$. As $\varepsilon \to 0$, $\tilde{\sigma} \sim -n\lambda\sigma'(0)$ or

$$\tilde{\sigma}(\varepsilon) \sim -n\sigma(\lambda(\varepsilon)) \tag{6.85}$$

where n is the order of the first non-zero derivative of λ with respect to ε. Note that this compares σ and $\tilde{\sigma}$ at the same value of λ, namely $\lambda(\varepsilon)$.

Proof: Since $\mathbf{u}(\varepsilon)$ is a steady bifurcating solution with $\lambda = \lambda(\varepsilon)$,

$$\mathbf{F}(\lambda(\varepsilon), \mathbf{u}(\varepsilon)) = \mathbf{0}$$

Differentiating with respect to ε

$$\lambda_\varepsilon \mathbf{F}_\lambda(\lambda, \mathbf{u}) + \mathbf{F}_\mathbf{u}(\lambda, \mathbf{u} | \mathbf{u}_\varepsilon) = \mathbf{0}$$

Taking the inner product with $\tilde{\boldsymbol{\phi}}^*$

$$\begin{aligned}\langle \mathbf{F}_\mathbf{u}(\lambda, \mathbf{u} | \mathbf{u}_\varepsilon), \tilde{\boldsymbol{\phi}}^* \rangle &= \langle \tilde{L}\mathbf{u}_\varepsilon, \tilde{\boldsymbol{\phi}}^* \rangle = \langle \mathbf{u}_\varepsilon, \tilde{L}^* \tilde{\boldsymbol{\phi}}^* \rangle \\ &= \tilde{\sigma} \langle \mathbf{u}_\varepsilon, \tilde{\boldsymbol{\phi}}^* \rangle = -\lambda_\varepsilon \langle \mathbf{F}_\lambda(\lambda, \mathbf{u}), \tilde{\boldsymbol{\phi}}^* \rangle \end{aligned} \tag{6.86}$$

which proves (6.84) if $\langle \mathbf{u}_\varepsilon, \boldsymbol{\phi}^* \rangle \neq 0$. As $\varepsilon \to 0$, $\mathbf{u} \to \varepsilon\boldsymbol{\phi}$, $\mathbf{u}_\varepsilon \to \boldsymbol{\phi}$, $\tilde{\boldsymbol{\phi}}^* \to \boldsymbol{\phi}^*$, $\mathbf{F}_\lambda(\lambda, \mathbf{u}) \to \mathbf{F}_{\lambda\mathbf{u}}(0, \mathbf{0} | \mathbf{u}) = L'(0)\mathbf{u} \to \varepsilon\sigma'(0)\boldsymbol{\phi}$, so that, from (6.84),

$$\tilde{\sigma} \sim -\varepsilon\lambda_\varepsilon\sigma'(0) \sim -n\lambda\sigma'(0)$$

comparing the Taylor expansions of λ_ε and λ. Since $\sigma(\lambda) = \lambda\sigma'(0) + O(\lambda^2)$, this gives $\tilde{\sigma} \sim -n\sigma$ as required. It follows that solutions which bifurcate super-critically (so that $\lambda > 0$ on the bifurcating solution) are stable, whereas those which bifurcate sub-critically are unstable. This result should be compared with the one-dimensional double point case (Theorem 6.29, see also Figs. 6.35, 6.38 and 6.41).

6.7 Bifurcation Analysis in a Finite One-dimensional Domain

This is the easiest case since the spectrum of L is discrete, and each function $\mathbf{u}_n$ may be found explicitly. Much work has been done on such problems since Turing (1952) postulated the combination of reaction and diffusion as the origin of spatial structure in certain biological systems. In particular, the work of Gierer and Meinhardt (1972), see also Gierer (1981), Meinhardt (1982), the Brussels school [see, for example, the book by Nicolis and Prigogine (1977)], Fife [see, for example, Fife (1979a)], and the surveys of Conway (1984) and Levin and Segel (1985) should be mentioned. As an example we shall consider a system arising in enzyme kinetics, namely

$$u_t = \lambda f(u, v) + D_1 \nabla^2 u, \qquad v_t = \lambda g(u, v) + D_2 \nabla^2 v \tag{6.87}$$

where

$$\begin{aligned} f(u, v) &= J_1 - u - \rho h(u, v) \\ g(u, v) &= \alpha(J_2 - v) - \rho h(u, v), \end{aligned} \tag{6.88}$$

and

$$h(u, v) = uv/(1 + u + Ku^2) \tag{6.89}$$

Here D_1, D_2, J_1, J_2, α, ρ and K are considered fixed, whereas λ is the variable, but positive, bifurcation parameter. Increasing λ corresponds to increasing the size of the domain. This system has been studied experimentally by Thomas (1976) and numerically by Kernevez *et al.* (1979); see also Kernevez (1980). It is qualitatively similar to a substrate inhibition model proposed by Seelig (1976) and studied by Mimura and Murray (1978b) and Britton and Murray (1979), and to other models in the literature. The system has been considered as a model for hydranth regeneration in the marine hydroid *Tubularia* by Britton *et al.* (1983); this work will be described later. It is easy to show that in a certain parameter range the system admits a single non-trivial steady state (u_0, v_0) which is stable to spatially homogeneous perturbation. We shall take the parameters to lie in this range and consider the stability of the steady state to non-homogeneous perturbations. In one dimension the system becomes

$$u_t = \lambda f(u, v) + D_1 u_{xx}, \qquad v_t = \lambda g(u, v) + D_2 v_{xx} \tag{6.90}$$

The finite domain may be taken to be $0 < x < \pi$ by rescaling the variables. We take zero-flux conditions at the boundaries:

$$u_x(0, t) = v_x(0, t) = 0, \qquad u_x(\pi, t) = v_x(\pi, t) = 0 \tag{6.91}$$

Dirichlet end conditions may also be treated by a similar analysis. The

Hilbert space is $L^2[0, \pi]$ with inner product

$$\langle \mathbf{w}_1, \mathbf{w}_2 \rangle = \int_0^\pi \mathbf{w}_1 \cdot \bar{\mathbf{w}}_2 \, dx \tag{6.92}$$

In this case, the linear operator L which we are considering is given by

$$L = \begin{bmatrix} D_1 \dfrac{d^2}{dx^2} + \lambda f_u & \lambda f_v \\ \lambda g_u & D_2 \dfrac{d^2}{dx^2} + \lambda g_v \end{bmatrix}$$

and the spectral problem is given by

$$L\boldsymbol{\phi} = \sigma \boldsymbol{\phi}$$

Since the eigenvalues of $-d^2/dx^2$ on the domain $(0, \pi)$ with zero-flux boundary conditions are given by k^2 for $k = 0, 1, 2, \ldots$, then the eigenvalues σ_k of L must satisfy

$$\begin{vmatrix} \sigma_k + D_1 k^2 - \lambda f_u & -\lambda f_v \\ -\lambda g_u & \sigma_k + D_2 k^2 - \lambda g_v \end{vmatrix} = 0 \tag{6.93}$$

Defining M by

$$M = \begin{bmatrix} f_u & f_v \\ g_u & g_v \end{bmatrix} \tag{6.94}$$

and the trace and determinant of M by

$$\operatorname{tr} M = f_u + g_v, \qquad \det M = f_u g_v - f_v g_u \tag{6.95}$$

Eq. (6.93) becomes

$$\sigma_k^2 + \{(D_1 + D_2)k^2 - \lambda \operatorname{tr} M\}\sigma_k + Q_k(\lambda) = 0 \tag{6.96}$$

where

$$Q_k(\lambda) = D_1 D_2 k^4 - \lambda k^2 (D_1 g_v + D_2 f_u) + \lambda^2 \det M \tag{6.97}$$

The Routh–Hurwitz criteria state that both the roots σ_k of (6.96) have negative real parts if and only if

$$(D_1 + D_2)k^2 - \lambda \operatorname{tr} M > 0 \tag{6.98}$$

$$Q_k(\lambda) > 0 \tag{6.99}$$

We have assumed that the steady state (u_0, v_0) is stable to spatially uniform perturbations. Such perturbations correspond to the zero eigenvalue of d^2/dx^2, so that $k = 0$. It follows that both roots σ_0 of (6.96) with $k = 0$ have negative real part, and so from (6.98), (6.99) and (6.97),

$$\operatorname{tr} M^0 < 0 \tag{6.100}$$

$$\det M^0 > 0 \tag{6.101}$$

where the superscript zero here and in the sequel denotes evaluation at the steady state. It follows that (6.98) holds at the steady state for all eigenvalues k^2, so that the only way that instability can set in is for (6.99) to be violated at the steady state. Since $\det M^0 > 0$, the equation $Q_k^0(\lambda) = 0$ has either no real positive roots or two, and it has two if

$$D_1 g_v^0 + D_2 f_u^0 > \sqrt{(4D_1 D_2 \det M^0)} > 0 \tag{6.102}$$

In this case the steady state becomes unstable for values of λ between the roots of $Q_k^0(\lambda) = 0$. This instability is caused entirely by diffusion (since if $D_1 = D_2 = 0$, the steady state is stable) and is called diffusional, or Turing, instability. For a comparison of the domains of Turing instability in different reaction–diffusion models see Murray (1982). For diffusional instability we must have $f_u^0 + g_v^0 < 0$, by (6.100), and $D_1 g_v^0 + D_2 f_u^0 > 0$. By (6.102). If $D_1 = D_2$, these cannot both hold, so it is impossible for scalar diffusion to destabilise a stable steady state. This was shown by the method of invariant sets and by spectral theory in Chapter 5. If $D_1 \neq D_2$, then for both conditions to hold we must have one of f_u^0 and g_v^0 positive and the other negative. Let us assume for definiteness that $f_u^0 > 0$, $g_v^0 < 0$ and $f_u^0 + g_v^0 < 0$. Then (6.102) holds, so that diffusional instability occurs, whenever D_2/D_1 is sufficiently large.

If (6.102) holds, then instability of the kth mode occurs at the value of λ given by the smaller positive solution of $Q_k^0(\lambda) = 0$, that is, at

$$\begin{aligned} \lambda &= k^2 \lambda_0 \\ &= k^2 \frac{D_1 g_v^0 + D_2 f_u^0 - \sqrt{(D_1 g_v^0 + D_2 f_u^0)^2 - 4D_1 D_2 \det M^0}}{2 \det M^0} \end{aligned} \tag{6.103}$$

which defines $\lambda_0 > 0$. Since the kth mode becomes unstable at $k^2 \lambda_0$, the first mode to become unstable as λ is increased is $k = 1$, and this occurs at $\lambda = \lambda_0$. We therefore carry out a bifurcation analysis about $\lambda = \lambda_0$, $u = u_0$ and $v = v_0$. Since $Q_1^0(\lambda_0) = 0$ (by the definition of λ_0), Eq. (6.96) implies that the bifurcation involves a simple real eigenvalue becoming equal to zero. It is easy to show that $Q_1'(\lambda_0) < 0$, so that

$$\sigma_1'(\lambda_0) = -\frac{Q_1'(\lambda_0)}{D_1 + D_2 - \lambda_0 \operatorname{tr} M^0} \tag{6.104}$$

and the transversality condition holds; we have strict loss of stability of the trivial solution, i.e. the steady state, as λ increases past λ_0. Thus the bifurcation analysis of the previous section is applicable.

We require the first eigenfunction of L and its adjoint at the bifurcation point. Since the only differential operator in L is d^2/dx^2 and since we have zero-flux boundary conditions at $x=0$ and $x=\pi$, the eigenfunction corresponding to $k=1$ must be of the form

$$\boldsymbol{\phi}(\lambda_0)=\begin{bmatrix} c_1 \\ c_2 \end{bmatrix} \cos x \tag{6.105}$$

where c_1 and c_2 are related by the equations

$$\begin{aligned}(-D_1+\lambda_0 f_u^0)c_1+\lambda_0 f_v^0 c_2&=0\\ \lambda_0 g_u^0 c_1+(-D_2+\lambda_0 g_v^0)c_2&=0\end{aligned} \tag{6.106}$$

By the definition of λ_0 these two equations are linearly dependent, and hence either of them may be used to give the ratio of c_1 to c_2. The adjoint operator L^* is given by

$$L^*=\begin{bmatrix} D_1\dfrac{d^2}{dx^2}+\lambda f_u & \lambda g_u \\ \lambda f_v & D_2\dfrac{d^2}{dx^2}+\lambda g_v \end{bmatrix}$$

so that its eigenfunction corresponding to $k=1$ at the bifurcation point is given by

$$\boldsymbol{\phi}^*(\lambda_0)=\begin{bmatrix} d_1 \\ d_2 \end{bmatrix} \cos x \tag{6.107}$$

where d_1 and d_2 are related by

$$\begin{aligned}(-D_1+\lambda_0 f_u^0)d_1+\lambda_0 g_u^0 d_2&=0\\ \lambda_0 f_v^0 d_1+(-D_2+\lambda_0 g_v^0)d_2&=0\end{aligned} \tag{6.108}$$

These equations are linearly dependent since those of (6.106) are, and either of them may be used to give the ratio of d_1 to d_2; c_1, c_2, d_1 and d_2 (taken to be real) must be chosen so that the normalisation condition $\langle\boldsymbol{\phi}, \boldsymbol{\phi}^*\rangle=1$ is satisfied at $\lambda=\lambda_0$, that is,

$$\int_0^\pi \mathbf{c}\cdot\mathbf{d}\cos^2 x\,dx=(c_1d_1+c_2d_2)\pi/2=1 \tag{6.109}$$

L cannot be a self-adjoint operator at the bifurcation point. To prove this, recall that f_u^0 and g_v^0 must be of opposite sign. Also $\det M^0 = f_u^0 g_v^0 - f_v^0 g_u^0 > 0$ (6.101), so f_v^0 and g_u^0 are of opposite sign. But L is only self-adjoint if $f_v^0 = g_u^0$. It follows that the linearised equations cannot derive from a potential, and this is an illustration of the general principle that the ordered structures we obtain through bifurcation analysis can only occur far from thermodynamic equilibrium. For detailed discussion on this point see Nicolis and Prigogine (1977).

We look for a solution in the form

$$\lambda = \sum_{n=0}^{\infty} \lambda_n \varepsilon^n, \qquad u = u(x) = \sum_{n=0}^{\infty} u_n(x)\varepsilon^n, \qquad v = v(x) = \sum_{n=0}^{\infty} v_n(x)\varepsilon^n \tag{6.110}$$

By substituting these into (6.87) and equating powers of ε, we obtain equations of the form (6.67)-(6.69). Here

$$\mathbf{F}(\lambda, \mathbf{u}) = \lambda \mathbf{f}(\mathbf{u}) + D\,\nabla^2\mathbf{u}$$

where $\mathbf{f} = (f, g)^{\mathrm{T}}$, $\mathbf{u} = (u, v)^{\mathrm{T}}$. Hence

$$\mathbf{F_u} = \lambda \mathbf{f_u} + D\,\nabla^2 = L, \qquad \mathbf{F}_{\mathbf{u}\lambda} = \mathbf{f_u} = M, \qquad \mathbf{F_{uu}} = \lambda \mathbf{f_{uu}}$$

so that our first two equations are

$$L^0\mathbf{u}_1 = 0 \tag{6.111}$$

$$L^0\mathbf{u}_2 = -\lambda_1 M^0\mathbf{u}_1 - \lambda_0(\mathbf{f}^0_{uu}u_1^2 + 2\mathbf{f}^0_{uv}u_1v_1 + \mathbf{f}^0_{vv}v_1^2)/2 \tag{6.112}$$

where $\mathbf{f}_{uu} = (f_{uu}, g_{uu})^{\mathrm{T}}$, etc. $\mathbf{u}_i = (u_i, v_i)^{\mathrm{T}}$ and L^0 and M^0 are evaluated at u_0, v_0 and λ_0. The solution of (6.111) is

$$\mathbf{u}_1 = \alpha\boldsymbol{\phi}(\lambda_0)$$

for any constant α. We require a normalisation condition analogous to (6.65), namely

$$\langle \mathbf{u} - \mathbf{u}_0, \boldsymbol{\phi}^*(\lambda_0)\rangle = \varepsilon \tag{6.113}$$

since we are carrying out a bifurcation analysis from the non-zero solution $\mathbf{u} = \mathbf{u}_0$. Thus ε is a measure of the amplitude of the perturbation from the steady state. It follows that

$$\langle \mathbf{u}_1, \boldsymbol{\phi}^*(\lambda_0)\rangle = 1$$

so that $\alpha = 1$, and $\mathbf{u}_1 = \boldsymbol{\phi}(\lambda_0)$. The Fredholm alternative (Theorem 6.71) applied to (6.112) requires that the inner product of the right-hand side with $\boldsymbol{\phi}^*(\lambda_0)$ be zero. However, the inner product of the nonlinear term with $\boldsymbol{\phi}^*(\lambda_0)$ is automatically zero since the quadratic terms u_1^2, u_1v_1 and v_1^2 do not involve first harmonics. Thus

$$0 = \langle \lambda_1 M^0\mathbf{u}_1, \boldsymbol{\phi}^*(\lambda_0)\rangle = \lambda_1\sigma_1'(\lambda_0)$$

so that, from (6.104), $\lambda_1 = 0$. Now $\mathbf{u}_2$ may be found by integrating (6.112). λ_2 may be found by applying the Fredholm orthogonality condition to the next perturbation equation, and may be either positive or negative depending on the other parameters of the system. If it is positive, we have a supercritical bifurcation and the bifurcating solution is stable; if it is negative, we have an unstable subcritical bifurcation. Figure 6.114 shows the bifurcation diagrams for each case.

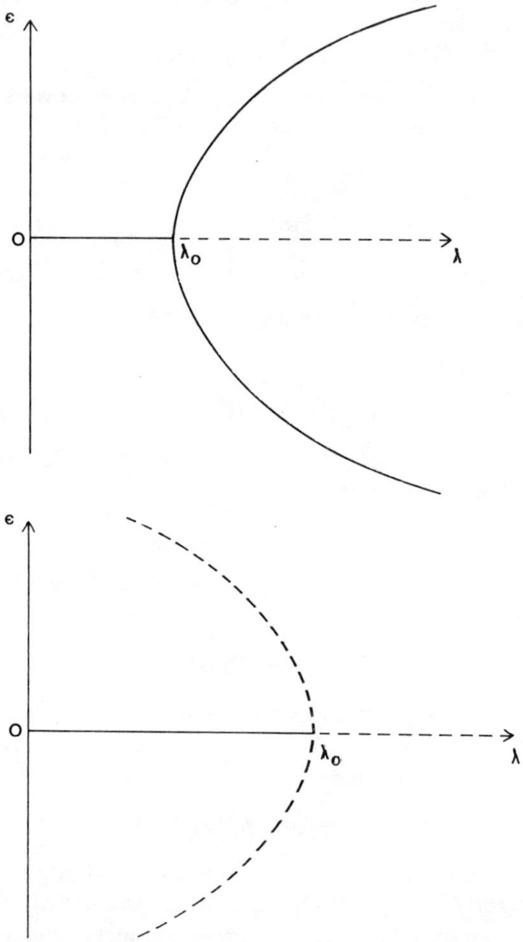

Fig. 6.114. Bifurcation diagrams for (6.90) with f and g given by (6.88) and (6.89) and zero-flux boundary conditions (6.91).

Note that the bifurcating solution has leading order term $\varepsilon \mathbf{u}_1$ given by $\varepsilon \mathbf{u}_1 = \varepsilon \boldsymbol{\phi}(\lambda_0) = \varepsilon \mathbf{c} \cos x$, and thus breaks the symmetry of the steady state; the domain attains a polarity. This symmetry-breaking is a common feature of bifurcating solutions, as we shall see later when discussing the Hopf bifurcation.

6.8 Bifurcation in a Finite Two-dimensional Domain: A Model for Hydranth Regeneration in *Tubularia*

The work in this section is based on a model for hydranth regeneration in the marine hydroid *Tubularia* (Britton *et al.*, 1983). In *Tubularia*, regeneration of an amputated hydranth takes place in about one and a half days. The sequence of clearly observable states is shown in Fig. 6.115. In (a)–(c) the shading and serrated lines represent reddish coloured pigment.

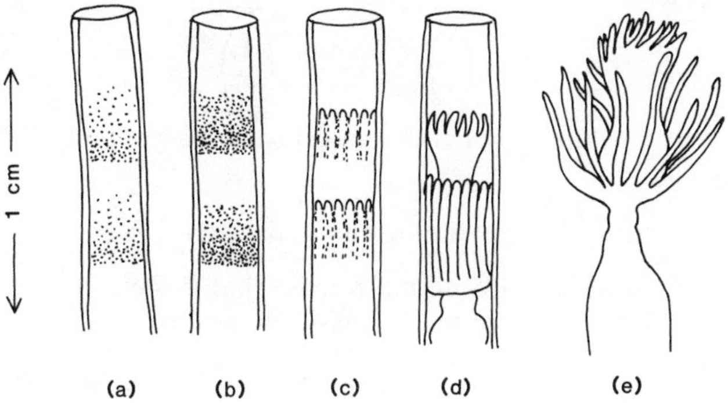

Fig. 6.115. Stages in hydranth regeneration in *Tubularia*.

The regular periodic configuration in (c) presages the tentacle distribution in (d) which evolves into the mature tentacle distribution in (e). There is an excellent film of the early stages of the process by J. Cooke and G. C. Webster, and the process is described in detail by Webster (1971) and for the similar *Obelia* by Beloussov *et al.* (1972). Work on hydroids has also been done by Lacalli (1980) and Lacalli and Harrison (1979). For a general discussion see Goodwin (1976).

It is proposed that the regular circumferential patterns seen during this development may be the result of a combination of reaction and diffusion. To produce such patterns a reaction–diffusion system would have to have a steady state stable to spatially homogeneous perturbations but unstable to spatially inhomogeneous ones (or at least more unstable to inhomogeneous than to homogeneous perturbations). Such a system is discussed in the previous section and we therefore use this as a model. It is not suggested that this particular system governs pattern formation in *Tubularia* but that it could be of this kind.

Tubularia is an approximately cylindrical animal. The main variations in pigment concentration occur in the radial and circumferential directions, and not in the axial direction. We therefore model the domain by a hollow cylinder, with internal radius r_1 and external radius r_2, with the usual cylindrical polar co-ordinates, and consider concentration profiles which vary with r and θ only, not with z. By non-dimensionalising r, we may take the inner radius of the hydroid to be unity. Thus the system becomes

$$u_t = \lambda f(u, v) + D_1 \nabla^2 u, \qquad v_t = \lambda g(u, v) + D_2 \nabla^2 v \tag{6.116}$$

where f and g are given by (6.88) and (6.89), $u = u(r, \theta, t)$ and $v = v(r, \theta, t)$ are in $C^{2,1}[\Omega \times (0, \infty)]$, i.e. they have continuous second-order spatial and first-order temporal derivatives, $\Omega = \{(r, \theta) | 1 < r < r_2/r_1 = \delta,\ 0 \leq \theta < 2\pi\}$, and

$$\nabla^2 \equiv \left(\frac{1}{r}\right)\frac{\partial}{\partial r}\left(r\frac{\partial}{\partial r}\right) + \left(\frac{1}{r^2}\right)\frac{\partial^2}{\partial \theta^2}$$

Thus δ is the non-dimensional outer radius of the hydroid. The boundary conditions are

$$\nabla u \cdot \mathbf{n} = \nabla v \cdot \mathbf{n} = 0 \qquad \text{on} \quad r = 1, \delta \tag{6.117}$$

where $\mathbf{n}$ is the unit outward normal vector, that is there is no flux of u or v into or out of the body of the hydroid. We also require

$$\begin{aligned} u(r, 0, t) &= u(r, 2\pi -, t), \qquad & v(r, 0, t) &= v(r, 2\pi -, t) \\ u_\theta(r, 0, t) &= u_\theta(r, 2\pi -, t), \qquad & v_\theta(r, 0, t) &= v_\theta(r, 2\pi -, t) \end{aligned} \tag{6.118}$$

that is, continuity of the concentrations and their derivatives on the line $\theta = 0$ or 2π.

The spectrum of L is again discrete in the two-dimensional case in a finite domain. Therefore a bifurcation analysis similar to that above may be carried out and is valid close to certain critical values λ_c (bifurcation points) of the parameter λ. In such an analysis, the form of the leading term $\mathbf{u}_1$ in the final solution is determined by a linear analysis; it satisfies $L^0\mathbf{u}_1 = 0$. Does a linear analysis give a good approximation to the form of the solution even if λ is not close to a bifurcation value, in the sense that the final steady solution of the fully nonlinear problem corresponds to the most unstable eigenfunction of the *linear* system? In one dimension numerical results suggest that it generally does. However, in two dimensions this is not true, as we shall show.

The linear problem involves analysis of

$$L\boldsymbol{\phi} = \sigma\boldsymbol{\phi} \tag{6.119}$$

where the linear operator $L(\lambda)$ is given by

$$L(\lambda) = \begin{pmatrix} D_1 \nabla^2 + \lambda f_u & \lambda f_v \\ \lambda g_u & D_2 \nabla^2 + \lambda g_v \end{pmatrix} = \lambda M + D \nabla^2 \tag{6.120}$$

If the eigenvalues of $-\nabla^2$ on the domain Ω with the given boundary and continuity conditions are given by k^2, then the spectral problem (6.119) has solutions or given by

$$\begin{vmatrix} \sigma + D_1k^2 - \lambda f_u & -\lambda f_v \\ -\lambda g_u & \sigma + D_2k^2 - \lambda g_v \end{vmatrix} = 0 \tag{6.121}$$

To find the possible values which k^2 may take we solve

$$\nabla^2\mathbf{u} + k^2\mathbf{u} = 0 \tag{6.122}$$

in Ω with

$$\mathbf{u}_r(1, \theta) = \mathbf{u}_r(\delta, \theta) = \mathbf{0} \tag{6.123}$$

and

$$\mathbf{u}(r, 0) = \mathbf{u}(r, 2\pi-), \qquad \mathbf{u}_\theta(r, 0) = \mathbf{u}_\theta(r, 2\pi-) \tag{6.124}$$

The continuity conditions (6.124) imply that we may look for a solution $\mathbf{u}(r, \theta)$ periodic of period 2π in θ, so that

$$\mathbf{u}(r, \theta) = \sum_{n=-\infty}^{\infty} c_n R_n(r)\, e^{in\theta} \tag{6.125}$$

Substituting this into (6.122) gives, by Fourier analysis,

$$R_n'' + (1/r)R_n' + (k^2 - n^2/r^2)R_n = 0 \tag{6.126}$$

where primes denote differentiation with respect to the argument. This is Bessel's equation, and has linearly independent solutions $J_n(kr)$ and $Y_n(kr)$. The linear combination

$$R_n(r) = J_n(kr)\,Y_n'(k) - J_n'(k)\,Y_n(kr) \tag{6.127}$$

satisfies the boundary condition at $r = 1$, and satisfies that at $r = \delta$ if $k = k_n$, where

$$J_n'(k_n\delta)\,Y_n'(k_n) - J_n'(k_n)\,Y_n'(k_n\delta) = 0 \tag{6.128}$$

For each n there is an infinity of solutions $k_{nj}, j = 1, 2, \ldots$, satisfying (6.128). These eigenvalues have been calculated for a range of δ by Bridge and Angrist (1962).

Each k_{nj} is an eigenvalue of minus the Laplacian on Ω with the given boundary and continuity conditions. The corresponding eigenvalues of the operator $L(\lambda)$ on Ω with the same conditions are obtained by substituting $k = k_{nj}$ into Eq. (6.121), so that for each eigenvalue k_{nj} of $-\nabla^2$ there are two eigenvalues σ_{nj} of $L(\lambda)$, given by the roots of

$$\sigma_{nj}^2 + \{(D_1 + D_2)k_{nj}^2 - \lambda \operatorname{tr} M\}\sigma_{nj} + Q_{nj}(\lambda) = 0 \tag{6.129}$$

where

$$Q_{nj}(\lambda) = \lambda^2 \det M - \lambda k_{nj}^2(D_2 f_u + D_1 g_v) + D_1 D_2 k_{nj}^4 \qquad (6.130)$$

As before we assume that the steady state is stable to spatially homogeneous perturbations, so that

$$\det M^0 > 0, \qquad \operatorname{tr} M^0 < 0$$

where the superscript zero again denotes evaluation at the steady state, and the only way for instability to occur is if $Q_{nj}^0(\lambda) < 0$, when one of the roots of (6.129) is positive. This can happen for at most a finite number of eigenvalues σ_{nj} [see Bridge and Angrist (1962)].

The eigenfunction corresponding to the eigenvalue σ_{nj} with $n \neq 0$ has sinusoidal variation in θ, with period $2\pi/|n|$, and hence has $|n|$ waves around the circumference. If $n = 0$, the eigenfunction has no θ dependence; but for the values of the parameters we took, all such eigenvalues had negative real parts. The second suffix j corresponds to the r variation of the eigenfunction; this has $j-1$ changes of sign between $r = 1$ and $r = \delta$ on any line $\theta = \text{const}$.

There are two questions which we now wish to address. First, for values of the parameters appropriate to the hydroid *Tubularia*, which are the linearly unstable modes? Second, what are the solutions of the fully nonlinear problem for these parameter values, and how well do they correspond to predictions based on the linear analysis? For $\delta = 1.2$, which is rather a low value for *Tubularia*, and for $\lambda = 25$, which corresponds to a small animal, the most unstable eigenvalue is σ_{51}. The corresponding eigenfunction has five waves around the circumference. The solution of the full nonlinear system, obtained numerically by the method of finite elements, is shown in Fig. 6.131a. It can be seen that the nonlinear system behaves in qualitatively the same way as the linear one. The reason for this is the quasi-one-dimensional nature of the domain; its width, 0.2, is much less than its circumference, 2π. The same is true of $\lambda = 100$, corresponding to an animal with twice the diameter. Here the most unstable eigenvalue is σ_{91}, and the numerical results are shown in Fig. 6.131b. In Fig. 6.131c, δ is taken to be 1.5, which is a more realistic value for *Tubularia.* When $\lambda = 25$, linear theory is still adequate since it gives the most unstable eigenvalue as σ_{51}. However, when $\lambda = 100$, so that the diameter is twice as large, the most unstable eigenvalues are $\sigma_{10,1}$, $\sigma_{11,1}$ and $\sigma_{6,2}$, all of which have very similar growth rates. The appearance of unstable modes with $j = 2$ can be thought of as a secondary bifurcation [see Auchmuty and Nicolis (1975), Herschkowitz-Kaufman (1975), Hiernaux and Erneux (1979)]. The final steady state, which is (numerically) stable, is shown in Fig. 6.131d. It can be seen that it is very close to that associated with $\sigma_{6,2}$ (i.e. one ring of six minima near the inner wall and another near the outer), especially in the bottom right

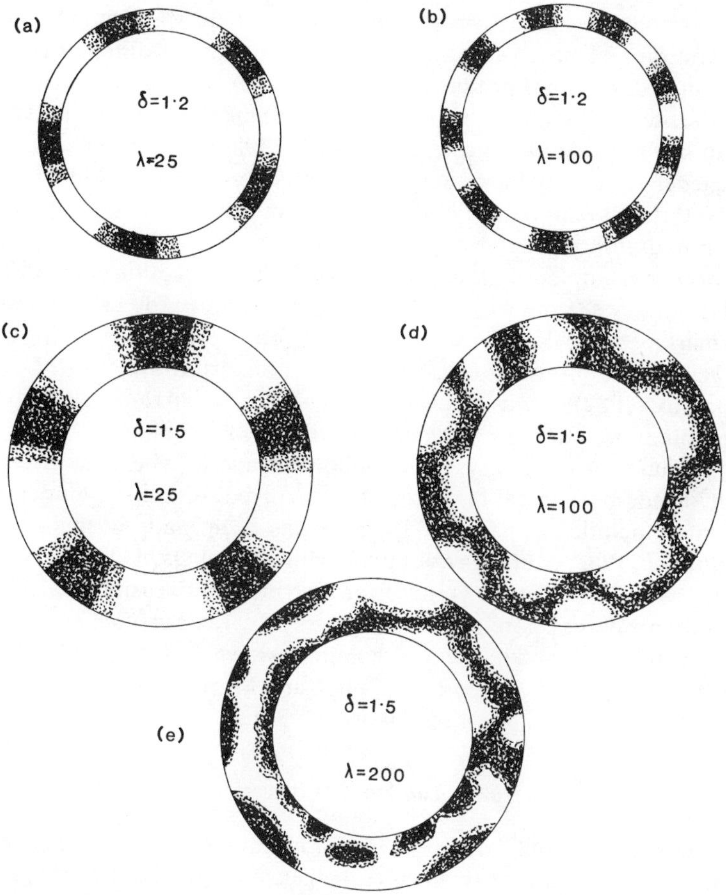

Fig. 6.131. Numerical solutions of the system (6.116) with (6.88) and (6.89). Dark regions represent areas where u is above the steady state and light regions where it is below it.

half. On the other hand, there are 11 patches below the steady state, so that $\sigma_{11,1}$ is also represented. In Fig. 6.131e, $\lambda = 200$ and the final steady state seems to bear no relation to the linear prediction (where the most unstable eigenvalues are $\sigma_{11,2}$ and several with $j = 1$). It seems that the number of excited modes is so large that it is difficult to decompose the pattern into its constituent parts.

Can reaction and diffusion then explain the mechanism of hydranth regeneration in *Tubularia*? The pattern in Fig. 6.131d, given by a reaction-diffusion system, corresponds very closely to the patterns of pigmentation seen in the regenerating animal (Fig. 6.115). However, the reaction–diffusion system is rather sensitive to changes in the parameters, whereas the *Tubularia* regeneration system is very robust. It is possible that a different reaction–diffusion system would produce good results over a wider parameter range, or on the other hand that there is some other mechanism at work in *Tubularia* to stabilise the required pattern. Kemmner (1984) has proposed a model for regeneration in hydra which involves reaction and diffusion with delay effects. It does seem plausible that reaction and diffusion do have a part to play in hydranth regeneration in *Tubularia.*

For other examples of pattern formation by reaction diffusion in developmental biology see the work of Murray (1981) and Bard (1981) on animal coat markings and Kauffman *et al.* (1978) on drosphila. Hunding (1983, 1984) considers bifurcation in spheroidal domains. Mimura *et al.* (1979), Kishimoto (1982), Fujii *et al.* (1982) and Ito (1984a) consider pattern formation in predator–prey systems. Murray (1982) compares Turing instability domains in various systems. In morphogenesis, where shape changes are involved, mechanical forces must be involved. A realistic model in this case must include such forces. Progress has been made in this direction (Odell *et al.*, 1981; Murray *et al.*, 1983; Murray and Oster, 1984; Oster and Odell, 1984). In ecology it is not clear that Fickian diffusion is always the most suitable description for animal movements. Taxis is often an important factor (Segel, 1984, and references therein; Kareiva, 1983), and higher-order derivatives may also be important (Cohen and Murray, 1981). For a general survey of ecological models see Okubo (1980).

6.9 The Hopf Bifurcation

So far we have only considered steady equilibrium solutions of $\mathbf{u}_t = \mathbf{F}(\lambda, \mathbf{u})$. Such solutions arise when real eigenvalues of the spectral problem cross the imaginary axis. If a pair of conjugate complex eigenvalues crosses this axis, then the bifurcating solutions may be periodic in t. Note that this is another example of symmetry-breaking, in the temporal rather than the spatial variable, since the symmetry of the forcing data, $\mathbf{F}(\lambda, \mathbf{u})$, which is steady, is broken in producing a time-periodic solution.

This bifurcation phenomenon was studied by Hopf (1942); for a translation of his original paper by L. N. Howard and N. Kopell see the important book of Marsden and McCracken (1976). It was also considered by Poincaré (1892–1899) and Andronov *et al.* and is sometimes called the Poincaré–Andronov–Hopf bifurcation. For a short account of Andronov's analysis see Minorsky (1974). A general reference on the subject is the book of Hassard *et al.* (1981). Early work on the Hopf bifurcation in reaction–diffusion systems was done by Auchmuty and Nicolis (1976) and others and stability is considered by Erneux and Herschkowitz-Kauffman (1979).

In the case of such a bifurcation in the kinetic system Cohen *et al.* (1977) have shown that the addition of diffusion may result in slow modulation of the amplitude in the bifurcating solution; see also Miura (1977). The connection between the Hopf bifurcation and spiral waves has been considered by Cohen, Neu and Rosales (1978), Duffy *et al.* (1980) and Hagan (1982). Hopf bifurcation in nerve conduction equations has been considered by Rinzel and Keener (1983).

Since we require two eigenvalues to cross the axis for such a bifurcation, the problem is essentially two-dimensional in its simplest form. This is the case we shall consider first. We therefore have

$$\dot{\mathbf{u}} = \mathbf{f}(\lambda, \mathbf{u}) \tag{6.132}$$

where $\mathbf{u} = (u_1, u_2)^{\mathrm{T}}$, and $\mathbf{u} \in B = C^1(\mathbb{R}, \mathbb{R}^2)$, $\mathbf{u}$ periodic. We consider bifurcation from the trivial steady state, so that

$$\mathbf{f}(\lambda, \mathbf{0}) = \mathbf{0} \tag{6.133}$$

for all λ, and define the linear operator $M(\lambda)$ and the nonlinear operator $\mathrm{N}(\lambda, \cdot)$ by

$$M(\lambda)\mathbf{u} = \mathbf{f}_{\mathbf{u}}(\lambda, \mathbf{0} | \mathbf{u}) \tag{6.134}$$

$$N(\lambda, \mathbf{u}) = \mathbf{f}(\lambda, \mathbf{u}) - M(\lambda)\mathbf{u} \tag{6.135}$$

so that

$$\dot{\mathbf{u}} = M(\lambda)\mathbf{u} + N(\lambda, \mathbf{u}) \tag{6.136}$$

M is a real 2×2 matrix. Its eigenvalues are complex if

$$(\mathrm{tr}\ \mathrm{M})^2 - 4 \det M < 0 \tag{6.137}$$

where tr M is the trace and det M the determinant of the matrix. We take this to be the case in some neighbourhood of the bifurcation point, which we take without loss of generality to be at $\lambda = 0$. Since M is a real 2×2 matrix, its eigenvalues are complex conjugates σ and $\bar{\sigma}$, where

$$\sigma(\lambda) = \xi(\lambda) + i\eta(\lambda), \qquad \bar{\sigma}(\lambda) = \xi(\lambda) - i\eta(\lambda) \tag{6.138}$$

Each eigenvalue is necessarily algebraically simple and has a corresponding eigenvector:

$$M\boldsymbol{\phi} = \sigma\boldsymbol{\phi}, \qquad M\bar{\boldsymbol{\phi}} = \bar{\sigma}\bar{\boldsymbol{\phi}}; \tag{6.139}$$

$\boldsymbol{\phi} = \boldsymbol{\phi}(\lambda) \in \mathbb{C}^2$, and we shall use the $\mathbb{C}^2$ inner product. The adjoint operator $M^* = M^*(\lambda)$ is equal to M^{T}, the transpose of the matrix M. It is easily seen that M^* has the same eigenvalues as M and corresponding eigenvectors given by $\boldsymbol{\phi}^* \subset \boldsymbol{\phi}^*(\lambda)$ and $\bar{\boldsymbol{\phi}}^*$. Normalising $\boldsymbol{\phi}$ appropriately, and using the fact that $\boldsymbol{\phi}$ and $\boldsymbol{\phi}^*$ are orthogonal since they are eigenvalues of M belonging to different eigenvalues, we have

$$\langle \boldsymbol{\phi}, \boldsymbol{\phi}^* \rangle = \boldsymbol{\phi} \cdot \bar{\boldsymbol{\phi}}^* = 1, \ \langle \boldsymbol{\phi}, \bar{\boldsymbol{\phi}}^* \rangle = \boldsymbol{\phi} \cdot \boldsymbol{\phi}^* = 0 \tag{6.140}$$

An easy computation shows that

$$M^* \boldsymbol{\phi}^* = \bar{\sigma} \boldsymbol{\phi}^*, \qquad M^* \bar{\boldsymbol{\phi}}^* = \sigma \bar{\boldsymbol{\phi}}^* \tag{6.141}$$

We assume that

$$\sigma(0) = i\omega_0 \neq 0, \qquad \operatorname{Re} \sigma'(0) = \xi'(0) \neq 0 \tag{6.142}$$

The first condition ensures that the eigenvalues of M are purely imaginary at the bifurcation point; the second is a transversality condition. Taking

$$\xi'(0) > 0 \tag{6.143}$$

for definiteness, it ensures strict loss of stability of the trivial solution as we pass through the bifurcation point.

We may now state

THEOREM 6.144 (Hopf Bifurcation Theorem in $\mathbb{R}^2$): Consider the two-dimensional bifurcation problem $\dot{\mathbf{u}} = \mathbf{f}(\lambda, \mathbf{u})$, where $\mathbf{f}$ is analytic and $\mathbf{f}(\lambda, \mathbf{0}) = \mathbf{0}$, and assume that the complex conjugate eigenvalues $\sigma(\lambda)$ and $\bar{\sigma}(\lambda)$ of the linearisation of $\mathbf{f}$ cross the imaginary axis from left to right at $\lambda = 0$, so that $\sigma(0) = i\omega_0 \neq 0$, $\operatorname{Re} \sigma'(0) > 0$. Then small amplitude periodic solutions bifurcate from the trivial solution as we pass through $\lambda = 0$.

REMARK 6.145: Again the analyticity condition may be weakened.

Proof: We shall give a constructive proof based on the treatment of Iooss and Joseph (1980); see also Minorsky (1974), or for a more rigorous version Chow and Hale (1982). For a different proof see Marsden and McCracken (1976).

Since $\boldsymbol{\phi}$ and $\bar{\boldsymbol{\phi}}$ are linearly independent, they span $\mathbb{C}^2$, so that we may look for a solution $\mathbf{u}$, which must be real, in the form

$$\mathbf{u}(t) = a(t)\boldsymbol{\phi} + \bar{a}(t)\bar{\boldsymbol{\phi}} \tag{6.146}$$

Substituting into (6.132) or (6.136),

$$\dot{a}\boldsymbol{\phi} + \dot{\bar{a}}\bar{\boldsymbol{\phi}} = M(a\boldsymbol{\phi} + \bar{a}\bar{\boldsymbol{\phi}}) + \mathbf{N}(\lambda, \mathbf{u}) = \sigma a \boldsymbol{\phi} + \bar{\sigma}\bar{a}\bar{\boldsymbol{\phi}} + \mathbf{N}(\lambda, \mathbf{u}) \tag{6.147}$$

Taking the inner product with $\boldsymbol{\phi}^*$ and using the normalisation conditions (6.140), we have

$$\dot{a} = \sigma a + \langle \mathbf{N}(\lambda, \mathbf{u}), \boldsymbol{\phi}^* \rangle = g(\lambda, a) = \sigma a + h(\lambda, a) = \sigma a + O(|a|^2) \tag{6.148}$$

so that $h(\lambda, a)$ represents the second- and higher-order terms in the equation. The linear equation at $\lambda = 0$ is

$$\dot{a} = \sigma(0)a = i\omega_0 a$$

so that $a = e^{i\omega_0 t} = e^{is_0}$, say, is a solution. We look for a solution which is a perturbation of this. The solution at criticality is 2π-periodic in $s_0 = \omega_0 t$, so we expect the solution near criticality to be 2π-periodic in $s = \omega t$, where ω is near ω_0. Thus we seek a solution in the form

$$a(t) = b(s;\, \varepsilon), \qquad s = \omega(\varepsilon)t, \qquad \omega(0) = \omega_0, \qquad \lambda = \lambda(\varepsilon) \tag{6.149}$$

where b is 2π-periodic in s and ε is a small parameter and is a measure of the amplitude of the bifurcating solution. Since we are working in the space of 2π-periodic functions, we introduce a corresponding inner product $[\cdot, \cdot]$, defined by

$$[p, q] = \frac{1}{2\pi} \int_0^{2\pi} p(s)\bar{q}(s)\, ds \tag{6.150}$$

Just as in the case of bifurcation from a real simple eigenvalue, we define ε in terms of the appropriate inner product. There we defined $\varepsilon = \langle \mathbf{u}, \boldsymbol{\phi}^*(0)\rangle$, so that the linearised equation at the bifurcation point, $L(0)\mathbf{u} = \mathbf{0}$, had solution $\mathbf{u} = \varepsilon\boldsymbol{\phi}(0)$. Here the linearised equation is $\dot{a} = \sigma a$, or using (6.149),

$$-\omega b' + \sigma b = Jb = 0 \tag{6.151}$$

where $J = J(\varepsilon) = -\omega(\varepsilon)(d/ds) + \sigma(\lambda(\varepsilon))$ and is the linearisation of $-\omega(d/ds) + g(\lambda, \cdot)$ about the trivial solution, and prime denotes differentiation with respect to s. At the bifurcation point $\varepsilon = 0$ this becomes

$$-\omega_0\, db/ds + i\omega_0 b = J_0 b = 0 \tag{6.152}$$

where $J_0 = J(0)$ and $\omega_0 = \omega(0)$, which has solution $b = \alpha\, e^{is}$ for some constant α. Since we wish the bifurcating solution to be $O(\varepsilon)$, we take $\alpha = \varepsilon$, so $b = \varepsilon e^{is}$. Then, taking the inner product with e^{is},

$$[b, e^{is}] = \frac{1}{2\pi} \int_0^{2\pi} \varepsilon e^{is}\, e^{-is}\, ds = \varepsilon \tag{6.153}$$

Thus we define $\varepsilon = [b, e^{is}]$. We then look for a solution in the form

$$b(s;\, \varepsilon) = \sum_{n=1}^{\infty} b_n(s)\varepsilon^n, \qquad \omega(\varepsilon) = \sum_{n=0}^{\infty} \omega_n \varepsilon^n, \qquad \lambda(\varepsilon) = \sum_{n=1}^{\infty} \lambda_n \varepsilon^n \tag{6.154}$$

Substituting into (6.148), we have

$$-\omega b' + g(\lambda, b) = -\omega b' + \sigma b + h(\lambda, b) = Jb + h(\lambda, b) = 0 \tag{6.155}$$

where

$$J(\varepsilon) = -\omega(\varepsilon)\frac{d}{ds} + \sigma(\lambda(\varepsilon))$$

$$= -\omega_0\frac{d}{ds} + \sigma(0) + \varepsilon\left(-\omega_1\frac{d}{ds} + \lambda_2\sigma'(0)\right)$$

$$+ \varepsilon^2\left(-\omega_2\frac{d}{ds} + \lambda_2\sigma'(0) + \frac{\lambda_1^{\ 2}}{2}\sigma''(0)\right) + \cdots$$

$$= J_0 + \varepsilon J_1 + \varepsilon^2 J_2 + \cdots \tag{6.156}$$

Equating powers of ε in (6.155), we have

$$J_0 b_1 = -\omega_0 b_1' + i\omega_0 b_1 = 0 \tag{6.157}$$

$$J_0 b_2 + J_1 b_1 + Q(b_1, \bar{b}_1)$$
$$= -\omega_0 b_2' + i\omega_0 b_2 - \omega_1 b_1' + \lambda_2\sigma'(0) b_1 + Q(b_1, \bar{b}_1) = 0 \tag{6.158}$$

where $Q(b_1, \bar{b}_1)$ denotes quadratic terms in b_1 and $\bar{b}_1$, and in general,

$$J_0 b_n + J_{n-1} b_1 + R_n = 0 \tag{6.159}$$

where R_n is a known function once the first $n-1$ equations have been solved. The normalisation condition (6.153) becomes

$$[b_1, e^{is}] = 1, \qquad [b_n, e^{is}] = 0, \qquad n > 1 \tag{6.160}$$

(6.167) is just the linear problem considered above and with the first of (6.160) has solution

$$b_1 = e^{is} \tag{6.161}$$

The operator J_0 with periodic boundary conditions is a Fredholm operator, so that the Fredholm alternative again holds. In this context it may be stated as follows.

THEOREM 6.162 (Fredholm Alternative for Periodic Equations): Equations of the form

$$J_0 b = f(s) = f(s + 2\pi)$$

are soluble for $b(s) = b(s + 2\pi)$ if and only if the Fourier expansion for $f(s)$ has no term proportional to e^{is}, that is, if

$$[f, e^{is}] = \frac{1}{2\pi}\int_0^{2\pi} f(s)\, e^{-is}\, ds = 0$$

This is again a Fredholm orthogonality condition. The solution is unique under the normalisation

$$[b, e^{is}] = \frac{1}{2\pi}\int_0^{2\pi} b(s)\, e^{-is}\, ds = k$$

for any constant k.

Proof: See Joseph and Sattinger (1972) or Sattinger (1973).

Applying this theorem to (6.158), it has a solution if and only if $[J_1 b_1, e^{is}] = 0$, since $[Q(b_1, \bar{b}_1), e^{is}] = 0$ because the quadratic terms in b_1 and $\bar{b}_1$ do not contain any terms in e^{is}. Thus, using (6.158) and (6.161),

$$[J_1 b_1, e^{is}] = \frac{1}{2\pi}\int_0^{2\pi} (-i\omega_1\, e^{is} + \lambda_1 \sigma'(0)\, e^{is})\, e^{-is}\, ds = -i\omega_1 + \lambda_1 \sigma'(0) = 0$$

It follows, from the assumption that $\xi'(0) = \operatorname{Re} \sigma'(0) \neq 0$, that $\lambda_1 = 0$, and so $\omega_1 = 0$. (6.158) may now be integrated to find b_2. Similarly, the unknown constants λ_{n-1} and ω_{n-1} in J_{n-1} may be found by applying the Fredholm orthogonality condition to (6.159), which gives $-i\omega_{n-1} + \lambda_{n-1}\sigma'(0)$ in terms of known functions; (6.159) may then be integrated to find b_n. It follows by induction that the perturbation equations are soluble at each order, so that the solutions may be found in the given form. Moreover, it may be shown that $\omega_i = \lambda_i = 0$ for all odd integers i, so that $\omega = \omega(\varepsilon)$ and $\lambda = \lambda(\varepsilon)$ are even functions of ε. Thus either both branches of the solutions bifurcate supercritically or both bifurcate subcritically—transcritical Hopf bifurcations do not exist. There is a third possibility: that $\lambda_i = 0$ for all $i > 0$ and $\lambda = \lambda_0 = 0$ gives a solution for all amplitudes ε. In this case the system at $\lambda = 0$ has a centre at the origin and an infinite family of periodic solutions.

It is important to determine whether a solution bifurcates subcritically or supercritically. As we shall see in the next section, this determines the stability of the bifurcating branches. Since the calculation often involves a large amount of algebra, we derive here a formula for λ_2 in the general case. If $\lambda_2 > 0$, then we have a supercritical bifurcation, and if $\lambda_2 < 0$, then it is subcritical. If $\lambda_2 = 0$, then higher-order terms must be considered. We assume that Eq. (6.136), namely

$$\dot{\mathbf{u}} = \mathbf{M}(\lambda)\mathbf{u} + \mathbf{N}(\lambda, \mathbf{u}),$$

is written in canonical form, i.e., such that

$$M(0) = \begin{bmatrix} 0 & \omega_0 \\ -\omega_0 & 0 \end{bmatrix}$$

We shall take $\omega_0 > 0$ without loss of generality. If the equation is not in this form it may be made so by a simple transformation of the form $\tilde{\mathbf{u}} = A\mathbf{u}$, where A is a constant matrix. $\mathbf{N}(\lambda, \mathbf{u})$ may be written in the form

$$\mathbf{N}(\lambda, \mathbf{u}) = (f_{uu}u^2/2 + f_{uv}uv + f_{vv}v^2/2 + f_{uuu}u^3/6 + f_{uuv}u^2v/2 + f_{uvv}uv^2/2 + f_{vvv}v^3/6 + \text{higher-order terms}, \cdots)^{\mathrm{T}}$$

where the dots represent the corresponding terms in g and the derivatives f_{uu}, etc., are evaluated at $(\lambda, \mathbf{0})$. The eigenvector $\boldsymbol{\phi}(\lambda)$ of $M(\lambda)$ may be found explicitly at $\lambda = 0$, and is given by $\boldsymbol{\phi}(0) = (1, i)^{\mathrm{T}}$, and since $\lambda_1 = 0$, then $\boldsymbol{\phi}(\lambda) = (1, i)^{\mathrm{T}} + O(\varepsilon^2)$. Similarly, $\boldsymbol{\phi}^*(0) = (1, i)^{\mathrm{T}}/2$, and $\boldsymbol{\phi}^*(\lambda) = (1, i)^{\mathrm{T}}/2 + O(\varepsilon^2)$. Also, $f_{uu}(\lambda, \mathbf{0}) = f_{uu}(0, \mathbf{0}) + O(\varepsilon^2) = f^0_{uu} + O(\varepsilon^2)$, say, with similar results for the other derivatives of f and g. Finally, $\mathbf{u} = b\boldsymbol{\phi} + \bar{b}\bar{\boldsymbol{\phi}}$, so that

$$\begin{aligned}
u &= (b + \bar{b})(1 + O(\varepsilon^2)) = \varepsilon(b_1 + \bar{b}_1) + \varepsilon^2(b_2 + \bar{b}_2) + O(\varepsilon^3)\\
&= \varepsilon u_1 + \varepsilon^2 u_2 + O(\varepsilon^3)\\
v &= (ib - i\bar{b})(1 + O(\varepsilon^2)) = \varepsilon(ib_1 - i\bar{b}_1) + \varepsilon^2(ib_2 - i\bar{b}_2) + O(\varepsilon^3)\\
&= \varepsilon v_1 + \varepsilon^2 v_2 + O(\varepsilon^3)
\end{aligned}$$

and

$$\begin{aligned}
h(\lambda, b) &= \langle \mathbf{N}(\lambda, b\boldsymbol{\phi} + \bar{b}\bar{\boldsymbol{\phi}}), \boldsymbol{\phi}^*\rangle\\
&= \{f^0_{uu}(b + \bar{b})^2/2 + f^0_{uv}(b + \bar{b})(ib - i\bar{b}) + f^0_{vv}(ib - i\bar{b})^2/2\\
&\quad + f^0_{uuu}(b + \bar{b})^3/6 + \cdots\}/2 - \{g^0_{uu}(b + \bar{b})^2/2 + \cdots\}i/2 + O(\varepsilon^4)
\end{aligned}$$

The equation to be solved is (6.155), or

$$-\omega b' + \sigma b + h(\lambda, b) = 0$$

As above, $O(\varepsilon)$ terms in this give $b_1 = e^{is}$. Using $\lambda_1 = \omega_1 = 0$, $O(\varepsilon^2)$ terms give

$$\begin{aligned}
-\omega_0 b_2' + i\omega_0 b_2 &= -\{f^0_{uu}(b_1 + \bar{b}_1)^2/2 + \cdots\}/2 - \{g^0_{uu}(b_1 + \bar{b}_1)^2/2 + \cdots\}i/2\\
&= -\{(f^0_{uu} - f^0_{vv} + 2g^0_{uv}) - i(-2f^0_{uv} + g^0_{uu} - g^0_{vv})\}\, e^{2is}/4\\
&\quad - \{(f^0_{uu} + f^0_{vv}) - i(g^0_{uu} + g^0_{vv})\}/2\\
&\quad - \{(f^0_{uu} - f^0_{vv} - 2g^0_{uv}) - i(2f^0_{uv} + g^0_{uu} - g^0_{vv})\}\, e^{-2is}/4\\
&= A\, e^{2is} + B + C\, e^{-2is}
\end{aligned}$$

This has a solution satisfying the normalisation condition $[b_2, e^{is}] = 0$ given by $b_2 = \alpha\, e^{2is} + \beta + \gamma\, e^{-2is}$, where

$$\begin{aligned}
(-2i\omega_0 + i\omega_0)\alpha &= -i\omega_0\alpha = A\\
i\omega_0\beta &= B\\
(2i\omega_0 + i\omega_0)\gamma &= 3i\omega_0\gamma = C
\end{aligned}$$

$O(\varepsilon^3)$ terms give

$$-\omega_0 b_3' + i\omega_0 b_3 - \omega_2 b_1' + \lambda_2 \sigma'(0) b_1$$
$$= -\{f^0_{uu} u_1 u_2 + f^0_{uv}(u_1 v_2 + u_2 v_1) + f^0_{vv} v_1 v_2 + f^0_{uuu} u_1^3/6 + \cdots\}/2$$
$$+ i\{g^0_{uu} u_1 u_2 + \cdots\}/2$$

By taking the inner product with e^{is}, using the fact that $[b_3, e^{is}] = 0$ (or, equivalently, by applying the Fredholm solvability condition), this gives

$$-i\omega_2 + \lambda_2 \sigma'(0) = -\{f^0_{uu}(\alpha + \bar{\gamma} + \beta + \bar{\beta}) + \cdots + f^0_{uuu}/2 + \cdots\}/2$$
$$+ i\{g^0_{uu}(\alpha + \bar{\gamma} + \beta + \bar{\beta}) + \cdots + g^0_{uuu}/2 + \cdots\}/2$$

Taking the real part,

$$\lambda_2 \operatorname{Re} \sigma'(0) = -\{f^0_{uu} \operatorname{Re}(\alpha + \bar{\gamma} + \beta + \bar{\beta}) + \cdots + f^0_{uuu}/2 + f^0_{uvv}/2\}/2$$
$$+ \{g^0_{uu} \operatorname{Im}(\alpha + \bar{\gamma} + \beta + \bar{\beta}) + \cdots + g^0_{uuv}/2 + g^0_{vvv}/2\}/2$$
$$= -\{f^0_{uu} g^0_{uu} - f^0_{vv} g^0_{vv} - f^0_{uv}(f^0_{uu} + f^0_{vv})$$
$$+ g^0_{uv}(g^0_{uu} + g^0_{vv})\}/(4\omega_0)$$
$$-(f^0_{uuu} + f^0_{uvv} + g^0_{uuv} + g^0_{vvv})/4$$
$$= D \qquad (6.163)$$

say, after some algebra. Since we have assumed that $\operatorname{Re} \sigma'(0) > 0$, then $\lambda_2 > 0$ if D is positive, and we then have a supercritical bifurcation. If D is negative, then $\lambda_2 < 0$ and the bifurcation is subcritical. If $D = 0$, then $\lambda_2 = 0$ and higher-order terms must be considered.

6.10 Stability of the Bifurcating Branches: The Factorisation Theorem for the Hopf Bifurcation

We are interested in the stability of the bifurcating branches. To investigate this, we consider a small disturbance of the solution $b(s; \varepsilon)$ which we have just shown how to construct, so that

$$a(t; \varepsilon) = b(s; \varepsilon) + z(t; \varepsilon) \qquad (6.164)$$

where a satisfies (6.148). By substituting this expression into (6.148) and linearising, we have

$$\dot{z}(t; \varepsilon) = g_a(\lambda(\varepsilon), b(s; \varepsilon)) z(t; \varepsilon) \qquad (6.165)$$

where the dot represents differentiation with respect to t. Here g_a is periodic of frequency $\omega(\varepsilon)$ in t, so that this equation is amenable to Floquet theory

(Iooss and Joseph, 1980). It follows that we may put

$$z(t;\,\varepsilon)=\exp(\gamma(\varepsilon)t)\,y(s;\,\varepsilon) \tag{6.166}$$

where $y(s;\,\varepsilon)=y(s+2\pi;\,\varepsilon)$. Then

$$\gamma y+\omega y'=g_a(\lambda,\,b)y \tag{6.167}$$

Defining the linear operator $\tilde{J}(\varepsilon)$ by

$$\tilde{J}y=-\omega y'+g_a(\lambda,\,b)y \tag{6.168}$$

this becomes

$$\tilde{J}y=\gamma y \tag{6.169}$$

where $\tilde{J}$ is the linearisation of $-\omega(d/ds)+g(\lambda,\cdot)$ about the bifurcating solution b, whereas J was the linearisation about the trivial solution. The problem thus reduces to the determination of the eigenvalues γ of $\tilde{J}$. By (6.166), if Re $\gamma>0$, then the perturbation $z\to\infty$ as $t\to\infty$ so that b is unstable; if Re $\gamma<0$, then $z\to 0$ and b is stable. If Re $\gamma=0$, then the stability is not determined by a linear analysis. At $\varepsilon=0$ we have

$$\tilde{J}(0)=\{-\omega_0(d/ds)+\sigma(0)\}=-\omega_0(d/ds)+i\omega(0)=J_0 \tag{6.170}$$

so that the eigenvalue problem becomes

$$\tilde{J}(0)y(s;\,0)=J_0y(s;\,0)=\gamma(0)y(s;\,0) \tag{6.171}$$

This has eigenvalues $\gamma_n(0)$ and corresponding eigenfunctions $y_n(s;\,0)$ 2π-periodic in s, given by

$$y_n(s;\,0)=e^{ins},\qquad \gamma_n(0)=-(n-1)i\omega_0 \tag{6.172}$$

Moreover, $\tilde{J}(0)$ has no further eigenvalues with 2π-periodic eigenfunctions. We may extend this result to values of $\varepsilon\neq 0$. If $\gamma(\varepsilon)$ is an eigenvalue of $\tilde{J}(\varepsilon)$ with eigenfunction $y(s;\,\varepsilon)$, then so is $\gamma(\varepsilon)+in\omega(\varepsilon)$, with eigenfunction $y(s;\,\varepsilon)\,e^{-ins}$, since

$$\begin{aligned}\tilde{J}(y\,e^{-ins})&=-\omega(y'\,e^{-ins}-iny\,e^{-ins})+g_a(\lambda,\,b)y\,e^{-ins}\\&=(\tilde{J}y)\,e^{-ins}+in\omega y\,e^{-ins}=(\gamma+in\omega)y\,e^{-ins}\end{aligned}$$

Thus if we can find one eigenvalue, $\gamma_1(\varepsilon)$ say, we may find eigenvalues of $\tilde{J}(\varepsilon)$ corresponding to each eigenvalue of $\tilde{J}(0)$, so that all the eigenvalues of $\tilde{J}$ are known. We thus investigate

$$\tilde{J}y_1=\gamma_1y_1 \tag{6.173}$$

where

$$y_1(s;\,0)=e^{is},\qquad \gamma_1(0)=0 \tag{6.174}$$

from (6.172). For simplicity of notation we drop the suffix 1, but still consider the eigenvalue such that $\gamma(0)=0$. Note that stability is determined by this eigenvalue alone, since the others differ from it by $in\omega$, a purely imaginary quantity.

As in the real simple case, we need to introduce the adjoint operator $\tilde{J}^*(\varepsilon)$ with respect to the inner product $[\cdot,\cdot]$. An easy computation gives

$$\tilde{J}^* = \omega\frac{d}{ds} + \overline{g_a(\lambda, b)} \tag{6.175}$$

Corresponding to the eigenvalue $\gamma(\varepsilon)$ of $\tilde{J}$, this operator has an eigenvalue $\bar{\gamma}(\varepsilon)$ and associated eigenfunction $y^*(s;\varepsilon)$,

$$\tilde{J}^* y^* = \bar{\gamma} y^* \tag{6.176}$$

At $\varepsilon=0$, the problem (6.176) reduces to the linearisation about the trivial solution. We have

$$\begin{aligned}\tilde{J}^*(0)y^*(s;0) &= \{\omega_0(d/ds)+\bar{\sigma}(0)\}y^*(s;0)\\ &= \{\omega_0(d/ds)-i\omega_0\}y^*(s;0)=\gamma(0)y^*(s;0)=0\end{aligned}$$

so that

$$y^*(s;0)=e^{is} \tag{6.177}$$

Before proving a factorisation theorem for the Hopf bifurcation, we need a preliminary result.

LEMMA 6.178: The inner product $[b', y^*]=0$.

Proof: Since b is a solution of (6.155),

$$\omega b' = g(\lambda, b) \tag{6.179}$$

Differentiating with respect to s,

$$-\omega b'' + g_a(\lambda, b)b' = \tilde{J}b' = 0 \tag{6.180}$$

Now, for any ε,

$$\gamma[b', y^*] = [b', \bar{\gamma}y^*] = [b', \tilde{J}^*y^*] = [\tilde{J}b', y^*] = 0$$

It follows that

$$[b', y^*] = 0 \tag{6.181}$$

as required.

THEOREM 6.182 (Factorisation Theorem for the Hopf Bifurcation in Two Dimensions): Consider the two-dimensional bifurcation problem

$$\mathbf{u}_t = \mathbf{f}(\lambda, \mathbf{u}) = \mathbf{0}, \qquad \mathbf{f}(\lambda, \mathbf{0}) = \mathbf{0} \tag{6.183}$$

where the linearisation of $\mathbf{f}(\lambda, \cdot)$ about the trivial solution has a pair of complex conjugate eigenvalues $\sigma(\lambda)$ and $\bar{\sigma}(\lambda)$ satisfying

$$\sigma(0) = i\omega_0 \neq 0, \qquad \text{Re}\ \sigma'(0) > 0 \tag{6.184}$$

Let $(\lambda, \mathbf{u}) = (\lambda, b\boldsymbol{\phi} + \bar{b}\bar{\boldsymbol{\phi}})$ be the non-trivial solution of (6.183) constructed above, so that

$$-\omega b' + g(\lambda, b) = 0$$

Let $\tilde{J}(\varepsilon)$ be the linearisation of $-\omega(d/ds) + g(\lambda, \cdot)$ about b, and let $\gamma(\varepsilon)$ be the eigenvalue of $\tilde{J}$ satisfying $\gamma(0) = 0$. Then $\mathbf{u} = b\boldsymbol{\phi} + \bar{b}\bar{\boldsymbol{\phi}}$ is stable if Re $\gamma < 0$ and is unstable if Re $\gamma > 0$, and γ is given by

$$\gamma(\varepsilon) = -[g_\lambda \lambda_\varepsilon, y^*]/[b_\varepsilon, y^*] \tag{6.185}$$

as long as $[b_\varepsilon, y^*] \neq 0$. As $\varepsilon \to 0$,

$$\gamma(\varepsilon) \sim -n\lambda\sigma'(0) \sim -n\sigma(\lambda(\varepsilon)) \tag{6.186}$$

where n is the order of the first non-zero derivative of λ with respect to ε. Note that this compares γ and σ at the same value of λ, namely $\lambda(\varepsilon)$. If n does not exist, i.e. all derivatives of λ are zero, then $\gamma \equiv 0$, the critical point is a centre and each periodic solution is orbitally stable.

Proof: Differentiating (6.179) with respect to ε, we have

$$-\omega_\varepsilon b' - \omega b'_\varepsilon + g_\lambda \lambda_\varepsilon + g_a b_\varepsilon = \tilde{J} b_\varepsilon - \omega_\varepsilon b' + g_\lambda \lambda_\varepsilon = 0 \tag{6.187}$$

Taking the inner product with y^* and using Lemma 6.178,

$$-[g_\lambda \lambda_\varepsilon, y^*] = [\tilde{J} b_\varepsilon, y^*] = [b_\varepsilon, \tilde{J}^* y^*] = \gamma[b_\varepsilon, y^*]$$

which gives (6.185) if $[b_\varepsilon, y^*] \neq 0$. As $\varepsilon \to 0$, $b \to \varepsilon\, e^{is}$, $b_\varepsilon \to e^{is}$, $y^* \to e^{is}$, $g \to \sigma b$, $g_\lambda \to \sigma'(0)\varepsilon\, e^{is}$, and since $[e^{is}, e^{is}] = 1$,

$$\gamma(\varepsilon) \sim -n\lambda\sigma'(0)$$

by comparing the Taylor series for λ and λ_ε. Since $\sigma(\lambda) = \lambda\sigma'(0) + O(\lambda^2)$, then $\gamma(\varepsilon) \sim -n\sigma(\lambda(\varepsilon))$, as required.

Since Re $\sigma'(0) > 0$ for strict loss of stability, (6.184), then branches which bifurcate supercritically (so that $\lambda > 0$) are stable, whereas those which

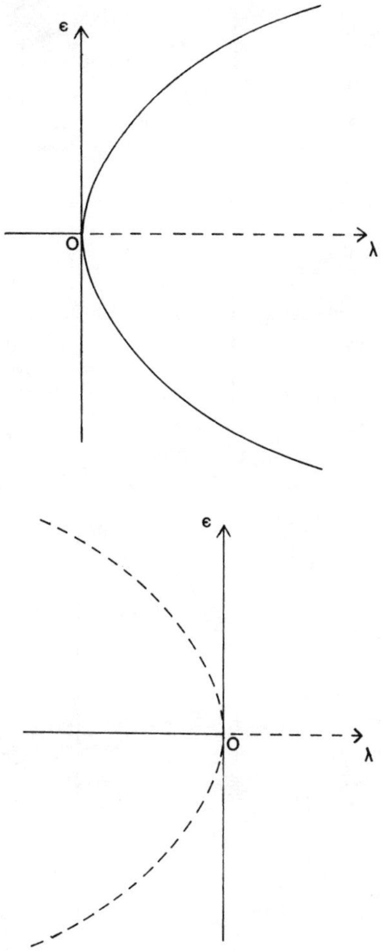

Fig. 6.188. Possible bifurcation diagrams for the Hopf bifurcation, Re $\sigma'(0) > 0$.

bifurcate subcritically ($\lambda < 0$) are unstable, as in the real simple case. This result holds more generally (Crandall and Rabinowitz, 1973; Sattinger, 1973). The possible bifurcation diagrams are shown in Fig. 6.188. The third possibility is that $\lambda = \lambda(\varepsilon) \equiv 0$, so that solutions of arbitrary amplitude exist for $\lambda = 0$. Then the critical point is a centre at $\lambda = 0$. A similar analysis shows that if Re $\sigma'(0) < 0$, then supercritical ($\lambda > 0$) bifurcations are unstable and subcritical ones are stable. The bifurcation diagrams are shown in Fig. 6.189.

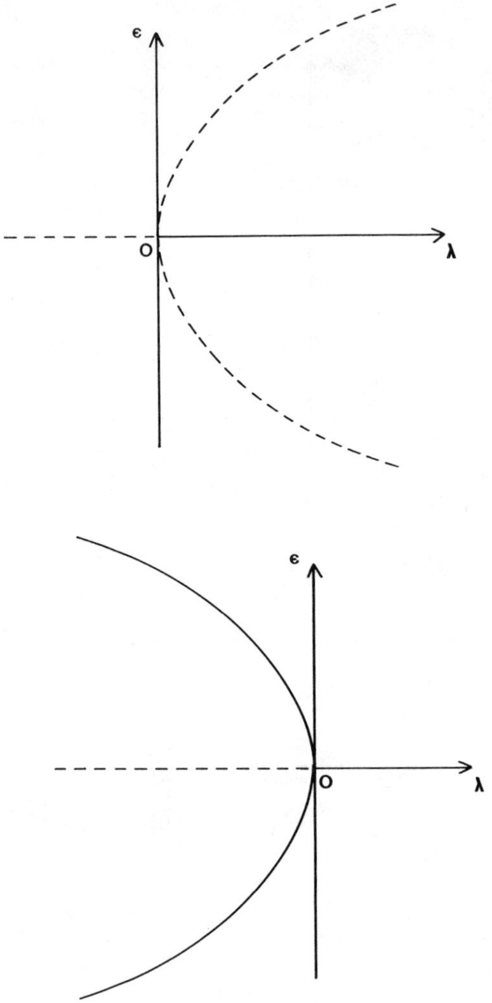

Fig. 6.189. Possible bifurcation diagrams for the Hopf bifurcation, Re $\sigma'(0) < 0$.

6.11 An Example of the Hopf Bifurcation in Two Dimensions

We shall consider the prey–predator system

$$\dot{\tilde{U}} = \tilde{U}^2(1-\tilde{U}) - \tilde{U}\tilde{V}, \qquad \dot{\tilde{V}} = a(\tilde{U}-\lambda)\tilde{V} \tag{6.190}$$

where $\tilde{U}$ and $\tilde{V}$ are dimensionless prey and predator populations, respectively, and λ is a parameter of the system [see Odell (1980)]. This differs from the Lotka–Volterra system with logistic effect in that the growth of the prey population at small values of $\tilde{U}$ is proportional to $\tilde{U}^2$ rather than to $\tilde{U}$. The justification for this is that reproduction depends on encounters between two members of the species rather than simply existence of one member, and such encounters are modelled by the law of mass action. The steady states of the system are given by $(\tilde{U}, \tilde{V}) = (0,0)$, $(1,0)$ and $(\lambda, \lambda(1-\lambda))$. The critical point at $(0,0)$ is a saddle-node, and that at $(1,0)$ is a saddle point. We shall analyse the critical point at $(\lambda, \lambda(1-\lambda))$ by defining

$$\tilde{u} = \tilde{U} - \lambda, \qquad \tilde{v} = \tilde{V} - \lambda(1-\lambda)$$

The system becomes

$$\begin{aligned} \dot{\tilde{u}} &= \lambda(1-2\lambda)\tilde{u} - \lambda\tilde{v} + (1-3\lambda)\tilde{u}^2 - \tilde{u}^3 - \tilde{u}\tilde{v} \\ \dot{\tilde{v}} &= a\lambda(1-\lambda)\tilde{u} - a\tilde{u}\tilde{v} \end{aligned} \tag{6.191}$$

which has eigenvalues σ given by

$$\sigma^2 - \lambda(1-2\lambda)\sigma + a\lambda^2(1-\lambda) = 0 \tag{6.192}$$

Thus the eigenvalues are purely imaginary if $\lambda = \lambda_0 = \frac{1}{2}$, and the critical point is an unstable focus if $\lambda < \lambda_0$ and $|\lambda - \lambda_0|$ is sufficiently small, and a stable focus if $\lambda > \lambda_0$ and $|\lambda - \lambda_0|$ is sufficiently small. Let us now attempt to draw the phase plane for this system from this information. For illustrative purposes we shall take $\lambda > \lambda_0$. Then the phase plane may be as in Fig. 6.193a, b or even as in c. There is no way we can decide which of these is correct simply from phase plane considerations. However, we may apply the Hopf bifurcation theorem to the system if the necessary transversality condition $\operatorname{Re} \sigma'(\lambda_0) \neq 0$ holds. But from (6.192),

$$\operatorname{Re} \sigma'(\lambda_0) = \frac{d}{d\lambda}\left\{\frac{\lambda(1-2\lambda)}{2}\right\}\Bigg|_{\lambda=\lambda_0} = -\frac{1}{2} < 0$$

so that the theorem obtains (Theorem 6.144, after the transformation $\lambda \to \lambda_0 - \lambda$). It follows immediately from the factorisation theorem (Theorem 6.162) that the phase plane cannot be as in Fig. 6.193c, at least not for λ sufficiently close to the bifurcation point, since supercritical bifurcations (such that $\lambda > \lambda_0$) are unstable, not semi-stable. To decide whether the bifurcation at $\lambda = \lambda_0$ is sub- or supercritical, so that we may decide between Fig. 6.193a and Fig. 6.193b, we need to find λ_2, since from Section 6.9 $\lambda = \lambda_0 + \varepsilon^2\lambda_2 + O(\varepsilon^3)$. To use Eq. (6.163) we must write the system (6.191) in canonical form, i.e. we must make a transformation $\mathbf{u} = A\tilde{\mathbf{u}}$ so that the linearisation of the system at the bifurcation point is in the form

$$\dot{\mathbf{u}} = M(\lambda_0)\mathbf{u}$$

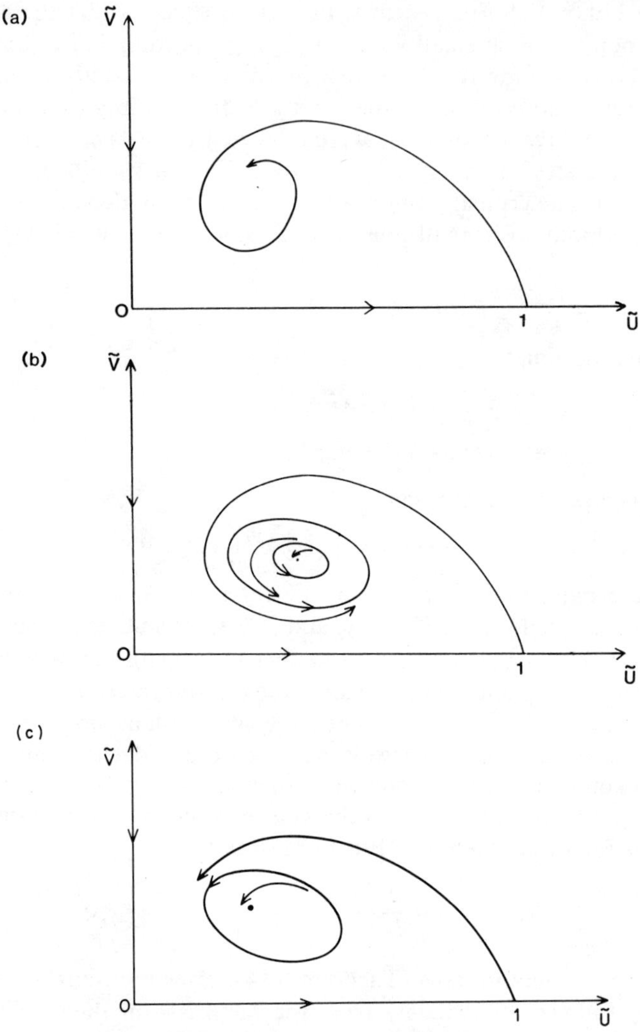

Fig. 6.193. Possible phase-plane diagrams for the system (6.190) with $\lambda > \lambda_0$, with (a) no periodic solutions, (b) one stable and one unstable limit cycle and (c) one semi-stable limit cycle.

where

$$M(\lambda_0) = \begin{bmatrix} 0 & \omega_0 \\ -\omega_0 & 0 \end{bmatrix}$$

and $\sigma(\lambda_0) = \sigma_0 = i\omega_0$. But from (6.192)

$$\sigma_0^2 + a\lambda_0^2(1-\lambda_0) = \sigma_0^2 + a/8 = 0$$

so that

$$\omega_0 = \sqrt{a/8}$$

The linearisation of (6.191) is

$$\dot{\tilde{\mathbf{u}}} = \tilde{M}(\lambda)\tilde{\mathbf{u}}$$

where

$$\tilde{M}(\lambda) = \begin{bmatrix} \lambda(1-2\lambda) & -\lambda \\ a\lambda(1-\lambda) & 0 \end{bmatrix}$$

so that

$$\tilde{M}(\lambda_0) = \tilde{M}(\tfrac{1}{2}) = \begin{bmatrix} 0 & -\frac{1}{2} \\ \frac{1}{4}a & 0 \end{bmatrix}$$

We make the transformation

$$u = \tilde{u}, \qquad v = -\tilde{v}/\sqrt{a/2} = -\tilde{v}/(2\omega_0)$$

and the system becomes

$$\begin{aligned} \dot{u} &= \lambda(1-2\lambda)u + 2\lambda\omega_0 v + (1-3\lambda)u^2 - u^3 - 2\omega_0 uv = f(\lambda, u, v) \\ \dot{v} &= -a\lambda(1-\lambda)u/(2\omega_0) + auv = g(\lambda, u, v) \end{aligned} \tag{6.194}$$

This is clearly in canonical form. In the notation of Section 6.9 we have

$$f^0_{uu} = 2(1-3\lambda_0) = -1, \qquad f^0_{uv} = -2\omega_0 = -\sqrt{a/2}, \qquad f^0_{vv} = 0,$$

$$f^0_{uuu} = -6, \qquad f^0_{uvv} = 0$$

$$g^0_{uu} = 0, \qquad g^0_{uv} = a, \qquad g^0_{vv} = 0, \qquad g^0_{uuv} = 0, \qquad g^0_{vvv} = 0$$

so that from (6.163)

$$\lambda_2 \operatorname{Re} \sigma'(\lambda_0) = -(-2\omega_0)/(4\omega_0) - (-6)/4 = 1/2 + 3/2 = 2$$

Thus $\lambda_2 < 0$, and the bifurcation is subcritical and therefore stable (since $\operatorname{Re} \sigma'(\lambda_0) < 0$). The bifurcation diagram is as in Fig. 6.195, at least for small amplitudes. Thus the phase plane for $\lambda > \lambda_0 = \frac{1}{2}$ has no (small amplitude) limit cycles and is probably as in Fig. 6.193a. In this simple two-dimensional case the Hopf bifurcation theorem may be supplemented by powerful phase plane techniques. For example, it is easy to show by using the Poincaré–Bendixson theorem that the system (6.190) has a stable limit cycle solution for all λ satisfying $0 < \lambda < \lambda_0$, a global result which cannot be derived from the Hopf bifurcation theorem, which only gives information close to the bifurcation point. The great advantage of the Hopf theorem is that it may also be applied to m-dimensional and to reaction–diffusion, or infinite-

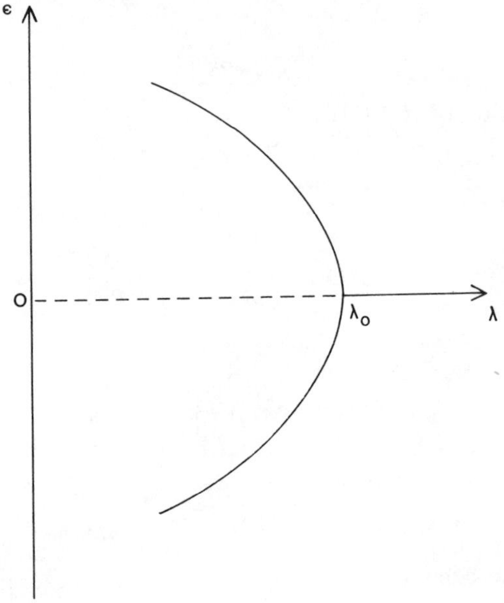

Fig. 6.195. The probable bifurcation diagram for the system (6.190).

dimensional, systems. Kishimoto *et al.* (1983) use the Hopf bifurcation theorem to prove the existence of stable spatio-temporal oscillations in an ecological system with three or more species.

6.12 The Hopf Bifurcation in *m* Dimensions and in Infinite Dimensions

Let us assume that in an m-dimensional or an infinite-dimensional problem there are two isolated complex conjugate simple eigenvalues with greatest real part which cross from the left to the right half of the complex plane. Then there occurs a Hopf bifurcation, which is an essentially two-dimensional phenomenon. To make this statement more precise and to see why it is true, we consider the reaction–diffusion system

$$\mathbf{u}_t = \mathbf{F}(\lambda, \mathbf{u}) = \mathbf{f}(\lambda, \mathbf{u}) + D\,\nabla^2\mathbf{u} \tag{6.196}$$

where $\mathbf{F}(\lambda, \mathbf{0}) = \mathbf{f}(\lambda, \mathbf{0}) = \mathbf{0}$, and define $L(\lambda)$ and $M(\lambda)$ as before,

$$L(\lambda) = \mathbf{F}_{\mathbf{u}}(\lambda, \mathbf{0}|\cdot) = M(\lambda) + D\,\nabla^2 = \mathbf{f}_{\mathbf{u}}(\lambda, \mathbf{0}|\cdot) + D\,\nabla^2 \tag{6.197}$$

Let $\sigma(\lambda)$ and $\bar{\sigma}(\lambda)$ be the eigenvalues with greatest real part, and let

$$L\boldsymbol{\phi}=\sigma\boldsymbol{\phi}, \qquad L\bar{\boldsymbol{\phi}}=\bar{\sigma}\bar{\boldsymbol{\phi}}, \qquad L^*\boldsymbol{\phi}^*=\bar{\sigma}\boldsymbol{\phi}^*, \qquad L^*\bar{\boldsymbol{\phi}}^*=\sigma\bar{\boldsymbol{\phi}}^* \tag{6.198}$$

where L^* is the adjoint operator of L, and

$$\langle\boldsymbol{\phi}, \boldsymbol{\phi}^*\rangle=1, \qquad \langle\boldsymbol{\phi}, \bar{\boldsymbol{\phi}}^*\rangle=0 \tag{6.199}$$

Let

$$\sigma(0)=i\omega_0\neq 0, \qquad \operatorname{Re}\sigma'(0)>0 \tag{6.200}$$

In the two-dimensional case $\mathbf{u}$ was written $\mathbf{u}(t)=a(t)\boldsymbol{\phi}+\bar{a}(t)\bar{\boldsymbol{\phi}}$. In this case we write

$$\mathbf{u}(t)=a(t)\boldsymbol{\phi}+\bar{a}(t)\bar{\boldsymbol{\phi}}+\mathbf{w}(t) \tag{6.201}$$

where $\mathbf{w}$ is real and

$$\langle\mathbf{w}, \boldsymbol{\phi}^*\rangle=\langle\mathbf{w}, \bar{\boldsymbol{\phi}}^*\rangle=0 \tag{6.202}$$

$\mathbf{w}$ is thus the component of $\mathbf{u}$ which is orthogonal to the subspace spanned by $\boldsymbol{\phi}$ and $\bar{\boldsymbol{\phi}}$. Substituting (6.201) into (6.196),

$$\begin{aligned}\dot{a}\boldsymbol{\phi}+\dot{\bar{a}}\bar{\boldsymbol{\phi}}+\dot{\mathbf{w}}&=\mathbf{F}(\lambda,\mathbf{u})=L(\lambda)\mathbf{u}+\mathbf{N}(\lambda,\mathbf{u})\\&=\sigma a\boldsymbol{\phi}+\bar{\sigma}\bar{a}\bar{\boldsymbol{\phi}}+L\mathbf{w}+\mathbf{N}(\lambda,\mathbf{u})\end{aligned} \tag{6.203}$$

where $\mathbf{N}(\lambda, \mathbf{u})$ is the nonlinear part of $\mathbf{F}(\lambda, \mathbf{u})$. By taking the inner product of this with $\boldsymbol{\phi}^*$ (or equivalently with $\bar{\boldsymbol{\phi}}^*$), we obtain

$$\dot{a}-\sigma a=\langle\mathbf{N}(\lambda,\mathbf{u}), \boldsymbol{\phi}^*\rangle \tag{6.204}$$

so that the equation for $\mathbf{w}$ is given by

$$\dot{\mathbf{w}}=L\mathbf{w}+\mathbf{N}(\lambda,\mathbf{u})-\langle\mathbf{N}(\lambda,\mathbf{u}), \boldsymbol{\phi}^*\rangle\boldsymbol{\phi}-\langle\mathbf{N}(\lambda,\mathbf{u}), \bar{\boldsymbol{\phi}}^*\rangle\bar{\boldsymbol{\phi}} \tag{6.205}$$

Let us assume that $|a|$ and $|\mathbf{w}|$ are initially $O(\varepsilon)$, where ε is a small parameter. Then the nonlinear terms in (6.205) are $O(\varepsilon^2)$ and the linear terms $\dot{\mathbf{w}}=L\mathbf{w}$ dominate. But the eigenvalues of L restricted to the space orthogonal to that spanned by $\boldsymbol{\phi}$ and $\bar{\boldsymbol{\phi}}$ all have negative real part, so that $|\mathbf{w}|$ decreases until the nonlinear terms become important. It follows that after some time $|\mathbf{w}|=O(|a|^2)=O(\varepsilon^2)$, so that to first-order $\mathbf{u}$ is of the same form and satisfies the same equation as in the two-dimensional case. This is what we mean by saying that the Hopf bifurcation is essentially two-dimensional. At higher orders the nonlinear coupling with $\mathbf{w}$ becomes important and this must be taken into account in finding a bifurcation solution. For a rigorous proof of the two-dimensional nature of Hopf bifurcations in higher-dimensional problems using the centre manifold theorem see, for example, Marsden and McCracken (1976).

The bifurcating solution $\mathbf{u}(t)$ in the m-dimensional or infinite-dimensional case may be constructed. Since we are looking for a periodic solution it is natural to set up a Hilbert space $P_{2\pi}$ of 2π-periodic functions sufficiently smooth for the relevant derivatives to be continuous, with inner product defined by

$$[\mathbf{p}, \mathbf{q}] = \frac{1}{2\pi} \int_0^{2\pi} \langle \mathbf{p}(s), \mathbf{q}(s) \rangle \, ds \tag{6.206}$$

We look for solutions of

$$\mathbf{u}_t = \mathbf{F}(\lambda, \mathbf{u})$$

in the form $\mathbf{u} = \mathbf{u}(s; \varepsilon)$ where $s = \omega t$. Defining

$$J(\varepsilon) = -\omega \frac{d}{ds} + \mathbf{F}_{\mathbf{u}}(\lambda, \mathbf{0} | \cdot) = -\omega \frac{d}{ds} + L(\lambda) \tag{6.207}$$

where L is the operator defined in (6.197), and

$$\mathbf{N}(\lambda, \mathbf{u}) = \mathbf{F}(\lambda, \mathbf{u}) - \mathbf{F}_{\mathbf{u}}(\lambda, \mathbf{0} | \mathbf{u}) \tag{6.208}$$

the equation becomes

$$-\omega \frac{d\mathbf{u}}{ds} + \mathbf{F}(\lambda, \mathbf{u}) = J\mathbf{u} + \mathbf{N}(\lambda, \mathbf{u}) = \mathbf{0} \tag{6.209}$$

At the bifurcation point $\varepsilon = 0$ the operator becomes

$$J(0) = J_0 = -\omega_0 \frac{d}{ds} + \mathbf{F}_{\mathbf{u}}(0, \mathbf{0} | \cdot) = -\omega_0 \frac{d}{ds} + L(0) \tag{6.210}$$

As in Section 6.9 we need to investigate the null-space and adjoint operator of the appropriate linear operator. Let $\boldsymbol{\phi}(\lambda)$ be the eigenvector of $L(\lambda)$ with eigenvalue $\sigma(\lambda)$, as before, and let $\boldsymbol{\phi}_0$ be its value at $\lambda = 0$, so that

$$L(0)\boldsymbol{\phi}_0 = \sigma(0)\boldsymbol{\phi}_0 = i\omega_0\boldsymbol{\phi}_0, \qquad L(0)\bar{\boldsymbol{\phi}}_0 = \bar{\sigma}(0)\bar{\boldsymbol{\phi}}_0 = -i\omega_0\bar{\boldsymbol{\phi}}_0 \tag{6.211}$$

It follows that

$$J_0\boldsymbol{\phi}_0 \, e^{is} = \mathbf{0} \tag{6.212}$$

and similarly

$$J_0\bar{\boldsymbol{\phi}}_0 \, e^{-is} = \mathbf{0} \tag{6.213}$$

Defining

$$\mathbf{z} = \boldsymbol{\phi}_0 \, e^{is} \tag{6.214}$$

then $\mathbf{z} \in P_{2\pi}$, and

$$J_0\mathbf{z} = J_0\bar{\mathbf{z}} = \mathbf{0} \tag{6.215}$$

so $\mathbf{z}$ and $\bar{\mathbf{z}}$ are in the null-space of J_0. Since $\sigma(0)$ and $\bar{\sigma}(0)$ are simple eigenvalues of $L(0)$, it follows that $\mathbf{z}$ and $\bar{\mathbf{z}}$ in fact span the null-space of J_0, so that any solution of $J_0\mathbf{v}=\mathbf{0}$ may be written as a linear combination of $\mathbf{z}$ and $\bar{\mathbf{z}}$. Defining the adjoint J_0^* in terms of the inner product $[\cdot, \cdot]$, by the requirement that

$$[J_0\mathbf{p}, \mathbf{q}] = [\mathbf{p}, J_0^*\mathbf{q}]$$

for all $\mathbf{p}(s)$ and $\mathbf{q}(s)$ in $P_{2\pi}$, we obtain

$$J_0^* = \omega_0 \frac{d}{ds} + L^*(0)$$

where $L^*(0)$ is defined as before, and $\boldsymbol{\phi}_0^*$ and $\bar{\boldsymbol{\phi}}_0^*$ are its eigenvalues at $\lambda = 0$. It follows that

$$J_0^*\mathbf{z}^* = J_0^*\bar{\mathbf{z}}^* = \mathbf{0} \tag{6.216}$$

where

$$\mathbf{z}^* = \boldsymbol{\phi}_0^* \, e^{is} \tag{6.217}$$

and $\mathbf{z}^* \in P_{2\pi}$.

We now seek solutions of $\mathbf{u}_t = \mathbf{F}(\lambda, \mathbf{u})$ in the form

$$\mathbf{u}(s; \varepsilon) = \sum_{n=1}^{\infty} \mathbf{u}_n(s)\varepsilon^n \tag{6.218}$$

with

$$\omega(\varepsilon) = \sum_{n=0}^{\infty} \omega_n\varepsilon^n, \qquad \lambda(\varepsilon) = \sum_{n=1}^{\infty} \lambda_n\varepsilon^n \tag{6.219}$$

where $s = \omega(\varepsilon)t$ and $\mathbf{u}$ is 2π-periodic in s. Here ε is a measure of the amplitude of $\mathbf{u}$ and may be defined by the normalising condition

$$\varepsilon = [\mathbf{u}, \mathbf{z}^*] \tag{6.220}$$

It follows, on substituting the power series representation for $\mathbf{u}$, that

$$[\mathbf{u}_1, \mathbf{z}^*] = 1, \qquad [\mathbf{u}_n, \mathbf{z}^*] = 0, \qquad n \geqslant 2 \tag{6.221}$$

The equation becomes

$$\omega\mathbf{u}' = \mathbf{F}(\lambda, \mathbf{u}) \tag{6.222}$$

By substituting (6.211) and (6.212) into this and equating powers of ε, we obtain

$$J_0\mathbf{u}_1 = 0 \tag{6.223}$$

$$J_0\mathbf{u}_2 - 2\omega_1\mathbf{u}_1' + 2\lambda_1\mathbf{F}_{\mathbf{u}\lambda}(0, \mathbf{0}|\mathbf{u}_1) + \mathbf{F}_{\mathbf{uu}}(0, \mathbf{0}|\mathbf{u}_1|\mathbf{u}_1) = \mathbf{0} \tag{6.224}$$

and in general

$$J_0\mathbf{u}_n - n\omega_{n-1}\mathbf{u}_1' + n\lambda_{n-1}\mathbf{F}_{\mathbf{u}\lambda}(0, \mathbf{0}|\mathbf{u}_1) + \mathbf{R}_{n-2} = \mathbf{0} \tag{6.225}$$

where $\mathbf{R}_{n-2}$ depends on terms of order $n-2$ or less. Equation (6.223) has general real solution

$$\mathbf{u}_1 = c\mathbf{z} + \bar{c}\bar{\mathbf{z}}$$

The normalisation condition (6.221) then gives $[\mathbf{u}_1, \mathbf{z}^*] = [c\mathbf{z} + \bar{c}\bar{\mathbf{z}}, \mathbf{z}^*] = c = 1$, so that

$$\mathbf{u}_1 = \mathbf{z} + \bar{\mathbf{z}} \tag{6.226}$$

To solve (6.224) we note that J_0 with periodic boundary conditions is again a Fredholm operator, so that the following version of the Fredholm alternative holds.

THEOREM 6.227: The equation

$$J_0\mathbf{u}_n = \mathbf{g}_{n-1}$$

where $\mathbf{g}_{n-1} \in P_{2\pi}$, is soluble for $\mathbf{u}_n \in P_{2\pi}$ if and only if

$$[\mathbf{g}_{n-1}, \mathbf{z}^*] = [\mathbf{g}_{n-1}, \bar{\mathbf{z}}^*] = 0 \tag{6.228}$$

This solution is unique (up to translation if $\mathbf{g}_{n-1}$ is translationally invariant) under the normalisation

$$[\mathbf{u}_n, \mathbf{z}^*] = k_n$$

for any given constant k_n, if we require that $\mathbf{u}_n$ be real.

Proof: See Sattinger (1973).

REMARK 6.229: If $\mathbf{g}_{n-1}$ is real, we need only apply the condition $[\mathbf{g}_{n-1}, \mathbf{z}^*] = 0$, since the second of (6.228) is then automatically satisfied.

Thus (6.224) may be solved under the condition that

$$[-2\omega_1\mathbf{u}_1' + 2\lambda_1\mathbf{F}_{\mathbf{u}\lambda}(0, \mathbf{0}|\mathbf{u}_1) + \mathbf{F}_{\mathbf{uu}}(0, \mathbf{0}|\mathbf{u}_1|\mathbf{u}_1), \mathbf{z}^*] = 0 \tag{6.230}$$

Note that

$$\mathbf{F}_{\mathbf{u}}(\lambda, \mathbf{0}|\boldsymbol{\phi}) = L(\lambda)\boldsymbol{\phi} = \sigma(\lambda)\boldsymbol{\phi}$$

so that

$$\mathbf{F}_{\mathbf{u}\lambda}(0, \mathbf{0}|\boldsymbol{\phi}_0) = \sigma'(0)\boldsymbol{\phi}_0$$

and

$$\begin{aligned}\mathbf{F}_{\mathbf{u}\lambda}(0,\mathbf{0}\,|\,\mathbf{u}_1) &= \mathbf{F}_{\mathbf{u}\lambda}(0,\mathbf{0}\,|\,\boldsymbol{\phi}_0\,e^{is}+\bar{\boldsymbol{\phi}}_0\,e^{-is})\\ &= \sigma'(0)\boldsymbol{\phi}_0\,e^{is}+\bar{\sigma}'(0)\bar{\boldsymbol{\phi}}_0\,e^{-is}\\ &= \sigma'(0)\mathbf{z}+\bar{\sigma}'(0)\bar{\mathbf{z}}\end{aligned}$$

It follows that $\langle \mathbf{F}_{\mathbf{u}\lambda}(0,\mathbf{0}\,|\,\mathbf{u}_1),\mathbf{z}^*\rangle=\sigma'(0)$, so that

$$[\mathbf{F}_{\mathbf{u}\lambda}(0,\mathbf{0}\,|\,\mathbf{u}_1),\mathbf{z}^*]=\sigma'(0) \tag{6.231}$$

Also $\mathbf{F}_{\mathbf{uu}}(0,\mathbf{0}\,|\,\mathbf{u}_1\,|\,\mathbf{u}_1)$ contains only quadratic terms, i.e. terms in e^{2is} and e^{-2is} and those independent of s. It follows that $[\mathbf{F}_{\mathbf{uu}}(0,\mathbf{0}\,|\,\mathbf{u}_1\,|\,\mathbf{u}_1),\mathbf{z}^*]=0$. Lastly, $[\mathbf{u}_1',\mathbf{z}^*]=i$, so that (6.230) becomes

$$-2i\omega_1+2\lambda_1\sigma'(0)=0 \tag{6.232}$$

The condition Re $\sigma'(0)\neq 0$ implies that $\lambda_1=0$, so that $\omega_1=0$. By induction it can be shown that $\lambda_{2n+1}=\omega_{2n+1}=0$ for all non-negative integers n, so that λ and ω are even functions of ε (Joseph and Sattinger, 1972). The functions $\mathbf{u}_n$ are found sequentially by solving the equations (6.225) subject to the normalisation conditions (6.221). Thus we have constructed the Hopf bifurcation solution in the general m-dimensional or infinite-dimensional case.

We are concerned with the stability of this solution. Floquet theory and an analysis entirely analogous to that in the two-dimensional case again shows that supercritical bifurcations are stable and subcritical ones unstable (Iooss and Joseph, 1980).

6.13 Plane-wave Solutions of Reaction–Diffusion Equations

It is well known that certain reacting and diffusing systems have travelling wave solutions. These may be seen experimentally, for example, in the Belousov–Zhabotinskii reaction (Winfree, 1974), where target patterns occur. Far from the origin of these patterns, the waves approximate to plane waves, and it is therefore of interest to investigate the possibility of plane-wave solutions of systems of reaction–diffusion equations. This was done using the Hopf bifurcation theorem by Kopell and Howard (1973), and we shall now describe their approach. Plane-wave solutions of the Belousov–Zhabotinskii reaction have been considered by Hastings (1976a).

Consider a system

$$\mathbf{u}_t=\mathbf{F}(\mathbf{u})=\mathbf{f}(\mathbf{u})+D\,\nabla^2\mathbf{u} \tag{6.233}$$

where for simplicity $\mathbf{u}$ is two-dimensional, $\mathbf{u}=(u_1,u_2)^{\mathrm{T}}$, and $\mathbf{u}$ represents deviation from a steady-state solution of the equations, so that

$$\mathbf{f}(\mathbf{0}) = \mathbf{0} \tag{6.234}$$

Let us assume that the steady state is oscillatorily unstable to small perturbations, that is, the eigenvalues of the matrix M defined by

$$M = \mathbf{f}_{\mathbf{u}}(\cdot) \tag{6.235}$$

are given by $p \pm iq$, where $p, q > 0$. A plane-wave solution of these equations is a periodic solution of the form

$$\mathbf{u} = \tilde{\mathbf{u}}(t - \mathbf{k} \cdot \mathbf{x}/\omega) \tag{6.236}$$

where ω is the angular frequency and $\mathbf{k}$ the wave-number vector of the plane wave. Substituting this form into (6.233), we obtain

$$\tilde{\mathbf{u}}' = \mathbf{f}(\tilde{\mathbf{u}}) + D(k^2/\omega^2)\tilde{\mathbf{u}}'' = \mathbf{f}(\tilde{\mathbf{u}}) + \lambda D\tilde{\mathbf{u}}'' \tag{6.237}$$

where $\lambda = k^2/\omega^2$. This is a fourth-order system of ordinary differential equations; we may write

$$\tilde{\mathbf{u}}' = \mathbf{v}, \qquad \mathbf{v}' = -(1/\lambda^2)D^{-1}\mathbf{f}(\tilde{\mathbf{u}}) + (1/\lambda^2)D^{-1}\mathbf{v} \tag{6.238}$$

where D^{-1} is the inverse of the diffusion matrix D. The eigenvalues of this system are given by the solutions of the fourth-order equation

$$\begin{vmatrix} -\sigma I & I \\ -\dfrac{1}{\lambda^2}D^{-1}M & \dfrac{1}{\lambda^2}D^{-1} - \sigma I \end{vmatrix} = 0 \tag{6.239}$$

where I is the 2×2 identity matrix. It can be shown (Kopell and Howard, 1973) that if the steady state is oscillatory unstable and if the diffusion matrix D is sufficiently close to a scalar matrix, then two complex conjugate eigenvalues of (6.239) cross the imaginary axis transversally at some point $\lambda = \lambda_0$. Taking λ as the bifurcation parameter, it follows from the Hopf bifurcation theorem that there exist periodic solutions of (6.237) of frequency ω, say. Defining $\hat{\mathbf{u}}$ by

$$\hat{\mathbf{u}}(\omega t - \mathbf{k} \cdot \mathbf{x}) = \tilde{\mathbf{u}}(t - \mathbf{k} \cdot \mathbf{x}/\omega) \tag{6.240}$$

we have solutions $\hat{\mathbf{u}}$ which are periodic of period 2π in their argument. These solutions may be constructed close to the bifurcation point in exactly the same way as was indicated for the general case in Section 6.10. However, in the case of a scalar diffusion matrix, $D = dI$, say, the special form of the equations allows a simpler construction.

Let the eigenvectors of the matrix M be given by $\boldsymbol{\psi}$ and $\bar{\boldsymbol{\psi}}$, where

$$M\boldsymbol{\psi} = \rho\boldsymbol{\psi} = (p + iq)\boldsymbol{\psi}, \qquad M\bar{\boldsymbol{\psi}} = \bar{\rho}\bar{\boldsymbol{\psi}} = (p - iq)\bar{\boldsymbol{\psi}} \tag{6.241}$$

These equations define ρ. The vectors $\boldsymbol{\psi}$ and $\bar{\boldsymbol{\psi}}$ are linearly independent, and so we must be able to express $\tilde{\mathbf{u}}$ as a linear combination of them, i.e. defining

$$\tau = t - \mathbf{k} \cdot \mathbf{x}/\omega \tag{6.242}$$

we must have

$$\tilde{\mathbf{u}}(\tau) = a(\tau)\boldsymbol{\psi} + \bar{a}(\tau)\bar{\boldsymbol{\psi}} \tag{6.243}$$

Then

$$\begin{aligned}\tilde{\mathbf{u}}' &= a'\boldsymbol{\psi} + \bar{a}'\bar{\boldsymbol{\psi}} = \mathbf{f}(\tilde{\mathbf{u}}) + d(k^2/\omega^2)\tilde{\mathbf{u}}'' \\ &= M\tilde{\mathbf{u}} + \mathbf{N}(\tilde{\mathbf{u}}) + d(k^2/\omega^2)\tilde{\mathbf{u}}'' \\ &= a\rho\boldsymbol{\psi} + \bar{a}\bar{\rho}\bar{\boldsymbol{\psi}} + \mathbf{N}(\tilde{\mathbf{u}}) + d(k^2/\omega^2)(a''\boldsymbol{\psi} + \bar{a}''\bar{\boldsymbol{\psi}})\end{aligned} \tag{6.244}$$

where $\mathbf{N}(\cdot)$ is the nonlinear part of the operator $\mathbf{f}(\cdot)$. Taking the inner product of this equation with the eigenvector $\boldsymbol{\psi}^*$ of the adjoint operator M^* we have

$$\begin{aligned}a' &= a\rho + d(k^2/\omega^2)a'' + \langle \mathbf{N}(\tilde{\mathbf{u}}), \boldsymbol{\psi}^*\rangle \\ &= a\rho + d(\mu/\omega^2)a'' + \langle \mathbf{N}(\tilde{\mathbf{u}}), \boldsymbol{\psi}^*\rangle = g(\mu, a)\end{aligned} \tag{6.245}$$

where we have defined $\mu = k^2 = \lambda\omega^2$. The linearised version of this equation has periodic solutions $a = e^{i\omega\tau}$ with $\omega = \omega_0$ if

$$i\omega_0 = \rho - \mu d$$

that is, if

$$\omega_0 = q, \qquad \mu = k^2 = p/d$$

This gives an explicit value for the bifurcation point, namely

$$\lambda = \lambda_0 + p/(dq^2) \qquad \text{or} \qquad \mu = \mu_0 = p/d \tag{6.246}$$

We now look for a solution which is a perturbation of this, i.e.

$$\begin{gathered}a(\tau) = b(s;\, \varepsilon), \qquad \omega = \omega(\varepsilon), \qquad \omega(0) = \omega_0, \qquad s = \omega(\varepsilon)\tau \\ \mu = \mu(\varepsilon), \qquad \mu(0) = \mu_0\end{gathered} \tag{6.247}$$

[cf. (6.149)]. We define

$$\varepsilon = \frac{1}{2\pi}\int_0^{2\pi} e^{-is} b(s;\, \varepsilon)\, ds = [b] \tag{6.248}$$

Substituting into (6.245), we have

$$\omega b' = g(\lambda, b) \tag{6.249}$$

where the prime now denotes differentiation with respect to s, with normalisation condition

$$[b] = \varepsilon$$

We look for a solution in the form

$$b(s;\varepsilon) = \sum_{n=1}^{\infty} b_n(s)\varepsilon^n, \qquad \omega(\varepsilon) = \sum_{n=0}^{\infty} \omega_n \varepsilon^n, \qquad \mu(\varepsilon) = \sum_{n=0}^{\infty} \mu_n \varepsilon^n \tag{6.250}$$

Substituting into (6.242) and equating powers of ε this becomes

$$\omega_0 b_1' - \rho b_1 - d\mu_0 b_1'' = K_0 b_1 = 0 \tag{6.251}$$

$$\omega_0 b_1' - \rho b_2 - d\mu_0 b_2'' = K_0 b_2 = -\omega_1 b_1' + d\mu_1 b_1'' + Q(b_1, \bar{b}_1) \tag{6.252}$$

and so on, where the linear operator K_0 is defined by

$$K_0 \equiv \omega_0 \frac{d}{ds} - \rho - d\mu_0 \frac{d^2}{ds^2} \tag{6.253}$$

and $Q(b_1, \bar{b}_1)$ denotes quadratic terms in b_1 and $\bar{b}_1$, with normalisation

$$[b_1] = 1, \qquad [b_n] = 0, \qquad n > 1 \tag{6.254}$$

Here K_0 with periodic boundary conditions is again a Fredholm operator. The first perturbation equation (6.251) with the above normalisation has periodic solution

$$b_1 = e^{is} \tag{6.255}$$

The second has a periodic solution if and only if the Fourier expansion of the right-hand side has no term in e^{is}. This is a Fredholm orthogonality condition. Since $Q(b_1, \bar{b}_1)$ has no term in e^{is}, this implies

$$-i\omega_1 - \mu_1 d = 0$$

so that $\omega_1 = \mu_1 = 0$. Then (6.252) can be solved for b_2, and the process may be continued for the higher order perturbation equations.

It follows that whenever a two-component reaction–diffusion system has an oscillatorily unstable steady state in the diffusionless case and the diffusion matrix is nearly scalar, then it has a plane-wave solution when diffusion is added. If the diffusion matrix is scalar the construction of this solution is particularly simple. Unfortunately, small amplitude plane waves are unstable, although it is possible that they attain stability through a turning point in the bifurcation diagram. See Kopell and Howard (1973) and Maginu (1981) for further discussion. Schneider (1983) has shown that if an m-component reaction–diffusion system has a non-degenerate limit cycle solution in the diffusionless case and the diffusion matrix has no eigenvalues on the imaginary axis, then it has a plane-wave solution when diffusion is added.

7. Asymptotic Methods for Oscillatory Systems

7.1 Introduction

A reaction–diffusion system contains certain parameters, both those appearing explicitly in the equations, such as rate constants and diffusion coefficients, and implicit ones such as the steady-state values of the quantities concerned. From the point of view of analysing the system, the most important parameters are non-dimensional, that is, they are pure numbers and do not depend on the system of units used. For example, the ratio of two diffusion coefficients is a non-dimensional parameter.

It may be that some of the non-dimensional parameters of a system are small, that is, they are very much less than one. In this case it may be possible to neglect the terms in the equation which involve the small parameter and obtain a good approximation to the solution of the problem. For example, if the ratio of two diffusion coefficients is small, it may be reasonable to treat the slowly diffusing substance or species as effectively non-diffusible, as in the model for rabies in Chapter 3. In general, however, it is not possible to do this. For example, we saw in Chapter 3 that the addition of a term $-\varepsilon u^2$ into the Lotka–Volterra equations, however small ε was, changed the nature of the solutions completely.

Even if one cannot neglect a term in a small parameter ε, it may be possible to obtain an approximate solution to the problem by using the fact that ε is small. This chapter is concerned with methods for constructing such approximations. These methods are of interest because they give the dependence of the solution of the problem on any given parameter, as opposed to numerical solutions which are obtained for particular sets of parameter values. Moreover, they can usually be refined to give any required degree of accuracy. It should be remembered that the equations themselves are only a representation of what is in reality a far more complex system, and that it is therefore impossible to obtain more than an approximation to the actual behaviour of the system.

Throughout this chapter we shall deal with non-dimensional systems. This is essential for a proper asymptotic analysis to be carried out, since otherwise it is impossible to tell which terms of an equation are small.

It is necessary to introduce some of the terminology of asymptotic analysis. We shall use the order notation.

DEFINITION 7.1: If ε is a small positive parameter, we say

$$f(\varepsilon)=O(g(\varepsilon)) \qquad \text{as} \quad \varepsilon \to 0$$

or just $f=O(g)$, if $|f|/|g|<K$ for some positive constant K whenever ε is sufficiently small. Similarly,

$$f(\varepsilon)=o(g(\varepsilon)) \qquad \text{as} \quad \varepsilon \to 0$$

or just $f=o(g)$, if $|f|/|g|\to 0$ as $\varepsilon \to 0$.

DEFINITION 7.2: A sequence of functions $\{f_n(\varepsilon)\}_{n\geq 0}$ is an *asymptotic sequence* as $\varepsilon \to 0$ if, for all n, $f_{n+1}=o(f_n)$ as $\varepsilon \to 0$. The most commonly used asymptotic sequence is that consisting of powers of ε where $f_n(\varepsilon)=\varepsilon^n$, that is the sequence $1, \varepsilon, \varepsilon^2, \ldots$.

DEFINITION 7.3: If $\{f_n(\varepsilon)\}$ is an asymptotic sequence then $\Sigma_{n=0}^{\infty} a_n f_n(\varepsilon)$, where the a_n's are constants, is an *asymptotic expansion* of the function $F(\varepsilon)$ as $\varepsilon \to 0$ if, for each $N\geq 0$,

$$F(\varepsilon)=\sum_{n=0}^{N} a_n f_n(\varepsilon)+o(f_N(\varepsilon))$$

as $\varepsilon \to 0$. In this case we write

$$F(\varepsilon)\approx \sum_{n=0}^{\infty} a_n f_n(\varepsilon)$$

In this chapter we shall come across several examples of functions $u(\mathbf{x}, t; \varepsilon)$ which can be written as asymptotic expansions. Since in this case u is a function of space and time as well as of ε the coefficients will also be functions of space and time and the expansion will be of the form

$$u(\mathbf{x}, t; \varepsilon)\approx \sum_{n=0}^{\infty} u_n(\mathbf{x}, t)\varepsilon^n$$

If $F(\varepsilon)$ has an asymptotic expansion in terms of an asymptotic sequence, then this expansion is unique; the coefficients a_n are obtained by successive applications of the limit process

$$a_0=\lim_{\varepsilon\to 0} F(\varepsilon)/f_0(\varepsilon)$$

$$a_m=\lim_{\varepsilon\to 0}\left\{F(\varepsilon)-\sum_{n=0}^{m-1} a_n f_n(\varepsilon)\right\}\Big/ f_m(\varepsilon)$$

DEFINITION 7.4: The first non-zero term in an asymptotic expansion is the *dominant*, or *leading order*, term. If $a_0 \neq 0$, then $a_0 f_0(\varepsilon)$ is the dominant term and we often write

$$F(\varepsilon) \sim a_0 f_0(\varepsilon)$$

as $\varepsilon \to 0$, or in our example,

$$u(\mathbf{x}, t; \varepsilon) \sim u_0(\mathbf{x}, t)$$

The error introduced in making such approximations may often be estimated, as discussed in detail in the book by Murray (1974a). We shall also concern ourselves with this point.

7.2 Regular and Singular Perturbations

The applications of asymptotic analysis are to perturbations, or small changes, in mathematical problems. These may be the addition of an extra term into the equations or a change in one of the parameters of the problem. Let ε be a measure of the size of the perturbation; for example, the additional term may have coefficient ε or a parameter may be changed by a fraction ε of its value. Then if the solution of the perturbed problem is close to that of the unperturbed one, the perturbation is *regular*; otherwise it is *singular*. We shall make this statement more precise.

DEFINITION 7.5: Let $\mathbf{u}(\varepsilon)$ satisfy the perturbed problem $P(\varepsilon)$ (where $P(\varepsilon)$ will normally consist of a system of differential equations in a spatio-temporal domain Q and some initial and boundary conditions to be satisfied by $\mathbf{u}$). Then the perturbation is *regular* if $\mathbf{u}$ is a continuous function of ε at $\varepsilon = 0$, i.e. if $\|\mathbf{u}(\varepsilon) - \mathbf{u}(0)\| \to 0$ as $\varepsilon \to 0$, where $\|\cdot\|$ is an appropriate norm, and *singular* otherwise.

Note that a perturbation may be regular with respect to one norm but singular with respect to another. For example, consider the problem $P(\varepsilon)$ consisting of a system of ordinary differential equations for $\mathbf{u}(t; \varepsilon)$,

$$\dot{\mathbf{u}} = \mathbf{f}(\mathbf{u}; \varepsilon)$$

with initial conditions

$$\mathbf{u}(0; \varepsilon) = \mathbf{u}_0(\varepsilon)$$

where $\mathbf{f}$ satisfies Lipschitz conditions in $\mathbf{u}$. Then if $\mathbf{f}$ and $\mathbf{u}_0$ are continuous functions of ε Picard's theorem (Theorem 2.2), this ensures that $\mathbf{u}$ also is. Hence this is a regular perturbation with respect to the norm $\|\cdot\|$ defined by $\|\cdot\| = \sup_{t\in[0,T]}|\cdot|$ for any finite constant T, for example. However, the solution of this problem does not necessarily remain close to the solution of the unperturbed problem for all time; it is necessary to restrict the range of t in defining the norm to be used. The perturbation of the Lotka–Volterra system described earlier is, in fact, regular with respect to this norm, but we know that the behaviour as $t\to\infty$ is completely changed, so that it is singular with respect to the norm $\|\cdot\| = \sup_{t\in[0,\infty)}|\cdot|$. It is therefore essential to be aware of the norm to be used when we define a perturbation as regular.

Solutions of regular perturbation problems may often be obtained as power series in ε. In the language of asymptotic analysis we look for an asymptotic expansion of $\mathbf{u}(\varepsilon)$ as $\varepsilon\to 0$ in terms of the asymptotic sequence $\{\varepsilon^n\}$. In cases in which t must be restricted, the straightforward asymptotic expansion in powers of ε may not be the most convenient asymptotic form in which to express the solution (unless we are only interested in the solutions over a finite time interval). This is particularly true in the case of perturbations of the harmonic oscillator which we shall discuss in this chapter.

7.3 Perturbations of the Harmonic Oscillator

The harmonic oscillator $\ddot{u}+u=0$, or equivalently

$$\dot{u} = -v, \qquad \dot{v} = u$$

has solutions $(u, v) = (A\cos(t+\alpha), A\sin(t+\alpha))$, where A and α are constants of integration. In many cases it is simpler to define $a = Ae^{i\alpha}/2$, so that the solutions become $(u, v) = (ae^{it}+\bar{a}e^{-it}, -iae^{it}+i\bar{a}e^{-it})$, since we are then dealing with only one (complex) unknown and we can often write down one complex differential equation to solve rather than two real ones. Since a is arbitrary, the harmonic oscillator has an infinite family of periodic solutions. It is a conservative system, since $u^2+v^2=A^2=4a\bar{a}$ is a constant of the motion, and the addition of small terms to such a system may change the nature of the solutions as $t\to\infty$ completely. An obvious example here is the equation $\ddot{u}+\varepsilon\dot{u}+u=0$, which instead of having periodic solutions has solutions which tend to zero as $t\to\infty$ if $\varepsilon>0$, and to infinity if $\varepsilon<0$. The perturbation $\varepsilon\dot{u}$ is regular in any finite time interval, since the solution depends analytically on ε by Picard's theorem, but because of the qualitative change in the solutions we cannot expect any asymptotic expansion of the solution depending on a regular perturbation procedure to be valid for all time. There are many techniques available to improve the asymptotic representation of such systems, which we shall consider particularly in relation to the equation

$$\ddot{u}-\varepsilon(1-u^2)\dot{u}+u=0$$

This is known as van der Pol's equation and models a spontaneously oscillating electrical circuit [see Minorsky (1974)]. It is used as an example since it is one of the simplest second-order differential equations to exhibit the behaviour we are interested in, that is, oscillations and specifically limit cycles. Note that it is a Liénard system (Definition 2.24) and therefore has a unique stable limit cycle (Theorem 2.23). It can also be thought of as the reaction system

$$\dot{u} = -v, \qquad \dot{v} = u + \varepsilon(1-u^2)v$$

or, by defining v differently,

$$\dot{u} = -v + \varepsilon(u - u^3/3), \qquad \dot{v} = u$$

These are both specific forms of the more general oscillatory reaction system

$$\dot{u} = -v + f(\varepsilon, u, v), \qquad \dot{v} = u + g(\varepsilon, u, v)$$

Let us define $\mathbf{u} = (u, v)^{\mathrm{T}}$, $\mathbf{f}(\varepsilon, \mathbf{u}) = (f(\varepsilon, u, v), g(\varepsilon, u, v))^{\mathrm{T}}$, and the linear operators L and M by

$$L\mathbf{u} = \frac{\partial \mathbf{u}}{\partial t} - M\mathbf{u} = \frac{\partial \mathbf{u}}{\partial t} - \begin{bmatrix} 0 & -1 \\ 1 & 0 \end{bmatrix}\mathbf{u} \tag{7.6}$$

Then the system becomes

$$L\mathbf{u} = \mathbf{f}(\varepsilon, \mathbf{u}) \tag{7.7}$$

where $\mathbf{f}(\varepsilon, \mathbf{0}) = \mathbf{f}(0, \mathbf{u}) = \mathbf{0}$. The essential feature of the matrix M from our point of view is that it has purely imaginary eigenvalues $\sigma = \pm i$. In fact the analysis that follows is easily extended to include any matrix M with purely imaginary eigenvalues, as we shall show.

Operators similar to L have already been discussed in Chapter 6, and the techniques for the construction of a Hopf bifurcating solution have their counterpart here. However, the periodic solutions constructed in Chapter 6 were of small amplitude, whereas here we consider solutions with amplitudes of $O(1)$. Moreover, we also construct non-periodic solutions of systems such as (7.7).

We review here some basic properties of the operator L.

PROPERTY 7.8: The system with $\varepsilon = 0$, $L\mathbf{u} = 0$, represents the harmonic oscillator and has solutions $u = ae^{it} + \bar{a}e^{-it}$, $v = -iae^{it} + i\bar{a}e^{-it}$, or in vector form

$$\mathbf{u} = a\boldsymbol{\phi} + \bar{a}\bar{\boldsymbol{\phi}} \tag{7.9}$$

where $\boldsymbol{\phi} = (1, -i)^{\mathrm{T}}e^{it}$ and $\bar{\boldsymbol{\phi}}$ is its complex conjugate and a is a complex constant; $\boldsymbol{\phi}$ and $\bar{\boldsymbol{\phi}}$ are the eigenfunctions of L with eigenvalue zero.

In the course of the perturbation procedures, equations of the form

$$L\mathbf{u} = \mathbf{g} \tag{7.10}$$

often occur, where $\mathbf{g}$ is a real periodic function of period 2π in t. Since L is a Fredholm operator on the space of 2π-periodic functions (cf. Chapter 6) it follows that

PROPERTY 7.11: Equation (7.10) has a 2π-periodic solution if and only if the following Fredholm orthogonality condition holds:

$$[\mathbf{g}, \boldsymbol{\phi}^*] = \frac{1}{2\pi}\int_0^{2\pi} \mathbf{g} \cdot \bar{\boldsymbol{\phi}}^* \, dt = 0 \tag{7.12}$$

where $\boldsymbol{\phi}^* = (1, -i)e^{it}/2$ is the adjoint eigenfunction of $\boldsymbol{\phi}$.

PROPERTY 7.13: If $[\mathbf{g}, \boldsymbol{\phi}^*] \neq 0$, then the solution of $L\mathbf{u} = \mathbf{g}$ is unbounded.

PROPERTY 7.14: If a solution $\mathbf{u}^p$ of $L\mathbf{u} = \mathbf{g}$ has been found, then any other solution is given by $\mathbf{u} = \mathbf{u}^p + a\boldsymbol{\phi} + \bar{a}\bar{\boldsymbol{\phi}}$ where a is an arbitrary complex constant. However, the solution may be made unique by applying a normalisation condition

$$[\mathbf{u}, \boldsymbol{\phi}^*] = k \tag{7.15}$$

where k is any constant.

Asymptotic techniques for systems close to the harmonic oscillator have been considered by many authors; see, for example, the books of Minorsky (1974), Jordan and Smith (1977), Kevorkian and Cole (1981) and Nayfeh (1973, 1981).

7.4 Failure of the Regular Perturbation Scheme

We shall consider first the reasons for the failure of the regular perturbation scheme for van der Pol's equation

$$\ddot{u} - \varepsilon(1 - u^2)\dot{u} + u = 0 \tag{7.16}$$

By substituting the form

$$u(t; \varepsilon) \approx u_0(t) + \varepsilon u_1(t) + \cdots \tag{7.17}$$

into this equation and equating powers of ε we obtain

$$\ddot{u}_0 + u_0 = 0, \qquad \ddot{u}_1 - (1 - u_0^2)\dot{u}_0 + u_1 = 0$$

$$\ddot{u}_2 - (1 - u_0^2)\dot{u}_1 + 2u_0u_1\dot{u}_0 + u_2 = 0$$

and so on. The first of these gives $u_0 = ae^{it} + \bar{a}e^{-it}$, which is of course the solution of the harmonic oscillator. The second then gives

$$\ddot{u}_1 + u_1 = (1-|a|^2)(iae^{it} - i\bar{a}e^{-it}) - ia^3e^{3it} + i\bar{a}^3e^{-3it}$$

which has general solution

$$u_1 = (1-|a|^2)(ae^{it} + \bar{a}e^{-it})t/2 + (ia^3e^{3it} - i\bar{a}^3e^{-3it})/8 + be^{it} + \bar{b}e^{-it}$$

It can be seen that u_1 is not uniformly $O(1)$ in general, because of the terms in $te^{\pm it}$. For the perturbation procedure to be valid we must have $\varepsilon u_1 \ll u_0$, i.e. $\varepsilon t \ll 1$, and the regular perturbation expansion is therefore only valid as long as $t \ll 1/\varepsilon$. As expected this region of validity is a finite interval, and to obtain a uniformly valid expansion for $t \in [0, \infty)$ another perturbation procedure must be used. Note that if $|a| = 1$, then u_1 *is* uniformly bounded. The van der Pol equation has a limit cycle solution whose amplitude $|a|$ is close to 1 (so that A is close to 2), and it may look as if the straightforward technique provides an acceptable approximation to this particular solution of the equation. However, it turns out that in this case the function u_2 has terms of the form $te^{\pm it}$ which cannot be made to disappear. Hence even when we are looking for the limit cycle solution, the representation obtained is only valid as long as $t \ll 1/\varepsilon^2$.

Let us consider this failure of the regular perturbation scheme in the context of the oscillatory reaction scheme

$$\dot{u} = -v + f(\varepsilon, u, v), \qquad \dot{v} = u + g(\varepsilon, u, v) \tag{7.18}$$

or

$$L\mathbf{u} = \mathbf{f}(\varepsilon, \mathbf{u}) \tag{7.19}$$

By substituting the form

$$\mathbf{u}(t; \varepsilon) \approx \mathbf{u}_0(t) + \varepsilon \mathbf{u}_1(t) + \cdots \tag{7.20}$$

and equating powers of ε, we obtain the recursive system of equations

$$L\mathbf{u}_0 = \mathbf{0} \tag{7.21}$$

$$L\mathbf{u}_1 = \mathbf{f}_\varepsilon(0, \mathbf{u}_0) \tag{7.22}$$

$$L\mathbf{u}_2 = \mathbf{f}_{\varepsilon\mathbf{u}}(0, \boldsymbol{u}_0 | \boldsymbol{u}_1) + \mathbf{f}_{\varepsilon\varepsilon}(0, \mathbf{u}_0)/2 \tag{7.23}$$

and so on. The first of these has solution

$$\mathbf{u}_0 = a\boldsymbol{\phi} + \bar{a}\bar{\boldsymbol{\phi}}$$

The second is of the form (7.10) considered in the introduction, since $\mathbf{u}_0$ is periodic of period 2π, and by Property 7.11 has a periodic solution if and only if the inner product

$$[\mathbf{f}_\varepsilon(0, \mathbf{u}_0), \boldsymbol{\phi}^*] = \frac{1}{2\pi}\int_0^{2\pi} \mathbf{f}_\varepsilon(0, \mathbf{u}_0) \cdot \bar{\boldsymbol{\phi}}^* \, dt = 0 \tag{7.24}$$

If $[\mathbf{f}_\varepsilon(0, \mathbf{u}_0), \boldsymbol{\phi}^*] \neq 0$, then by Property 7.13 any solution is unbounded as $t \to \infty$. The result generalises the situation discussed before for the van der Pol equation. In the case of the limit cycle solution the modulus of the constant a is such that condition (7.24) is satisfied, but it can then be shown that (7.23) has no bounded solution.

7.5 The Method of Renormalisation

The method of renormalisation is a technique for obtaining an asymptotic expansion of a periodic solution of a perturbation problem. The reason for the breakdown of the straightforward method for finding the periodic solution of van der Pol's equation is that this solution has a period very close to, but not equal to, 2π. Such a solution cannot be well approximated indefinitely by an asymptotic expansion whose first term has period 2π. The perturbation $-\varepsilon(1-u^2)\dot{u}$ of the harmonic oscillator in this case has changed its frequency from 1 to ω, say, and the asymptotic scheme used must take this into account. Since we expect the frequency perturbation to be small, $O(\varepsilon)$ in fact, we write

$$\omega \approx 1 + \varepsilon\omega_1 + \varepsilon^2\omega_2 + \cdots \tag{7.25}$$

and define a new time variable τ by

$$\tau = \omega t \approx (1 + \varepsilon\omega_1 + \varepsilon^2\omega_2 + \cdots)t \tag{7.26}$$

We then look for a solution u as a function of the new time variable τ and ε, with asymptotic expansion

$$u(\tau; \varepsilon) \approx u_0(\tau) + \varepsilon u_1(\tau) + \cdots \tag{7.27}$$

This is known as the method of renormalisation. Note that the functions u_0, u_1, etc., are different from those defined in the last section. In terms of τ the van der Pol equation becomes

$$\omega^2 u'' - \omega\varepsilon(1-u^2)u' + u = 0$$

where prime denotes differentiation with respect to τ. By substituting for u and ω their asymptotic expansions (7.25) and (7.27), respectively, and equating powers of ε, we obtain

$$u_0''+u_0=0 \tag{7.28}$$

$$u_1''+2\omega_1 u_0''-(1-u_0^2)u_0'+u_1=0 \tag{7.29}$$

$$u_2''+2\omega_1 u_1''+(\omega_1^2+2\omega_2)u_0''-(1-u_0^2)u_1'+2u_0u_1u_0'+u_2=0 \tag{7.30}$$

and so on. These equations as they stand do not have unique solutions, since each will have an arbitrary constant of integration $a_n e^{i\tau}+\bar{a}_n e^{-i\tau}$, with period $2\pi/\omega$ in terms of the original time variable t. However, we may eliminate this indeterminacy by requiring that the component of u with this frequency be included entirely in the first approximation u_0. That is, defining a complex constant a by

$$a=[u, e^{i\tau}]=\frac{1}{2\pi}\int_0^{2\pi} ue^{-i\tau}\,d\tau \tag{7.31}$$

we require that

$$[u_0, e^{i\tau}]=a, \qquad [u_n, e^{i\tau}]=0, \qquad n>0 \tag{7.32}$$

These equations may be compared with (6.153) and (6.160); $[\cdot,\cdot]$ is in fact an inner product on the space of 2π-periodic functions. Equation (7.28) has general solution $u_0=a_0e^{i\tau}+\bar{a}_0e^{-i\tau}$. By applying the first condition of (7.32), this becomes

$$u_0=ae^{i\tau}+\bar{a}e^{-i\tau} \tag{7.33}$$

Equation (7.29) then gives

$$\begin{aligned} u_1''+u_1 &= 2\omega_1(ae^{i\tau}+\bar{a}e^{-i\tau}) \\ &\quad +(1-a^2e^{2i\tau}-2a\bar{a}-\bar{a}^2e^{-2i\tau})(iae^{i\tau}-i\bar{a}e^{-i\tau}) \\ &= 2\omega_1(ae^{i\tau}+\bar{a}e^{-i\tau})+(1-|a|^2)(iae^{i\tau}-i\bar{a}e^{-i\tau}) \\ &\quad -ia^3e^{3i\tau}+i\bar{a}^3e^{-3i\tau} \end{aligned} \tag{7.34}$$

which has general solution

$$\begin{aligned} u_1 &= \omega_1\tau(-iae^{i\tau}+i\bar{a}e^{-i\tau})+(1-|a|^2)(ae^{i\tau}+\bar{a}e^{-i\tau})\tau/2 \\ &\quad +(ia^3e^{3i\tau}-i\bar{a}^3e^{-3i\tau})/8+be^{i\tau}+\bar{b}e^{-i\tau} \end{aligned} \tag{7.35}$$

where b is an arbitrary complex constant. For this to be uniformly bounded we must have $|a|=1$ and $\omega_1=0$. We do not expect to determine arg a, since the time origin is arbitrary. The condition $[u_1, e^{i\tau}]=0$ implies that $b=0$, so that u_1 is given uniquely by

$$u_1=(ia^3e^{3i\tau}-i\bar{a}^3e^{-3i\tau})/8$$

The equation for u_2 is then

$$u_2''+u_2=2\omega_2(ae^{i\tau}+\bar{a}e^{-i\tau})$$
$$-3(1-a^2e^{2i\tau}-2-\bar{a}^2e^{-2i\tau})(a^3e^{3i\tau}+\bar{a}^3e^{-3i\tau})/8$$
$$-2(ia^2e^{2i\tau}-i\bar{a}^2e^{-2i\tau})(ia^3e^{3i\tau}-i\bar{a}^3e^{-3i\tau})/8 \qquad (7.36)$$

This will produce unbounded solutions u_2 if there are any terms in $e^{\pm i\tau}$, known as *secular* terms, on the right-hand side of the equation. But the coefficient of $e^{i\tau}$ is given by $2\omega_2 a+3a/8-a/4$, so that we require

$$\omega_2=-\tfrac{1}{16} \qquad (7.37)$$

to eliminate secular terms. Now u_2 may be found from (7.36) under the condition $[u_2, e^{i\tau}]=0$.

The subsequent equations are considered consecutively using the requirements that their solutions are bounded, or that the equations contain no secular terms, and that $[u_n, e^{i\tau}]=0$. The first condition determines ω_n and the second ensures that the solution u_n is unique. Thus a representation for the limit cycle solution of the van der Pol oscillator is derived. It is given by

$$u\approx ae^{i\tau}+\bar{a}e^{-i\tau}+\varepsilon(ia^3e^{3i\tau}-i\bar{a}^3e^{-3i\tau})+\cdots \qquad (7.38)$$

where $|a|=1$, or in terms of trigonometric functions,

$$u\approx 2\cos(\tau+\alpha)+\varepsilon\sin 3(\tau+\alpha)+\cdots \qquad (7.39)$$

where $\tau=\omega t$ and

$$\omega\approx 1-\varepsilon^2/16+\cdots \qquad (7.40)$$

Let us now consider the method of renormalisation in the context of the oscillatory reaction scheme given by (7.19),

$$L\mathbf{u}=\mathbf{f}(\varepsilon,\mathbf{u}) \qquad (7.41)$$

where

$$L=\frac{\partial}{\partial t}-M=\frac{\partial}{\partial t}-\begin{bmatrix}0 & -1\\ 1 & 0\end{bmatrix}$$

To apply the method of renormalisation we make the change of variable

$$\tau=\omega t\approx(1+\varepsilon\omega_1+\varepsilon^2\omega_2+\cdots)t \qquad (7.42)$$

Then (7.41) becomes

$$L\mathbf{u}=\left(\omega\frac{\partial}{\partial\tau}-M\right)\mathbf{u}=\mathbf{f}(\varepsilon,\mathbf{u}) \qquad (7.43)$$

or equivalently

$$\omega\left(\frac{\partial}{\partial\tau}-M\right)\mathbf{u}=\omega L_0\mathbf{u}=(1-\omega)M\mathbf{u}+\mathbf{f}(\varepsilon,\mathbf{u})$$

where $L_0\equiv\partial/\partial\tau-M$ and is a Fredholm operator. Let $\boldsymbol{\phi}_0=(1,-i)^{\mathrm{T}}e^{i\tau}$ and $\bar{\boldsymbol{\phi}}_0$ be the eigenfunctions of L_0 with eigenvalue zero and $\boldsymbol{\phi}_0^*=(1,-i)^{\mathrm{T}}e^{i\tau}/2$ be the adjoint eigenfunction of $\boldsymbol{\phi}_0$. Let us define a complex constant a, the amplitude of the solution, by

$$a=[\mathbf{u},\boldsymbol{\phi}_0^*] \tag{7.44}$$

We look for an asymptotic representation of $\mathbf{u}$ in the form

$$\mathbf{u}(\tau;\varepsilon)\approx\mathbf{u}_0(\tau)+\varepsilon\mathbf{u}_1(\tau)+\cdots \tag{7.45}$$

By substituting into (7.43) and equating powers of ε, we obtain the following sequence of equations:

$$L_0\mathbf{u}_0=\mathbf{0} \tag{7.46}$$

$$L_0\mathbf{u}_1=-\omega_1\mathbf{u}_0'+\mathbf{f}_\varepsilon(0,\mathbf{u}_0) \tag{7.47}$$

$$L_0\mathbf{u}_2=-\omega_2\mathbf{u}_0'-\omega_1\mathbf{u}_1'+\mathbf{f}_{\varepsilon\mathbf{u}}(0,\mathbf{u}_0|\mathbf{u}_1)+\mathbf{f}_{\varepsilon\varepsilon}(0,\mathbf{u}_0)/2 \tag{7.48}$$

and so on, where prime again denotes differentiation with respect to τ. Again these equations do not have unique solutions, since each has an arbitrary constant of integration $a_n\boldsymbol{\phi}_0+\bar{a}_n\bar{\boldsymbol{\phi}}_0$. From (7.44) the normalising conditions are

$$[\mathbf{u}_0,\boldsymbol{\phi}_0^*]=a,\qquad[\mathbf{u}_n,\boldsymbol{\phi}_0^*]=0,\qquad n>0 \tag{7.49}$$

Equation (7.46) has general solution $\mathbf{u}_0=a_0\boldsymbol{\phi}_0+\bar{a}_0\bar{\boldsymbol{\phi}}_0$, which with the normalisation $[\mathbf{u}_0,\boldsymbol{\phi}_0^*]=a$ becomes

$$\mathbf{u}_0=a\boldsymbol{\phi}_0+\bar{a}\bar{\boldsymbol{\phi}}_0 \tag{7.50}$$

Equation (7.49) has a periodic solution if and only if

$$[-\omega_1\mathbf{u}_0'+\mathbf{f}_\varepsilon(0,\mathbf{u}_0),\boldsymbol{\phi}_0^*]=[-\omega_1\mathbf{u}_0'+\mathbf{f}_\varepsilon(0,\mathbf{u}_0),\bar{\boldsymbol{\phi}}_0^*]=0 \tag{7.51}$$

or equivalently, since $\mathbf{u}_0$ and $\mathbf{f}$ are real,

$$\omega_1[\mathbf{u}_0',\boldsymbol{\phi}_0^*+\bar{\boldsymbol{\phi}}_0^*]=[\mathbf{f}_\varepsilon(0,\mathbf{u}_0),\boldsymbol{\phi}_0^*+\bar{\boldsymbol{\phi}}_0^*]$$

If each component of $\mathbf{f}_\varepsilon(0,\mathbf{u}_0)=\mathbf{f}_\varepsilon(0,a\boldsymbol{\phi}_0+\bar{a}\bar{\boldsymbol{\phi}}_0)$ may be written as a power series in u_0 and v_0, which we assume to be the case, then the terms in $e^{i\tau}$ will be of the form $a^{n+1}\bar{a}^ne^{i\tau}=a|a|^{2n}e^{i\tau}$, where n is a non-negative integer. Since the inner product with $\boldsymbol{\phi}_0^*$ picks out these terms, it follows that $[\mathbf{f}_\varepsilon(0,\mathbf{u}_0),\boldsymbol{\phi}_0^*]$ may be written in the form

$$[\mathbf{f}_\varepsilon(0,\mathbf{u}_0),\boldsymbol{\phi}_0^*]=a\Lambda(|a|^2) \tag{7.52}$$

where Λ has a power series expansion with complex coefficients. We also have

$$[\mathbf{u}_0', \boldsymbol{\phi}_0^*] = [ia\boldsymbol{\phi}_0 - i\bar{a}\bar{\boldsymbol{\phi}}_0, \boldsymbol{\phi}_0^*] = ia \tag{7.53}$$

so that (7.51) becomes

$$i\omega_1 a = a\Lambda(|a|^2) \tag{7.54}$$

and we have

$$0 = \text{Re}\,\Lambda(|a|^2), \qquad \omega_1 = \text{Im}\,\Lambda(|a|^2)$$

We assume that the first of these has at least one non-trivial solution for $|a|$. Again arg a is arbitrary since the time origin is. Equation (7.47) then has a periodic solution which is unique under the normalisation $[\mathbf{u}_1, \boldsymbol{\phi}_0^*] = 0$. Equation (7.48) now has a periodic solution if and only if

$$[-\omega_2\mathbf{u}_0' - \omega_1\mathbf{u}_1' + \mathbf{f}_{\varepsilon\mathbf{u}}(0, \mathbf{u}_0|\mathbf{u}_1) + \mathbf{f}_{\varepsilon\varepsilon}(0, \mathbf{u}_0)/2, \boldsymbol{\phi}_0^* + \bar{\boldsymbol{\phi}}_0^*] = 0$$

or since $[\mathbf{u}_0', \boldsymbol{\phi}_0^* + \bar{\boldsymbol{\phi}}_0^*] = -2\,\text{Im}\,a[\mathbf{u}_1', \boldsymbol{\phi}_0^* + \bar{\boldsymbol{\phi}}_0^*] = 0$,

$$\omega_2 = [\mathbf{f}_{\varepsilon\mathbf{u}}(0, \mathbf{u}_0|\mathbf{u}_1) + \mathbf{f}_{\varepsilon\varepsilon}(0, \mathbf{u}_0)/2, \boldsymbol{\phi}_0^* + \bar{\boldsymbol{\phi}}_0^*]/(-2\,\text{Im}\,a) \tag{7.55}$$

and is clearly real. With this choice of ω_2, Eq. (7.50) has a unique periodic solution satisfying $[\mathbf{u}_2, \boldsymbol{\phi}_0^*] = 0$. Similarly, secular terms at higher orders may be eliminated by choosing ω_n appropriately and the resulting equations have unique solutions $\mathbf{u}_n$; thus a periodic solution of the system may be built up.

7.6 The Two-timing Method

The method of renormalisation thus gives an asymptotic representation of a periodic solution of the required system. The van der Pol oscillator has only one periodic solution which is a stable limit cycle, and all other solutions (except the trivial one $u \equiv 0$) spiral onto this solution. How can we obtain asymptotic representations of these other solutions?

The problem with the straightforward regular perturbation technique was the appearance in the expansion of terms such as εte^{it}. The solution thus depends on the combination εt as well as on t itself, but the nature of the dependence on εt is not clear from the expansion. As a simpler example, consider the equation

$$\ddot{u} + 2\varepsilon\dot{u} + u = 0$$

A regular perturbation expansion of the equation of the form

$$u(t;\,\varepsilon)\approx u_0(t)+\varepsilon u_1(t)+\cdots$$

gives

$$u(t;\,\varepsilon)\approx ae^{it}+\bar{a}e^{-it}-\varepsilon t(ae^{it}+\bar{a}e^{-it})+\cdots$$

which looks as if it is unbounded as $t\to\infty$. However, the exact solution can be found as the equation is linear, and is given by

$$u(t;\,\varepsilon)=e^{-\varepsilon t}\{a\exp(it\sqrt{1-\varepsilon^2})+\bar{a}\exp(-it\sqrt{1-\varepsilon^2})\}$$

This, of course, agrees with the regular expansion on the assumption that εt is small, since $e^{-\varepsilon t}$ may then be approximated by $1-\varepsilon t+O(\varepsilon^2)$. Note that u is periodic in a fast time scale $t\sqrt{1-\varepsilon^2}$ but is modulated on a slow time scale εt.

This leads to the idea that the dependence on a slow time scale should be considered explicitly. Thus we define

$$T_1=\Omega t\approx(\varepsilon+\varepsilon^2\Omega_2+\varepsilon^3\Omega_3+\cdots)t \tag{7.56}$$

and consider u as a function of T_1 and a slow time variable T_0, which are considered to be independent. To deal with time dependence such as $\exp(it\sqrt{1-\varepsilon^2})$ above we define

$$T_0=\omega t\approx(1+\varepsilon^2\omega_2+\varepsilon^3\omega_3+\cdots)t \tag{7.57}$$

Note that there is no $\varepsilon\omega_1$ term in this definition. This is because such time dependence may always be dealt with by the $T_1=\Omega t$ variable, as we shall see. We then put $u(t;\,\varepsilon)=u(T_0,\,T_1;\,\varepsilon)$ and look for an asymptotic expansion in the form

$$u(T_0,\,T_1;\,\varepsilon)\approx u_0(T_0,\,T_1)+\varepsilon u_1(T_0,\,T_1)+\cdots \tag{7.58}$$

This is known as the method of double time scales or the two-timing method. Here T_1 is a slow time scale, since when T_0 or t changes by $O(1)$, then T_1 changes by $O(\varepsilon)$. By the chain rule we have

$$\frac{\partial}{\partial t}=\omega\frac{\partial}{\partial T_0}+\Omega\frac{\partial}{\partial T_1} \tag{7.59}$$

$$\frac{\partial^2}{\partial t^2}=\omega^2\frac{\partial^2}{\partial T_0^2}+2\omega\Omega\frac{\partial^2}{\partial T_0\,\partial T_1}+\Omega^2\frac{\partial^2}{\partial T_1^2} \tag{7.60}$$

Hence the van der Pol equation becomes

$$\omega^2\frac{\partial^2 u}{\partial T_0^2}+2\Omega\omega\frac{\partial^2 u}{\partial T_0\,\partial T_1}+\Omega^2\frac{\partial^2 u}{\partial T_1^2}-\varepsilon(1-u^2)\left(\omega\frac{\partial u}{\partial T_0}+\Omega\frac{\partial u}{\partial T_1}\right)+u=0 \tag{7.61}$$

By substituting in the expansion (7.58) for u and equating powers of ε, we obtain

$$\frac{\partial^2 u_0}{\partial T_0^2} + u_0 = 0 \tag{7.62}$$

$$\frac{\partial^2 u_1}{\partial T_0^2} + 2\frac{\partial^2 u_0}{\partial T_0\,\partial T_1} - (1-u_0^2)\frac{\partial u_0}{\partial T_0} + u_1 = 0 \tag{7.63}$$

$$\frac{\partial^2 u_2}{\partial T_0^2} + 2\omega_2\frac{\partial^2 u_0}{\partial T_0^2} + 2\frac{\partial^2 u_1}{\partial T_0\partial T_1} + 2\Omega_2\frac{\partial^2 u_0}{\partial T_0\,\partial T_1} + \frac{\partial^2 u_0}{\partial T_1^2}$$
$$- (1-u_0^2)\left(\frac{\partial u_1}{\partial T_0} + \frac{\partial u_0}{\partial T_1}\right) + 2u_0 u_1\frac{\partial u_0}{\partial T_0} + u_2 = 0 \tag{7.64}$$

and so on. Note that in this case we do not apply any normalisation conditions corresponding to (7.32). This is because we need the flexibility afforded by allowing u_n for $n > 0$ to have first harmonics in order to eliminate secular terms in the higher-order equations.

Since we are looking for a solution of a second-order equation, we need to apply two initial conditions. Let us take

$$u(0) = U, \qquad u_t(0) = -V \tag{7.65}$$

The first of these becomes, in terms of $u_0, u_1, \ldots,$

$$u_0(0,0) = U, \qquad u_n(0,0) = 0, \qquad n > 1 \tag{7.66}$$

By using (7.58) and (7.59), we have

$$\frac{\partial u}{\partial t} \approx \omega\frac{\partial u_0}{\partial T_0} + \Omega\frac{\partial u_0}{\partial T_1} + \varepsilon\left(\omega\frac{\partial u_1}{\partial T_0} + \Omega\frac{\partial u_1}{\partial T_1}\right) + \cdots$$
$$\approx \frac{\partial u_0}{\partial T_0} + \varepsilon\left(\frac{\partial u_0}{\partial T_1} + \frac{\partial u_1}{\partial T_0}\right) + \varepsilon^2\left(\frac{\partial u_1}{\partial T_1} + \omega_2\frac{\partial u_0}{\partial T_0} + \Omega_2\frac{\partial u_0}{\partial T_1} + \frac{\partial u_2}{\partial T_0}\right) + O(\varepsilon^3) \tag{7.67}$$

By equating powers of ε, the second initial condition becomes

$$\frac{\partial u_0}{\partial T_0}(0,0) = -V$$

$$\frac{\partial u_n}{\partial T_0}(0,0) = -\frac{\partial u_{n-1}}{\partial T_1}(0,0) - \sum_{r=2}^{n}\omega_r\frac{\partial u_{n-r}}{\partial T_0}(0,0) \tag{7.68}$$
$$- \sum_{r=2}^{n}\Omega_r\frac{\partial u_{n-r}}{\partial T_1}(0,0), \qquad n > 0$$

The equation for u_0, (7.62), gives

$$u_0(T_0, T_1) = a(T_1)e^{iT_0} + \bar{a}(T_i)e^{-iT_0} \tag{7.69}$$

The initial conditions (7.66) and (7.68) give

$$a(0) + \bar{a}(0) = U, \qquad ia(0) - i\bar{a}(0) = -V$$

so that

$$a(0) = (U + iV)/2 = a_0 \tag{7.70}$$

say. Equation (7.63) then gives

$$\begin{aligned}\frac{\partial^2 u_1}{\partial T_0^2} + u_1 &= (1 - a^2e^{2iT_0} - 2a\bar{a} - \bar{a}^2e^{-2iT_0})(iae^{iT_0} - i\bar{a}e^{-iT_0}) \\ &\quad - 2(ia'e^{iT_0} - i\bar{a}'e^{-iT_0}) \\ &= (1 - a\bar{a})(iae^{iT_0} - i\bar{a}e^{-iT_0}) - ia^3e^{3iT_0} + i\bar{a}^3e^{-3iT_0} \\ &\quad - 2(ia'e^{iT_0} - i\bar{a}'e^{-iT_0})\end{aligned} \tag{7.71}$$

where primes denote differentiation with respect to the argument T_1. As usual we want u_1 to be bounded, and for this to be so the inner product of the right-hand side of (7.71) with e^{iT_0} must be zero, i.e. there must be no terms in e^{iT_0} on the right-hand side. Hence

$$(1 - |a|^2)a - 2a' = 0 \tag{7.72}$$

Defining $\rho = |a|^2$, we obtain

$$\rho' = (1 - \rho)\rho \tag{7.73}$$

We thus have a differential equation for the square of the amplitude $|a|$. If $\rho_0 = \rho(0) = |a_0|^2 = 1$, the solution is $\rho \equiv 1$, the limit cycle solution. If $\rho_0 \neq 1$, separation of variables gives

$$dT_1 = \frac{d\rho}{\rho(1-\rho)} = \left(\frac{1}{\rho} + \frac{1}{1-\rho}\right) d\rho$$

Integrating,

$$T_1 = \log[\rho(1 - \rho_0)/\rho_0(1 - \rho)]$$

or, rearranging,

$$\rho = \frac{\rho_0 e^{T_1}}{1 - \rho_0(1 - e^{T_1})} \tag{7.74}$$

It can be seen that for $\rho_0 > 0$, $\rho \to 1$ as $T_1 \to \infty$; in other words, the limit cycle is stable.

With this expression for $\rho=|a|^2$, we may solve (7.72) for a. With the initial condition $a(0)=a_0$ we obtain

$$a^2=a_0^2\frac{e^{T_1}}{1-\rho_0+\rho_0e^{T_1}} \tag{7.75}$$

u_1 is now obtained from (7.71). Recalling that a satisfies (7.72), this becomes

$$\frac{\partial^2u_1}{\partial T_0^2}+u_1=-ia^3e^{3iT_0}+i\bar{a}^3e^{-3iT_0}$$

which gives

$$u_1=(ia^3e^{3iT_0}-i\bar{a}^3e^{-3iT_0})/8+be^{iT_0}+\bar{b}e^{-iT_0} \tag{7.76}$$

where $b=b(T_1)$. Substituting this expression into the Eq. (7.64) for u_2 and eliminating secular terms gives a first-order linear differential equation for $b(T_1)$. In the general case this may also have a secular solution; for example, if $a(T_1)$ includes exponential terms e^{T_1} or e^{-T_1}, these will be of the form $T_1e^{T_1}$ or $T_1e^{-T_1}$. We then have $|u_1/u_0|=O(T_1)$, which tends to infinity as $T_1\to\infty$, so that the natural requirement that $|u_1/u_0|$ be bounded is violated. Such secular terms must therefore be eliminated, and this may generally be done by choosing ω_2 and Ω_2 appropriately, as we shall show. Similarly, the elimination of secular terms in the differential equation for u_n determines a first-order linear differential equation for the arbitrary function in u_{n-1}, and elimination of any secular terms in this differential equation determines ω_n and Ω_n. The first-order differential equations may be solved subject to the initial conditions (7.66) and (7.68). Each u_n and u_n/u_0 is then bounded, and the expression

$$u\approx u_0+\varepsilon u_1+\cdots+\varepsilon^n u_n+O(\varepsilon^{n+1})$$

is a uniformly valid solution of the differential equation, and satisfies the initial conditions to the required order.

In this derivation we have assumed that it is possible to eliminate secular terms in the equations by requiring that the constants, or more strictly functions, of integration satisfy certain first-order differential equations and by choosing the coefficients ω_n and Ω_n appropriately. We shall prove that this is the case as long as the problem is non-degenerate, in a sense to be made precise later, for the more general class of Eqs. (7.19), or

$$L\mathbf{u}=\mathbf{f}(\varepsilon,\mathbf{u}) \tag{7.77}$$

where $\mathbf{f}(0,\mathbf{u})=\mathbf{f}(\varepsilon,\mathbf{0})=\mathbf{0}$. In this case we look for a solution u as a function of the two variables $T_0=\omega t$ and $T_1=\Omega t$ and seek expansions in the form

$$\mathbf{u}(T_0,T_1;\varepsilon)\approx\mathbf{u}_0(T_0,T_1)+\varepsilon\mathbf{u}_1(T_0,T_1)+\cdots \tag{7.78}$$

Initial conditions are given as

$$\mathbf{u}(0) = \mathbf{U} = (U, V)^{\mathrm{T}} \tag{7.79}$$

or in terms of the $\mathbf{u}_n$'s,

$$\mathbf{u}_0(0, 0) = \mathbf{U}, \qquad \mathbf{u}_n(0, 0) = \mathbf{0}, \qquad n > 0 \tag{7.80}$$

The equations become

$$\omega\frac{\partial \mathbf{u}}{\partial T_0} + \Omega\frac{\partial \mathbf{u}}{\partial T_1} = M\mathbf{u} + \mathbf{f}(\varepsilon, \mathbf{u}) \tag{7.81}$$

or equivalently,

$$\omega\left(\frac{\partial}{\partial T_0} - M\right)\mathbf{u} = \omega L_0\mathbf{u} = -\Omega\frac{\partial \mathbf{u}}{\partial T_1} - (\omega - 1)M\mathbf{u} + \mathbf{f}(\varepsilon, \mathbf{u})$$

where

$$L_0 = \frac{\partial}{\partial T_0} - M = \frac{\partial}{\partial T_0} - \begin{bmatrix} 0 & -1 \\ 1 & 0 \end{bmatrix}.$$

Substituting the expansions for $\mathbf{u}$ into this and equating powers of ε, we have

$$L_0\mathbf{u}_0 = \mathbf{0} \tag{7.82}$$

$$L_0\mathbf{u}_1 + \frac{\partial \mathbf{u}_0}{\partial T_1} = \mathbf{f}_\varepsilon(0, \mathbf{u}_0) \tag{7.83}$$

$$L_0\mathbf{u}_2 + \omega_2\frac{\partial \mathbf{u}_0}{\partial T_0} + \Omega_2\frac{\partial \mathbf{u}_0}{\partial T_1} + \frac{\partial \mathbf{u}_1}{\partial T_1} = \mathbf{f}_{\varepsilon\mathbf{u}}(0, \mathbf{u}_0|\mathbf{u}_1) + \frac{\mathbf{f}_{\varepsilon\varepsilon}(0, \mathbf{u}_0)}{2} \tag{7.84}$$

and so on. The first of these has solution

$$\mathbf{u}_0 = a(T_1)\boldsymbol{\phi}_0 + \bar{a}(T_1)\bar{\boldsymbol{\phi}}_0 \tag{7.85}$$

where $\boldsymbol{\phi}_0 = (1, -i)^T e^{iT_0}$ is an eigenfunction of L_0 with zero eigenvalue. The second, by properties 7.11 and 7.13 of L_0, has a solution bounded as a function of T_0 (which is 2π-periodic in T_0) if and only if

$$\left[-\frac{\partial \mathbf{u}_0}{\partial T_1} + \mathbf{f}_\varepsilon(0, \mathbf{u}_0), \boldsymbol{\phi}_0^*\right] = \left[-\frac{\partial \mathbf{u}_0}{\partial T_1} + \mathbf{f}_\varepsilon(0, \mathbf{u}_0), \bar{\boldsymbol{\phi}}_0^*\right] = 0 \tag{7.86}$$

where $\boldsymbol{\phi}_0^* = \frac{1}{2}(1, -i)^{\mathrm{T}} e^{iT_0}$ is the adjoint of $\boldsymbol{\phi}_0$. Since $\mathbf{u}_0$ and $\mathbf{f}$ are real, these conditions are equivalent. This is a first-order ordinary differential equation for $a(T_1)$, which is to be solved subject to the initial condition, from (7.80) and (7.85),

$$a(0)\begin{pmatrix}1\\-i\end{pmatrix}+\bar{a}(0)\begin{pmatrix}1\\i\end{pmatrix}=\mathbf{U}=\begin{pmatrix}U\\V\end{pmatrix} \tag{7.87}$$

or, eliminating $\bar{a}(0)$,

$$a(0)=(U+iV)/2 \tag{7.88}$$

As in (7.53) we have

$$[\mathbf{f}_\varepsilon(0,\mathbf{u}_0),\boldsymbol{\phi}_0^*]=a\Lambda(|a|^2) \tag{7.89}$$

where $\Lambda(\cdot)$ is some function, not necessarily real. Thus

$$a'=a\Lambda(|a|^2) \tag{7.90}$$

Defining $a=re^{i\theta}$, this becomes

$$r'=r\,\mathrm{Re}\,\Lambda(r^2) \tag{7.91}$$

$$\theta'=\mathrm{Im}\,\Lambda(r^2) \tag{7.92}$$

subject to initial conditions

$$r(0)=\sqrt{a(0)\bar{a}(0)}=\sqrt{(U^2+V^2)/4}$$

$$\theta(0)=\tan^{-1}(V/U)$$

(7.91) may be solved by quadrature, and hence (7.92) may be solved on substituting in for r.

We have thus chosen a so that (7.87) is satisfied, and therefore the equation for $\mathbf{u}_1$, (7.83), has a solution bounded as a function of T_0 and 2π-periodic in T_0 which may be written in the form

$$\mathbf{u}_1=\mathbf{u}_1^p+b\boldsymbol{\phi}_0+\bar{b}\bar{\boldsymbol{\phi}}_0=\mathbf{u}_1^p+\mathbf{u}_1^c \tag{7.93}$$

where $b=b(T_1)$, $\mathbf{u}_1^p=\mathbf{u}_1^p(T_0,a)$ is a particular real solution 2π-periodic in T_0, and $\mathbf{u}_1^c=b\boldsymbol{\phi}_0+\bar{b}\bar{\boldsymbol{\phi}}_0$ is the complementary function. We may without loss of generality take

$$[\mathbf{u}_1^p,\boldsymbol{\phi}_0^*]=[\mathbf{u}_1^p,\bar{\boldsymbol{\phi}}_0^*]=0 \tag{7.94}$$

which determines $\mathbf{u}_1^p$ uniquely. Then the equation for $\mathbf{u}_2$, (7.84), has a solution bounded as a function of T_0, which is 2π-periodic in T_0, if and only if

$$\left[-\omega_2\frac{\partial\mathbf{u}_0}{\partial T_0}-\Omega_2\frac{\partial\mathbf{u}_0}{\partial T_1}-\frac{\partial\mathbf{u}_1}{\partial T_1}+\mathbf{f}_{\varepsilon\mathbf{u}}(0,\mathbf{u}_0|\mathbf{u}_1)+\frac{\mathbf{f}_{\varepsilon\varepsilon}(0,\mathbf{u}_0)}{2},\boldsymbol{\phi}_0^*\right]=0 \tag{7.95}$$

Now

$$\left[\frac{\partial \mathbf{u}_0}{\partial T_0}, \boldsymbol{\phi}_0^*\right] = [ia\boldsymbol{\phi}_0 - i\bar{a}\bar{\boldsymbol{\phi}}_0, \boldsymbol{\phi}_0^*] = ia, \tag{7.96}$$

$$\left[\frac{\partial \mathbf{u}_0}{\partial T_1}, \boldsymbol{\phi}_0^*\right] = [a'\boldsymbol{\phi}_0 + \bar{a}'\bar{\boldsymbol{\phi}}_0, \boldsymbol{\phi}_0^*] = a' = a\Lambda \tag{7.97}$$

and since, from (7.94), we have $[\partial \mathbf{u}_1^p/\partial T_1, \boldsymbol{\phi}_0^*] = 0$, then (7.93) implies

$$\left[\frac{\partial \mathbf{u}_1}{\partial T_1}, \boldsymbol{\phi}_0^*\right] = b' \tag{7.98}$$

It remains to consider $[\mathbf{f}_{\varepsilon\mathbf{u}}(0, \mathbf{u}_0|\mathbf{u}_1) + \mathbf{f}_{\varepsilon\varepsilon}(0, \mathbf{u}_0)/2, \boldsymbol{\phi}_0^*]$. But $\mathbf{u}_0$ and $\mathbf{u}_1^p$ are known in terms of a, so that $[\mathbf{f}_{\varepsilon\mathbf{u}}(0, \mathbf{u}_0|\mathbf{u}_1^p) + \mathbf{f}_{\varepsilon\varepsilon}(0, \mathbf{u}_0)/2, \boldsymbol{\phi}_0^*]$ is a known function of a, and is of the form $aP(|a|^2)$ since terms in e^{iT_0} are again picked out by the inner product. Thus

$$[\mathbf{f}_{\varepsilon\mathbf{u}}(0, \mathbf{u}_0|\mathbf{u}_1^p) + \mathbf{f}_{\varepsilon\varepsilon}(0, \mathbf{u}_0)/2, \boldsymbol{\phi}_0^*] = aP(|a|^2) \tag{7.99}$$

Finally,

$$\begin{aligned}
[\mathbf{f}_{\varepsilon\mathbf{u}}(0, \mathbf{u}_0|\mathbf{u}_1^c), \boldsymbol{\phi}_0^*] &= \frac{d}{d\varepsilon}[\mathbf{f}_{\varepsilon}(0, \mathbf{u}_0 + \varepsilon\mathbf{u}_1^c), \boldsymbol{\phi}_0^*]|_{\varepsilon=0} \\
&= \frac{d}{d\varepsilon}(a + \varepsilon b)\Lambda(|a + \varepsilon b|^2)|_{\varepsilon=0} \\
&= b\Lambda(|a|^2) + a\Lambda'(|a|^2)(\bar{a}b + a\bar{b})
\end{aligned} \tag{7.100}$$

From (7.95) with (7.96) to (7.100), we have

$$b' = b\Lambda + a(a\bar{b} + \bar{a}b)\Lambda' + aP - i\omega_2 a - \Omega_2 a\Lambda \tag{7.101}$$

The form of this equation leads us to consider the combination $a\bar{b} + \bar{a}b$. Differentiating this we obtain

$$\begin{aligned}
(a\bar{b} + \bar{a}b)' &= a\bar{b}' + a'\bar{b} + \bar{a}b' + \bar{a}'b \\
&= a(\bar{b}\bar{\Lambda} + \bar{a}(\bar{a}b + a\bar{b})\bar{\Lambda}' + \bar{a}\bar{P} + i\omega_2\bar{a} - \Omega_2\bar{a}\bar{\Lambda}) + a\Lambda\bar{b} \\
&\quad + \bar{a}(b\Lambda + a(a\bar{b} + \bar{a}b)\Lambda' + aP - i\omega_2 a - \Omega_2 a\Lambda) + \bar{a}\bar{\Lambda}b \\
&= (a\bar{b} + \bar{a}b)\{(\Lambda + \bar{\Lambda}) + |a|^2(\Lambda' + \bar{\Lambda}')\} + |a|^2\{(P + \bar{P}) - \Omega_2(\Lambda + \bar{\Lambda})\}
\end{aligned}$$

where we have used the fact that $a' = a\Lambda$. This is a real first-order differential equation for the real quantity $a\bar{b} + \bar{a}b$. Putting $y = a\bar{b} + \bar{a}b$, the equation is of the form

$$y' = Q(r^2)y + R(r^2) \tag{7.102}$$

where $r=|a|$, $Q(r^2)=2\text{ Re }\Lambda(r^2)+2r^2\text{ Re }\Lambda'(r^2)$, and $R(r^2)=2r^2\text{ Re }P(r^2)-2\Omega_2 r^2\text{ Re }\Lambda(r^2)$. This linear equation may be integrated to obtain $y=y(r^2(T_1))$, where

$$\frac{y(r^2(T_1))}{r^2(T_1)\text{ Re }\Lambda(r^2(T_1))}-\frac{y(r^2(0))}{r^2(0)\text{ Re }\Lambda(r^2(0))}$$
$$=\int_{r^2(0)}^{r^2(T_1)}\frac{\text{Re }P(\rho)-\Omega_2\text{ Re }\Lambda(\rho)}{\rho(\text{Re }\Lambda(\rho))^2}\,d\rho \tag{7.103}$$

Using $\Lambda=a'/a$, (7.101) becomes

$$(b/a)'=y\Lambda'+P-i\omega_2-\Omega_2\Lambda \tag{7.104}$$

We shall consider two cases. First, let there be a stable limit cycle at $r=r_c\neq 0$, and let $\text{Re }\Lambda(r_c^2)=\text{Re }\Lambda_c=0$ (from (7.91)) and $\text{Re }\Lambda'(r_c^2)=\text{Re }\Lambda_c'<0$ for stability; $\text{Re }\Lambda_c'\neq 0$ is the required non-degeneracy condition. Then from (7.103) y tends to a finite limit y_c for arbitrary values of Ω_2, since $\text{Re }\Lambda(\rho)$ has a simple zero at $\rho=r_c^2$. From (7.102) we have

$$(\text{Re }\Lambda_c+r_c^2\text{ Re }\Lambda_c')y_c+r_c^2(\text{Re }P_c-\Omega_2\text{ Re }\Lambda_c)=0$$

where $\text{Re }P_c=\text{Re }P(r_c^2)$. But since $\text{Re }\Lambda_c=0$, $r_c^2\neq 0$,

$$\text{Re }\Lambda_c'\,y_c+\text{Re }P_c=0$$

Now integrating (7.104) using (7.91), we have

$$\frac{b(T_1)}{a(T_1)}-\frac{b(0)}{a(0)}=\int_{r^2(0)}^{r^2(T_1)}\frac{y(\rho)\Lambda'(\rho)+P(\rho)-i\omega_2-\Omega_2\Lambda(\rho)}{2\rho\Lambda(\rho)}\,d\rho \tag{7.105}$$

The integrand has a simple pole at $\rho=r_c^2$, so that b/a is finite if ω_2 and Ω_2 may be chosen in such a way that the numerator is zero at $r=r_c$, that is if

$$\begin{aligned}y_c\Lambda_c'+P_c-i\omega_2-\Omega_2\Lambda_c&=y_c\text{ Re }\Lambda_c'+\text{Re }P_c-\Omega_2\text{ Re }\Lambda_c\\&\quad+i(y_c\text{ Im }\Lambda_c'+\text{Im }P_c-\omega_2-\Omega_2\text{ Im }\Lambda_c)\\&=i(y_c\text{ Im }\Lambda_c'+\text{Im }P_c-\omega_2-\Omega_2\text{ Im }\Lambda_c)=0.\end{aligned}$$

There is some arbitrariness in the choice of ω_2 and Ω_2 here, but a bounded solution may always be obtained by taking $\Omega_2=0$, $\omega_2=y_c\text{ Im }\Lambda_c'+\text{Im }P_c$.

Let us now consider the case where $r\to r_c=0$ as $T_1\to\infty$, and $\text{Re }\Lambda_c\neq 0$ for non-degeneracy, and let $\text{Re }P_c-\Omega_2\text{ Re }\lambda_c=0$. Then from (7.103) $y(\rho)=O(\rho^2)$ as $T_1\to\infty$, and the integrand in (7.105) has a simple pole at $\rho=0$. Thus b/a is finite if ω_2 and Ω_2 are chosen such that the numerator is zero at $r=r_c$, that is if

$$P_c-i\omega_2-\Omega_2\Lambda_c=\text{Re }P_c-\Omega_2\text{ Re }\Lambda_c+i(\text{Im }P_c-\omega_2-\Omega_2\text{ Im }\Lambda_c)=0$$

This confirms our choice of Ω_2 and also gives ω_2; we have

$$\Omega_2 = \text{Re } P_c/\text{Re } \Lambda_c$$

$$\omega_2 = \text{Im } P_c - \Omega_2 \text{ Im } \Lambda_c$$

In either case it follows that equation (7.104) may be integrated to give a suitable function b which satisfies the boundedness and initial conditions required. This function is defined by (7.105) and (7.103) assuming that $b(0)$ and $y(r^2(0))$ are known. But the function $\mathbf{u}_1$ satisfies the initial conditions $\mathbf{u}_1(0,0)=\mathbf{0}$, from (7.80). From (7.93),

$$\mathbf{u}_1(0,0) = \mathbf{u}_1^p(0,0) + b(0)(1,-i)^{\mathrm{T}} + \bar{b}(0)(1,i)^{\mathrm{T}}$$

so that

$$b(0) = -(u_1^p(0,0) + iv_1^p(0,0))/2$$

where $\mathbf{u}_1^p(0,0) = (u_1^p(0,0), v_1^p(0,0))^T$. Then the initial value of y is given by

$$y(r^2(0)) = a(0)\bar{b}(0) + \bar{a}(0)b(0)$$

The function $\mathbf{u}_1$ which we have constructed then satisfies the boundedness conditions and initial conditions required. Higher order equations are of the same form as (7.101) so that $\mathbf{u}_2$, $\mathbf{u}_3$ and so on satisfying the boundedness and initial conditions required may be constructed and an asymptotic representation for $\mathbf{u}$ found, given by (7.78). In the case of a solution tending to a non-degenerate stable limit cycle such a representation may be found with $\Omega_n = 0$ for all n, so that $\Omega = \varepsilon$ and $T_1 = \varepsilon t$.

7.7 The Two-timing Method for Reaction–Diffusion Systems

The two-timing method may be extended to reaction–diffusion systems, although in this case the theory is much less developed. The system

$$\begin{aligned} u_t &= -v + f(\varepsilon, u, v) + \varepsilon D_{11}\nabla^2 u + \varepsilon D_{12}\nabla^2 v \\ v_t &= u + g(\varepsilon, u, v) + \varepsilon D_{21}\nabla^2 u + \varepsilon D_{22}\nabla^2 v \end{aligned} \tag{7.106}$$

or

$$L\mathbf{u} = \mathbf{f}(\varepsilon, \mathbf{u}) + \varepsilon D\,\nabla^2\mathbf{u} \tag{7.107}$$

where D is the diffusion matrix (assumed independent of ε for simplicity only), and L is the operator defined in (7.6), is again close to the harmonic oscillator $L\mathbf{u}=\mathbf{0}$, so that we might expect two-timing to be appropriate. We therefore define $T_0 = \omega t$, $T_1 = \Omega t$, where $\omega = 1 + \varepsilon^2\omega_2 + \varepsilon^3\omega_3 + \cdots$, $\Omega = \varepsilon + \varepsilon^2\Omega_2 + \varepsilon^3\Omega_3 + \cdots$, as before, and look for solutions as functions of the two-time variables T_0 and T_1 and the space variables $\mathbf{x}$, with asymptotic expansions

$$\mathbf{u}(T_0, T_1, \mathbf{x}; \varepsilon) \approx \mathbf{u}_0(T_0, T_1, \mathbf{x}) + \varepsilon \mathbf{u}_1(T_0, T_1, \mathbf{x}) + \cdots \tag{7.108}$$

Substituting this into the system above and equating powers of ε gives

$$\frac{\partial u_0}{\partial T_0} = -v_0, \qquad \frac{\partial v_0}{\partial T_0} = u_0$$

$$\frac{\partial u_1}{\partial T_0} + \frac{\partial u_0}{\partial T_1} = -v_1 + f_\varepsilon(0, u_0, v_0) + D_{11}\,\nabla^2 u_0 + D_{12}\,\nabla^2 v_0$$

$$\frac{\partial v_1}{\partial T_0} + \frac{\partial v_0}{\partial T_1} = u_1 + g_\varepsilon(0, u_0, v_0) + D_{21}\,\nabla^2 u_0 + D_{22}\,\nabla^2 v_0$$

and so on, or in operator notation,

$$L_0 \mathbf{u}_0 = \mathbf{0} \tag{7.109}$$

$$L_0 \mathbf{u}_1 = -\frac{\partial \mathbf{u}_0}{\partial T_1} + \mathbf{f}_\varepsilon(0, \mathbf{u}_0) + D\,\nabla^2 \mathbf{u}_0 \tag{7.110}$$

and so on, where

$$L_0 = \frac{\partial}{\partial T_0} - M = \frac{\partial}{\partial T_0} - \begin{pmatrix} -1 & 0 \\ 0 & 1 \end{pmatrix}$$

as before. The first of these gives

$$\mathbf{u}_0 = a(T_1, \mathbf{x})\boldsymbol{\phi}_0 + \bar{a}(T_1, \mathbf{x})\bar{\boldsymbol{\phi}}_0 \tag{7.111}$$

analogous to (7.84) in the spatially uniform case, but with a now a function of $\mathbf{x}$ as well as the slow time variable T_1. The initial and boundary conditions are easily derived and we shall not discuss them here. The equation for $\mathbf{u}_1$, (7.110), has a solution if and only if

$$\left[\frac{\partial \mathbf{u}_0}{\partial T_1}, \boldsymbol{\phi}_0^*\right] = \frac{\partial a}{\partial T_1} = [\mathbf{f}_\varepsilon(0, \mathbf{u}_0) + D\nabla^2 \mathbf{u}_0, \boldsymbol{\phi}_0^*] \tag{7.112}$$

A simple computation shows that

$$[D\nabla^2 \mathbf{u}_0, \boldsymbol{\phi}_0^*] = (D_{11} - iD_{12} + iD_{21} + D_{22})\nabla^2 a = \delta\,\nabla^2 a \tag{7.113}$$

say, so that the equation becomes

$$\frac{\partial a}{\partial T_1} = [\mathbf{f}_\varepsilon(0, \mathbf{u}_0), \boldsymbol{\phi}_0^*] + \delta\,\nabla^2 a = a\Lambda(|a|^2) + \delta\,\nabla^2 a \tag{7.114}$$

where Λ is as in the previous section. This is a parabolic partial differential equation for a. Note that it cannot in general be written as a single equation for $\rho = |a|^2$, in contrast to the situation in the diffusionless case, and is therefore very much more difficult to solve. A particular form of this equation will be discussed in Section 7.11. Higher-order approximations will not be considered.

7.8 Averaging Methods

Another class of asymptotic methods useful in dealing with oscillatory ordinary differential equations and the associated reaction–diffusion equations is known as the method of averaging. It is similar to the two-timing method in that the amplitude and phase of the variable are considered to be slowly varying functions of time, but the approach is quite different.

Consider again the van der Pol equation

$$\ddot{u} - \varepsilon(1-u^2)\dot{u} + u = 0 \tag{7.115}$$

When $\varepsilon = 0$, this has solution

$$u = ae^{it} + \bar{a}e^{-it}$$

so that

$$\dot{u} = iae^{it} - i\bar{a}e^{-it}$$

where a is a complex constant. We have already seen that the effect of the nonlinear term in the van der Pol equation is to make a vary with time, but that this variation is over times which are long compared to the period of oscillation 2π. The method of averaging makes use of the slowness of these variations. We first define a time-varying function $a(t)$ in terms of the function $u(t)$ by the equations

$$u(t) = a(t)e^{it} + \bar{a}(t)e^{-it} \tag{7.116}$$

$$\dot{u}(t) = ia(t)e^{it} - i\bar{a}(t)e^{-it} \tag{7.117}$$

where $u(t)$ satisfies van der Pol's equation and the required initial conditions. Note that this imposes a condition on a, since we require that the right-hand side of (7.117) be the differential of the right-hand side of (7.116), i.e.

$$\dot{a}e^{it} + iae^{it} + \dot{\bar{a}}e^{-it} - i\bar{a}e^{-it} = iae^{it} - i\bar{a}e^{-it}$$

or

$$\dot{a}e^{it} + \dot{\bar{a}}e^{-it} = 0 \tag{7.118}$$

We also have, differentiating (7.117),

$$\ddot{u} = i\dot{a}e^{it} - ae^{it} - i\dot{\bar{a}}e^{-it} - \bar{a}e^{-it}$$

so the requirement that u satisfy the differential equation (7.115) leads to

$$i\dot{a}e^{it} - i\dot{\bar{a}}e^{-it} = \varepsilon(1 - a^2e^{2it} - 2a\bar{a} - \bar{a}^2e^{-2it})(iae^{it} - i\bar{a}e^{-it}) \qquad (7.119)$$

We therefore have two coupled real differential equations (7.118) and (7.119) for the complex unknown a. Eliminating $\dot{\bar{a}}$ gives

$$2i\dot{a}e^{it} = \varepsilon(1 - a^2e^{2it} - 2a\bar{a} - \bar{a}^2e^{-2it})(iae^{it} - i\bar{a}e^{-it})$$

or

$$\dot{a} = \tfrac{1}{2}\varepsilon(1 - a^2e^{2it} - 2a\bar{a} - \bar{a}^2e^{-2it})(a - \bar{a}e^{-2it}) \qquad (7.120)$$

This is an *exact* differential equation for the function a defined by (7.116) and (7.117). It is, of course, nonlinear and intractable but does show, as expected, that a is a slowly varying function of t, i.e. that $\dot{a}$ is $O(\varepsilon)$. Hence a and its derivatives are constant to first order over the period π of the periodic functions. Taking an average of Eq. (7.120) over this period, we have

$$\frac{1}{\pi}\int_0^{\pi} \dot{a}dt = \frac{\varepsilon}{2\pi}\int_0^{\pi} [(1 - |a|^2)(a - \bar{a}e^{-2it}) - a^3e^{2it} + \bar{a}^3e^{-4it}]\, dt$$

or using the fact that a and $\dot{a}$ are approximately constant over a period

$$\dot{a} \simeq \tfrac{1}{2}\varepsilon a(1 - |a|^2) \qquad (7.121)$$

This result is correct up to terms of $O(\varepsilon)$ and agrees with the result obtained by the two-timing method (equation 7.72).

Consider now the oscillatory reaction scheme

$$\dot{u} = -v + f(\varepsilon, u, v), \qquad \dot{v} = u + g(\varepsilon, u, v)$$

or

$$L\mathbf{u} = \mathbf{f}(\varepsilon, \mathbf{u}) \qquad (7.122)$$

The system with $\varepsilon = 0$ has solution

$$\mathbf{u} = a\boldsymbol{\phi}_0 + \bar{a}\bar{\boldsymbol{\phi}}_0$$

where a is a constant and $\boldsymbol{\phi}_0 = (1, -i)^T e^{it}$. For $\varepsilon \neq 0$ we make the transformation

$$\mathbf{u} = a(t)\boldsymbol{\phi}_0 + \bar{a}(t)\bar{\boldsymbol{\phi}}_0 \qquad (7.123)$$

where a is now a function of t. Substituting this form into (7.122) it follows that

$$\dot{a}\boldsymbol{\phi}_0 + \dot{\bar{a}}\bar{\boldsymbol{\phi}}_0 = \mathbf{f}(\varepsilon, a\boldsymbol{\phi}_0 + \bar{a}\bar{\boldsymbol{\phi}}_0) \qquad (7.124)$$

This vector differential equation for the complex unknown a corresponds to the two scalar differential equations (7.118) and (7.119) obtained for the van der Pol equation. Then taking the inner product of (7.124) with $\boldsymbol{\phi}_0^*$ and using the fact that a and $\dot{a}$ are approximately constant over a period, we have

$$\dot{a} \simeq [\mathbf{f}(\varepsilon, a\boldsymbol{\phi}_0 + \bar{a}\bar{\boldsymbol{\phi}}_0), \boldsymbol{\phi}_0^*] \tag{7.125}$$

or, to first order in ε,

$$\dot{a} \simeq [\varepsilon \mathbf{f}_\varepsilon(0, a\boldsymbol{\phi}_0 + \bar{a}\bar{\boldsymbol{\phi}}_0), \phi_0^*] = \varepsilon a \Lambda(|a|^2) \tag{7.126}$$

agreeing to this order with the two-timing method.

7.9 The Krylov–Bogoliubov–Mitropolsky Method

The averaging method is capable of extension to higher orders of accuracy, and is then known as the Krylov-Bogoliubov-Mitropolsky method [see Minorsky (1974)]. The idea on which this extension is based is that since a is a slowly varying function of time, its derivatives may be expressed as functions of a slow variable. Rather than introduce a slow time scale, like $T_1 = \Omega t$ in the two-timing method, the Krylov-Bogoliubov-Mitropolsky method uses a itself, or more accurately an approximation to a, as the slow variable.

The starting point is the exact differential equation (7.124) for a, where $\mathbf{u} = a\boldsymbol{\phi}_0 + \bar{a}\bar{\boldsymbol{\phi}}_0$, given by

$$\dot{a}\boldsymbol{\phi}_0 + \dot{\bar{a}}\bar{\boldsymbol{\phi}}_0 = \mathbf{f}(\varepsilon, a\boldsymbol{\phi}_0 + \bar{a}\bar{\boldsymbol{\phi}}_0)$$

We now look for a in the form

$$a(t; \varepsilon) \approx b(t) + \varepsilon b_1(b, t) + \varepsilon^2 b_2(b, t) + \cdots \tag{7.127}$$

where b and t are considered to be *independent* variables and the b_n's are 2π-periodic functions of t. We expect b to be slowly varying, so that $\dot{b}$ has an asymptotic expansion in terms of ε given by

$$\dot{b} \approx \varepsilon g_1(b) + \varepsilon^2 g_2(b) + \cdots \tag{7.128}$$

where the function g_n are independent of t. Note that

$$\mathbf{u} \approx \mathbf{u}_0 + \varepsilon \mathbf{u}_1 + \varepsilon^2 \mathbf{u}_2 + \cdots \tag{7.129}$$

where

$$\mathbf{u}_0 = b\boldsymbol{\phi}_0 + \bar{b}\bar{\boldsymbol{\phi}}_0, \qquad \mathbf{u}_n = b_n\boldsymbol{\phi}_0 + \bar{b}_n\bar{\boldsymbol{\phi}}_0, \qquad n > 0 \tag{7.130}$$

Just as in the two-timing method, we are concerned with an initial value problem. If

$$a(0) = a_0 = (U + iV)/2$$

as in (7.88), the initial conditions for b and the b_n's will be

$$b(0)=a_0, \qquad b_n(a_0,0)=0, \qquad n>0 \tag{7.131}$$

Now

$$\dot{a}\approx(\varepsilon g_1+\varepsilon^2 g_2+\cdots)+\varepsilon(b_{1b}(\varepsilon g_1+\cdots)+b_{1t})+\varepsilon^2(\cdots+b_{2t})+\cdots \tag{7.132}$$

where subscripts denote partial differentiation with respect to b and t, and

$$\mathbf{f}(\varepsilon, a\boldsymbol{\phi}_0+\bar{a}\bar{\boldsymbol{\phi}}_0)=\mathbf{f}(\varepsilon,\mathbf{u})\approx\varepsilon\mathbf{f}_\varepsilon(0,\mathbf{u}_0)+\varepsilon\mathbf{f}_{\varepsilon\mathbf{u}}(0,\mathbf{u}_0|\,\varepsilon\mathbf{u}_1+\cdots)+\cdots \tag{7.133}$$

Let us denote the ordinary inner product of two complex vectors by $\langle\cdot,\cdot\rangle$, so that

$$\langle\mathbf{u},\mathbf{v}\rangle=\mathbf{u}\cdot\bar{\mathbf{v}}$$

for any vectors $\mathbf{u},\mathbf{v}$. The square bracket inner product is thus the average of this over a period. Taking the ordinary inner product of (7.124) with $\boldsymbol{\phi}_0^*$, we obtain

$$\dot{a}=\langle\mathbf{f}(\varepsilon, a\boldsymbol{\phi}_0+\bar{a}\bar{\boldsymbol{\phi}}_0),\boldsymbol{\phi}_0^*\rangle \tag{7.134}$$

Substituting the expressions (7.132) and (7.133) into this equation and equating powers of ε, we obtain

$$g_1+b_{1t}=\langle\mathbf{f}_\varepsilon(0,\mathbf{u}_0),\boldsymbol{\phi}_0^*\rangle \tag{7.135}$$

$$g_2+b_{1b}g_1+b_{2t}=\langle\mathbf{f}_{\varepsilon\mathbf{u}}(0,\mathbf{u}_0|\mathbf{u}_1)+\mathbf{f}_{\varepsilon\varepsilon}(0,\mathbf{u}_0)/2,\boldsymbol{\phi}_0^*\rangle \tag{7.136}$$

and so on. We shall take the average of these equations over a period in t, recalling that b and t are being considered as independent variables. Note that since b_n is 2π-periodic in t, the average of b_{nt} over a period is

$$[b_{nt},1]=\frac{1}{2\pi}\int_0^{2\pi}b_{nt}\,dt=\frac{1}{2\pi}[b_n]_0^{2\pi}=0 \tag{7.137}$$

Taking the average of (7.135) over a period and noting that g_1 is independent of t, we have

$$g_1=[\mathbf{f}_\varepsilon(0,\mathbf{u}_0),\boldsymbol{\phi}_0^*]$$

Thus g_1 is determined and is given by $b\Lambda(|b|^2)$, so that to leading order $b'=\varepsilon b\Lambda(|b|^2)$ from (7.128), cf. (7.90). Note that this is an exact equation for the approximation b, as opposed to (7.126) which is an approximate equation for the exact a. Now, from (7.135), b_1 satisfies

$$b_{1t}=\langle\mathbf{f}_\varepsilon(0,\mathbf{u}_0),\boldsymbol{\phi}_0^*\rangle-[\mathbf{f}_\varepsilon(0,\mathbf{u}_0),\boldsymbol{\phi}_0^*] \tag{7.138}$$

But $\partial/\partial t$ is a Fredholm operator on the space of 2π-periodic functions with constant eigenvector, so that (7.138) has a bounded solution, which is

2π-periodic, if and only if the average of the right-hand side is zero. But this follows immediately, so that we may find such a solution. Then the general solution is given by

$$b_1(b, t) = b_1^p(b, t) + B_1(b) \tag{7.139}$$

where b_1^p is a particular solution and B_1 is a function of integration. b_1 must satisfy the initial condition (7.131), so that

$$b_1(a_0, 0) = b_1^p(a_0, 0) + B_1(a_0) = 0$$

or

$$B_1(a_0) = -b_1^p(a_0, 0)$$

This must hold for any initial condition a_0, so that $B_1(b) = -b_1^p(b, 0)$, and

$$b_1(b, t) = b_1^p(b, t) - b_1^p(b, 0) \tag{7.140}$$

Thus b_1, and hence $\mathbf{u}_1$, is determined. To find b_2, we take the average of (7.136) over a period, using (7.137), to obtain

$$g_2 + g_1[b_{1b}, 1] = [\mathbf{f}_{\varepsilon\mathbf{u}}(0, \mathbf{u}_0|\mathbf{u}_1) + \mathbf{f}_{\varepsilon\varepsilon}(0, \mathbf{u}_0)/2, \boldsymbol{\phi}_0^*]$$

which determines g_2, so that (7.136) becomes

$$b_{2t} = \langle \mathbf{f}_{\varepsilon\mathbf{u}}(0, \mathbf{u}_0|\mathbf{u}_1), \boldsymbol{\phi}_0^* \rangle - [\mathbf{f}_{\varepsilon\mathbf{u}}(0, \mathbf{u}_0|\mathbf{u}_1), \boldsymbol{\phi}_0^*] - g_1\{b_{1b} - [b_{1b}, 1]\} \tag{7.141}$$

This has a bounded solution which is 2π-periodic if the average of the right-hand side over a period is zero, and again this follows immediately. The initial conditions for b_2 may be satisfied as before, so that b_2, and hence $\mathbf{u}_2$, is determined. Higher-order terms may be found similarly, and an asymptotic expansion for $\mathbf{u}$ built up. The only problem is that b is not given explicitly, but in the form of an asymptotic expansion for $\dot{b}$. It may be shown under fairly general assumptions (Bogoliubov and Mitropolsky, 1961) that the asymptotic expansion for b to nth order may be obtained by integrating the asymptotic expansion for $\dot{b}$ to nth order, that is

$$\dot{b} = \varepsilon g_1(b) + \varepsilon^2 g_2(b) + \cdots + \varepsilon^n g_n(b)$$

with initial conditions $b(0) = a_0$.

It may be shown (Morrison, 1966) that the two-timing method and the Krylov-Bogoliubov-Mitropolsky method for nearly linear ordinary differential equations give equivalent asymptotic expansions. The two-timing method is usually simpler to apply, but the Krylov-Bogoliubov-Mitropolsky method is more general and may be used when it is not obvious what slow time scale is to be used.

7.10 Averaging Methods for Reaction–Diffusion Systems

Averaging methods may also be applied to reaction–diffusion systems. Consider the system

$$u_t = -v + f(\varepsilon, u, v) + \varepsilon D_{11}\,\nabla^2 u + \varepsilon D_{12}\,\nabla^2 v$$
$$v_t = u + g(\varepsilon, u, v) + \varepsilon D_{21}\,\nabla^2 u + \varepsilon D_{22}\,\nabla^2 v$$

or in operator notation

$$L\mathbf{u} = \mathbf{f}(\varepsilon, \mathbf{u}) + \varepsilon D\,\nabla^2\mathbf{u} \tag{7.142}$$

where L is the linear operator defined above. This system has solution $(u, v) = (A\cos(t+\alpha),\ A\sin(t+\alpha))$ or $\mathbf{u} = a\boldsymbol{\phi}_0 + \bar{a}\bar{\boldsymbol{\phi}}_0$ if $\varepsilon = 0$. We assume that if $\varepsilon \neq 0$, the solution is of the form

$$\mathbf{u}(t, \mathbf{x}; \varepsilon) = a(t, \mathbf{x}; \varepsilon)\boldsymbol{\phi}_0 + \bar{a}(t, \mathbf{x}; \varepsilon)\bar{\boldsymbol{\phi}}_0$$

This is, of course, the straightforward averaging method with a now a function of $\mathbf{x}$ as well as t. The equations become

$$a_t\boldsymbol{\phi}_0 + ia\boldsymbol{\phi}_0 + \bar{a}_t\bar{\boldsymbol{\phi}}_0 - i\bar{a}\bar{\boldsymbol{\phi}}_0 = ia\boldsymbol{\phi}_0 - i\bar{a}\bar{\boldsymbol{\phi}}_0 + \mathbf{f}(\varepsilon, \mathbf{u}) + \varepsilon D\,\nabla^2 a\boldsymbol{\phi}_0 + \varepsilon D\,\nabla^2\bar{a}\bar{\boldsymbol{\phi}}_0$$

or

$$a_t\boldsymbol{\phi}_0 + \bar{a}_t\bar{\boldsymbol{\phi}}_0 = \mathbf{f}(\varepsilon, \mathbf{u}) + \varepsilon D\,\nabla^2\mathbf{u} \tag{7.143}$$

This is an exact real vector differential equation for the complex unknown a, and is therefore in principle sufficient to determine a (subject to initial and boundary conditions). As in the spatially uniform case, we simplify this equation by using the assumed slow variation of a. We take the square brackets inner product of (7.143) with $\boldsymbol{\phi}_0^* = (1, -i)^{\mathrm{T}}e^{it}/2$, and use the fact that a is slowly varying so that it and its derivatives may be considered constant over the period 2π of the eigenfunctions $\boldsymbol{\phi}_0$ and $\bar{\boldsymbol{\phi}}_0$. Since $[\boldsymbol{\phi}_0, \boldsymbol{\phi}_0^*] = 1$ and $[\bar{\boldsymbol{\phi}}_0, \boldsymbol{\phi}_0^*] = 0$, the equation becomes

$$a_t = [\mathbf{f}(\varepsilon, \mathbf{u}), \boldsymbol{\phi}_0^*] + \varepsilon[D\,\nabla^2\mathbf{u}, \boldsymbol{\phi}_0^*]$$

or to first order

$$a_t \simeq \varepsilon a\Lambda(|a|^2) + \varepsilon\delta\,\nabla^2 a \tag{7.144}$$

where $\delta = (D_{11} - iD_{12} + iD_{21} + D_{22})$ and

$$a\Lambda(|a|^2) = [\mathbf{f}_\varepsilon(0, a\boldsymbol{\phi}_0 + \bar{a}\bar{\boldsymbol{\phi}}_0), \boldsymbol{\phi}_0^*]$$

where the average is taken considering a constant. The equation is an approximate one for the *exact* solution $\mathbf{u} = a\boldsymbol{\phi}_0 + \bar{a}\bar{\boldsymbol{\phi}}_0$, since we have used the slowly-varying approximation to derive. To continue the analysis to higher order a Krylov–Bogoliubov–Mitropolsky technique must be used, and we look for a in the form

$$a(t, \mathbf{x}; \varepsilon) \approx b(t, \mathbf{x}) + \varepsilon b_1(b, t) + \cdots \tag{7.145}$$

where

$$b_t(t, \mathbf{x}) \approx \varepsilon g_1(b) + \varepsilon^2 g_2(b) + \cdots \tag{7.146}$$

and $\mathbf{u}$ is as in (7.129) and (7.130). An analysis analogous to the spatially uniform case gives the differential equation for b as

$$b_t = \varepsilon\{b\Lambda(|b|^2) + \delta\, \nabla^2 b\} \tag{7.147}$$

just as in the two-timing method.

7.11 Perturbations of Strongly Nonlinear Oscillatory Systems

Reaction–diffusion oscillatory kinetics are very often not simply perturbations of a linear oscillator but strongly nonlinear, so that the techniques above must be modified. We shall assume in this section that the mth order nonlinear kinetic system

$$\mathbf{u}_t = \mathbf{f}(0, \mathbf{u})$$

has a globally asymptotically stable 2π-periodic limit cycle solution $\mathbf{u} = \mathbf{a}(t)$, and look for solutions of a perturbed problem

$$\mathbf{u}_t = \mathbf{f}(\varepsilon, \mathbf{u}) + \varepsilon D\, \nabla^2 \mathbf{u} \tag{7.148}$$

which are everywhere close to this solution. We shall consider two-timing and averaging methods for such equations. Two-timing methods in similar situations have been considered by Hagan (1981) and Neu (1979). The behaviour of the unperturbed problem is fundamentally different from the linear oscillator in that the most general form of the limit cycle solution is just

$$\mathbf{u} = \mathbf{a}(t + \psi)$$

where ψ is a constant, whereas the linear problem $\mathbf{u}_t = L\mathbf{u}$ has solution

$$\mathbf{u} = a\boldsymbol{\phi} + \bar{a}\bar{\boldsymbol{\phi}}$$

where a is a complex constant and $\boldsymbol{\phi}$ and $\bar{\boldsymbol{\phi}}$ are the eigenfunctions of L corresponding to the purely imaginary eigenvalues. Thus the nonlinear oscillator has one arbitrary constant, the phase ψ, whereas the linear oscillator has two, the real and imaginary parts of the complex amplitude a.

As usual in the two-timing technique we define fast and slow time scales T_0 and T_1 by $T_0=\omega t$, $T_1=\Omega t$, where $\omega\approx 1+\varepsilon^2\omega_2+\varepsilon^3\omega_3+\cdots$, $\Omega\approx\varepsilon+\varepsilon^2\Omega_2+\varepsilon^3\Omega_3+\cdots$. The equation becomes

$$\omega\frac{\partial\mathbf{u}}{\partial T_0}+\Omega\frac{\partial\mathbf{u}}{\partial T_1}=\mathbf{f}(\varepsilon,\mathbf{u})+\varepsilon D\,\nabla^2\mathbf{u}$$

We look for a solution in the form

$$\mathbf{u}=\mathbf{u}(T_0,T_1,\mathbf{x})\approx\mathbf{u}_0(T_0,T_1,\mathbf{x})+\varepsilon\mathbf{u}_1(T_0,T_1,\mathbf{x})+\cdots$$

Substituting this form into the equation and equating powers of ε we have

$$\frac{\partial\mathbf{u}_0}{\partial T_0}=\mathbf{f}(0,\mathbf{u}_0)$$

$$\frac{\partial\mathbf{u}_1}{\partial T_0}+\frac{\partial\mathbf{u}_0}{\partial T_1}=\mathbf{f}_{\mathbf{u}}(0,\mathbf{u}_0|\mathbf{u}_1)+\mathbf{f}_\varepsilon(0,\mathbf{u}_0)+D\,\nabla^2\mathbf{u}_0$$

and so on, or

$$\frac{\partial\mathbf{u}_0}{\partial T_0}=\mathbf{f}(0,\mathbf{u}_0)$$

$$L\mathbf{u}_1\equiv\frac{\partial\mathbf{u}_1}{\partial T_0}-\mathbf{f}_{\mathbf{u}}(0,\mathbf{u}_0|\mathbf{u}_1)$$

$$=-\frac{\partial\mathbf{u}_0}{\partial T_1}+\mathbf{f}_\varepsilon(0,\mathbf{u}_0)+D\,\nabla^2\mathbf{u}_0 \qquad (7.149)$$

and so on. The first of these, by assumption, has a solution

$$\mathbf{u}_0=\mathbf{a}(T_0+\psi(T_1,\mathbf{x}))$$

2π-periodic in T_0, where ψ is as yet undetermined. This is the limiting behaviour of the general solution, since the limit cycle is assumed to be globally asymptotically stable. We shall not consider the initial value problem, but shall assume that the solution is close to the limit cycle at any point. To solve (7.149), we must consider the linear operator $L\equiv(\partial/\partial T_0)-\mathbf{f}_{\mathbf{u}}(0,\mathbf{u}_0|\cdot)$. Since $\mathbf{u}_0$ is periodic in T_0 this is amenable to treatment by Floquet theory. The homogeneous problem

$$L\mathbf{v}\equiv\frac{\partial\mathbf{v}}{\partial T_0}-\mathbf{f}_{\mathbf{u}}(0,\mathbf{u}_0|\mathbf{v})=0$$

has a 2π-periodic solution

$$\mathbf{v}=\mathbf{v}(T_0,T_1,\mathbf{x})=\mathbf{a}'(T_0+\psi(T_1,\mathbf{x}))$$

and $m-1$ linearly independent solutions, where m is the order of the system.

To ensure stability of the limit cycle solution we assume that all linearly independent solutions decay exponentially. Let $L^* \equiv (\partial/\partial T_0) + \mathbf{f}_\mathbf{u}^T(0, \mathbf{u}_0|\cdot)$ be the adjoint operator of L, and let $\mathbf{v}^* = \mathbf{v}^*(T_0 + \psi)$ be the adjoint eigenfunction of $\mathbf{v}$ satisfying $L^*\mathbf{v}^* = 0$ and the normalisation condition

$$[\mathbf{a}', \mathbf{v}^*] = 1 \tag{7.150}$$

Now equation (7.149) has a solution which is 2π-periodic in T_0 if and only if

$$\left[-\frac{\partial \mathbf{u}_0}{\partial T_1} + \mathbf{f}_\varepsilon(0, \mathbf{u}_0) + D\,\nabla^2 \mathbf{u}_0, \mathbf{v}^*\right] = 0$$

or, substituting $\mathbf{u}_0 = \mathbf{a}(T_0 + \psi)$,

$$\left[-\mathbf{a}'\frac{\partial \psi}{\partial T_1} + \mathbf{f}_\varepsilon(0, \mathbf{a}) + D(\mathbf{a}''|\nabla\psi|^2 + \mathbf{a}'\,\nabla^2\psi), \mathbf{v}^*\right] = 0$$

Using $[\mathbf{a}, \mathbf{v}^*] = 1$ and defining

$$\begin{aligned} d &= \frac{1}{2\pi}\int_0^{2\pi} D\mathbf{a}'(s)\cdot\mathbf{v}^*(s)\,ds = [D\mathbf{a}', \mathbf{v}^*] \\ d\gamma &= \frac{1}{2\pi}\int_0^{2\pi} D\mathbf{a}''(s)\cdot\mathbf{v}^*(s)\,ds = [D\mathbf{a}'', \mathbf{v}^*] \\ \alpha &= \frac{1}{2\pi}\int_0^{2\pi} \mathbf{f}_\varepsilon(0, \mathbf{a}(s))\cdot\mathbf{v}^*(s)\,ds = [\mathbf{f}_\varepsilon(0, \mathbf{a}), \mathbf{v}^*] \end{aligned} \tag{7.151}$$

we have

$$\frac{\partial \psi}{\partial T_1} = \alpha + d(\nabla^2\psi + \gamma|\nabla\psi|^2)$$

Using the Cole–Hopf transformation $z = \exp \gamma(\psi - \alpha T_1)$ this may be simplified to

$$\frac{\partial z}{\partial T_1} = d\,\nabla^2 z$$

so that z satisfies the heat equation and may be found by standard techniques. If this equation is satisfied then (7.149) may be solved for $\mathbf{u}_1$, and the analysis may be continued to any order.

In averaging techniques for perturbations of the harmonic oscillator, the complex amplitude a is taken to be a slowly varying function of time, and expansions for the solution $\mathbf{u}$ are considered as functions of two-time variables, the time t representing the fast variation due to the linear oscillator and an approximation b to the amplitude representing the slow variation due to the perturbation. In the nonlinear case the phase ψ must be the slow variable, and we take $\mathbf{u}$ to be a function of t and ψ. Thus we look for a solution of (7.148) in the form

$$\mathbf{u} = \mathbf{u}(t, \psi) \approx \mathbf{u}_0(t, \psi) + \varepsilon \mathbf{u}_1(t, \psi) + \cdots$$

where $\psi = \psi(t, \mathbf{x})$ is a slowly varying function of time. Thus the equation becomes

$$\mathbf{u}_t + \mathbf{u}_\psi \psi_t = \mathbf{f}(\varepsilon, \mathbf{u}) + \varepsilon D \nabla^2 \mathbf{u}$$

$O(1)$ terms give

$$\partial \mathbf{u}_0 / \partial t = \mathbf{f}(\mathbf{u}_0)$$

so that

$$\mathbf{u}_0 = \mathbf{a}(t + \psi)$$

where $\mathbf{a}(\cdot)$ is the stable limit cycle, since we are looking for a solution close to this. The next order terms give

$$\varepsilon \frac{\partial \mathbf{u}_1}{\partial t} - \varepsilon \mathbf{f}_{\mathbf{u}}(0, \mathbf{u}_0 | \mathbf{u}_1) = -\frac{\partial \mathbf{u}_0}{\partial \psi} \psi_t + \varepsilon \mathbf{f}_\varepsilon(0, \mathbf{u}_0) + \varepsilon D \nabla^2 \mathbf{u}_0$$

Just as in the two-timing method this problem has a bounded solution, which is 2π-periodic, if and only if:

$$\left[-\frac{\partial \mathbf{u}_0}{\partial \psi} \psi_t + \varepsilon \mathbf{f}_\varepsilon(0, \mathbf{u}_0) + \varepsilon D \nabla^2 \mathbf{u}_0, \mathbf{v}^* \right] = 0$$

Now

$$\frac{\partial \mathbf{u}_0}{\partial \psi} = \mathbf{a}'(t + \psi)$$

and

$$D \nabla^2 \mathbf{u}_0 = D(\mathbf{a}'' |\nabla \psi|^2 + \mathbf{a}' \nabla^2 \psi)$$

so that this condition becomes

$$\frac{1}{2\pi} \int_0^{2\pi} (-\mathbf{a}' \psi_t + \varepsilon \mathbf{f}_\varepsilon(0, \mathbf{a}) + \varepsilon D \mathbf{a}'' |\nabla \psi|^2 + \varepsilon D \mathbf{a}' \nabla^2 \psi) \cdot \mathbf{v}^* \, ds = 0$$

where the arguments of $\mathbf{a}$ and its derivatives are $t + \psi$. We now use the hypothesis of slow time variation of ψ and its derivatives, so that these may be considered constant over a period, to approximate the solvability condition by

$$\psi_t = \varepsilon d(\nabla^2 \psi + \gamma |\nabla \psi|^2) + \varepsilon \alpha \tag{7.152}$$

using (7.150) and (7.151). This is a partial differential equation for the perturbed phase ψ, and is essentially the same equation as was obtained by the two-timing method.

Hagan (1981) applied a two-timing analysis similar to ours to the system

$$\mathbf{u}_t = \mathbf{f}(\mathbf{u}) + \varepsilon \mathbf{g}(\mathbf{x}, \mathbf{u}) + \varepsilon D\, \nabla^2 \mathbf{u}$$

where the term $\varepsilon \mathbf{g}(\mathbf{x}, \mathbf{u})$ represents inhomogeneities in the system. In this case the phase ψ satisfies

$$\frac{\partial \psi}{\partial T_1} = \alpha(\mathbf{x}) + d(\nabla^2 \psi + \gamma |\nabla \psi|^2)$$

and the transformation $z = e^{\gamma\psi}$ leads to

$$\frac{\partial z}{\partial T_1} = \alpha(\mathbf{x}) z + d\, \nabla^2 z,$$

a linear partial differential equation for z. Hagan shows that stable target patterns of the original system with bounded initial conditions may arise, but only if $\mathbf{g}(\mathbf{x}, \mathbf{u}) \neq \mathbf{0}$. Such target patterns arise experimentally in the Belousov–Zhabotinkii reaction, but may be virtually eliminated by careful filtering of the chemical components (resulting in the term $\mathbf{g}(\mathbf{x}, \mathbf{u})$ which represents inhomogeneities becoming equal to zero). Hagan also shows that higher frequency target patterns eventually engulf neighbouring lower frequency patterns, again in line with experimental observation.

7.12 The Hopf Bifurcation and Λ–Ω Systems

Much work has been done on a particular kind of reaction–diffusion system known as a Λ–Ω system (Kopell and Howard, 1973; Cohen *et al.*, 1978; Greenberg, 1978). Although these systems have been thought of as unrealistic, close to an ordinary supercritical Hopf bifurcation of the diffusionless system any reaction–diffusion system may be approximated by a Λ–Ω system. We show here how this result is obtained by the method of averaging. It may also be proved by two-timing methods (Duffy *et al.*, 1980; Hagan, 1981). A general reaction–diffusion system may be written

$$\mathbf{u}_t = \mathbf{f}(\lambda, \mathbf{u}) + D\, \nabla^2 \mathbf{u} \tag{7.153}$$

where $\mathbf{u} = (u_1, u_2, \ldots, u_m)^{\mathrm{T}}$ is an m-vector of, e.g. concentrations, $\mathbf{f}(\lambda, \mathbf{u})$ is the vector of the reaction terms, D is the matrix of diffusion coefficients (which we take to be constant for simplicity only) and λ is a bifurcation parameter. All variables and parameters are as usual non-dimensional. We assume that the spatial variation is slow, so that to a first approximation $\mathbf{u}$

satisfies the kinetic equations

$$\mathbf{u}_t = \mathbf{f}(\lambda, \mathbf{u}) \tag{7.154}$$

We assume without loss of generality that $\mathbf{u} = \mathbf{0}$ is a steady-state solution of these equations for all λ, that it is stable for $\lambda < 0$, and that it becomes unstable *via* an ordinary Hopf bifurcation as λ increases past 0. Thus if $\mathbf{f}$ is split into its linear and non-linear parts

$$\mathbf{f}(\lambda, \mathbf{u}) = \mathbf{f}_{\mathbf{u}}(\lambda, \mathbf{0} | \mathbf{u}) + \mathbf{N}(\lambda, \mathbf{u}) = M(\lambda)\mathbf{u} + \mathbf{N}(\lambda, \mathbf{u}) \tag{7.155}$$

then all the eigenvalues of the matrix $M(\lambda)$ have negative real part if $\lambda < 0$, and two complex conjugate eigenvalues cross the imaginary axis at $\lambda = 0$. Let these two eigenvalues be $\sigma_1(\lambda) = \sigma(\lambda)$ and $\sigma_2(\lambda) = \bar{\sigma}(\lambda)$ and the remaining ones be $\sigma_3(\lambda), \ldots, \sigma_m(\lambda)$. Then

$$\sigma(\lambda) = \xi(\lambda) + i\eta(\lambda)$$

where

$$\xi(0) = 0, \qquad \eta(0) = \omega_0 \neq 0, \qquad \xi'(0) > 0 \tag{7.156}$$

The first two of these ensure that σ and $\bar{\sigma}$ are purely imaginary at the bifurcation point, and the third is a transversality condition as required for the Hopf bifurcation theorem (Theorem 6.144).

We may decompose any $\mathbf{u} \in \mathbb{R}^m$ as

$$\mathbf{u}^{\mathrm{T}} = (\mathbf{u}_P^{\mathrm{T}}, \mathbf{u}_Q^{\mathrm{T}})$$

according to the spectrum of $M(\lambda)$, so that $\mathbf{u}_P \in P$, where P is the eigenspace corresponding to the eigenvalues σ and $\bar{\sigma}$, and $\mathbf{u}_Q \in Q$, where Q corresponds to $\sigma_3, \ldots, \sigma_m$. Then we may assume that

$$M(\lambda) = \begin{bmatrix} M_P(\lambda) & O(\lambda) \\ O(\lambda) & M_Q(\lambda) \end{bmatrix} \tag{7.157}$$

where

$$M_P(\lambda) = \begin{bmatrix} \xi(\lambda) & -\eta(\lambda) \\ \eta(\lambda) & \xi(\lambda) \end{bmatrix}$$

and the eigenvalues of $M_Q(\lambda)$ all have negative real part. Since we are considering a bifurcation from the zero solution, then all elements of $\mathbf{u}_Q$ will remain exponentially small and we may neglect them in any asymptotic solution. Thus the kinetic system reduces to one with two components,

$$\frac{\partial \mathbf{u}_P}{\partial t} = M_P(\lambda)\mathbf{u}_P + \mathbf{N}_P(\lambda, \mathbf{u}_P)$$

where $\mathbf{N}_P$ represents the nonlinear terms. Hopf's bifurcation theorem (6.144) ensures the existence close to $\lambda = 0$ of small amplitude oscillatory solutions of the system of equations whose amplitude is $O(\sqrt{|\lambda|})$. We therefore define a small parameter ε by

$$\varepsilon^2 = |\lambda|$$

Since we are assuming that the bifurcation is supercritical, then the oscillatory solutions guaranteed by the Hopf bifurcation theorem exist for $\lambda > 0$ and are stable. Thus we may take

$$\varepsilon^2 = \lambda$$

Since $\mathbf{u}_P$ is $O(\varepsilon)$, we look for a solution

$$\mathbf{u}_P = \varepsilon \hat{\mathbf{u}}_P$$

Since we have assumed slow spatial variations, as in previous cases, we must also re-scale the space variable. Thus we define

$$\mathbf{x} = \varepsilon \hat{\mathbf{x}}$$

so that ∇^2 with respect to $\mathbf{x}$ becomes $\varepsilon^2 \hat{\nabla}^2$ with respect to $\hat{\mathbf{x}}$. We also define D_P to be the 2×2 matrix of diffusion coefficients of $\mathbf{u}_P$.

We now drop the subscript P and the hats for simplicity of notation to obtain the system

$$\mathbf{u}_t = M(\lambda)\mathbf{u} + \varepsilon^{-1} N(\lambda, \mathbf{u}) + \varepsilon^2 D\,\nabla^2 \mathbf{u} \tag{7.158}$$

where

$$M(\lambda) = \begin{bmatrix} \xi(\lambda) & -\eta(\lambda) \\ \eta(\lambda) & \xi(\lambda) \end{bmatrix} \tag{7.159}$$

Expanding $\mathbf{N}$ about $(0, \mathbf{0})$ we have

$$\mathbf{N}(\lambda, \mathbf{u}) = \mathbf{N}_{\mathbf{uu}}(0, \mathbf{0}|\mathbf{u}|\mathbf{u})\varepsilon^2/2 + \mathbf{N}_{\mathbf{uuu}}(0, \mathbf{0}|\mathbf{u}|\mathbf{u}|\mathbf{u})\varepsilon^3/6 + O(\varepsilon^4) \tag{7.160}$$

The equations become

$$\begin{aligned} \mathbf{u}_t = {} & M(0)\mathbf{u} + \varepsilon \mathbf{N}_{\mathbf{uu}}(0, \mathbf{0}|\mathbf{u}|\mathbf{u})/2 \\ & + \varepsilon^2\{M'(0)\mathbf{u} + \mathbf{N}_{\mathbf{uuu}}(0, \mathbf{0}|\mathbf{u}|\mathbf{u}|\mathbf{u})/6 + D\,\nabla^2\mathbf{u}\} + O(\varepsilon^3) \end{aligned} \tag{7.161}$$

where

$$M(0) = \begin{bmatrix} \xi(0) & -\eta(0) \\ \eta(0) & \xi(0) \end{bmatrix} = \begin{bmatrix} 0 & -\omega_0 \\ \omega_0 & 0 \end{bmatrix} \tag{7.162}$$

and

$$M'(0) = \begin{bmatrix} \xi'(0) & -\eta'(0) \\ \eta'(0) & \xi'(0) \end{bmatrix}$$

The system with $\varepsilon=0$ is

$$\mathbf{u}_t = M(0)\mathbf{u} \tag{7.163}$$

which has solution

$$\mathbf{u} = a\boldsymbol{\phi}_0 + \bar{a}\bar{\boldsymbol{\phi}}_0 \tag{7.164}$$

where $\boldsymbol{\phi}_0 = e^{i\omega_0 t}(1, -i)^{\mathrm{T}}$, and a is a complex constant. We therefore look for solutions of the same form with $a(t, \mathbf{x})$ a slowly varying function of t. The slow spatial variation has been built into the equations by the re-scaling of $\mathbf{x}$. The equations become

$$\begin{aligned} a_t\boldsymbol{\phi}_0 + i\omega_0 a\boldsymbol{\phi}_0 + \bar{a}_t\bar{\boldsymbol{\phi}}_0 - i\omega_0\bar{a}\bar{\boldsymbol{\phi}}_0 &= i\omega_0 a\boldsymbol{\phi}_0 - i\omega_0\bar{a}\bar{\boldsymbol{\phi}}_0 + \varepsilon\mathbf{N}_{\mathbf{uu}}(0, \mathbf{0}|\mathbf{u}|\mathbf{u})/2 \\ &+ \varepsilon^2\{\sigma'(0)(a\boldsymbol{\phi}_0 + \bar{a}\bar{\boldsymbol{\phi}}_0) + \mathbf{N}_{\mathbf{uuu}}(0, \mathbf{0}|\mathbf{u}|\mathbf{u}|\mathbf{u})/6 + D\,\nabla^2\mathbf{u}\} + O(\varepsilon^3) \end{aligned}$$

or

$$\begin{aligned} a_t\boldsymbol{\phi}_0 + \bar{a}_t\bar{\boldsymbol{\phi}}_0 &= \varepsilon\mathbf{N}_{\mathbf{uu}}(0, \mathbf{0}|\mathbf{u}|\mathbf{u})/2 + \varepsilon^2\{\sigma'(0)(a\boldsymbol{\phi}_0 + \bar{a}\bar{\boldsymbol{\phi}}_0) \\ &+ \mathbf{N}_{\mathbf{uuu}}(0, \mathbf{0}|\mathbf{u}|\mathbf{u}|\mathbf{u})/6 + D\nabla^2\mathbf{u}\} + O(\varepsilon^3) \end{aligned} \tag{7.165}$$

Taking the scalar product with $\boldsymbol{\phi}_0^* = \frac{1}{2}e^{i\omega_0 t}(1, -i)^{\mathrm{T}}$, this becomes

$$\begin{aligned} a_t &= \varepsilon\langle\mathbf{N}_{\mathbf{uu}}(0, \mathbf{0}|\mathbf{u}|\mathbf{u}), \boldsymbol{\phi}_0^*\rangle/2 + \varepsilon^2\{\sigma'(0)a \\ &+ \langle\mathbf{N}_{\mathbf{uuu}}(0, \mathbf{0}|\mathbf{u}|\mathbf{u}|\mathbf{u}), \boldsymbol{\phi}_0^*\rangle/6 + \langle D\,\nabla^2\mathbf{u},\ \boldsymbol{\phi}_0^*\rangle\} + O(\varepsilon^3) \end{aligned} \tag{7.166}$$

where

$$\begin{aligned} \langle D\,\nabla^2\mathbf{u}, \boldsymbol{\phi}_0^*\rangle &= (D_{11} - iD_{12} + iD_{21} + D_{22})\,\nabla^2 a/2 \\ &+ (D_{11} + iD_{12} - iD_{21} - D_{22})\,\nabla^2 a e^{-2i\omega_0 t}/2 \end{aligned} \tag{7.167}$$

We now apply the averaging technique. That is, we take an average of the equations over a period $2\pi/\omega_0$ and use the assumption that a is a slowly varying function of t. It follows that a and its derivatives will not change significantly over a period, so that we may take them to be effectively constants (considering a constant) so that $\langle\mathbf{N}_{\mathbf{uu}}(0, \mathbf{0}|\mathbf{u}|\mathbf{u}), \boldsymbol{\phi}_0^*\rangle$ consists of The expression $\mathbf{N}_{\mathbf{uu}}(0, \mathbf{0}|\mathbf{u}|\mathbf{u})$ consists of terms in $e^{2i\omega_0 t}$ and $e^{-2i\omega_0 t}$ and constants (considering a constant) so that $\langle\mathbf{N}_{\mathbf{uu}}(0, \mathbf{0}|\mathbf{u}|\mathbf{u}) \cdot \boldsymbol{\phi}_0^*\rangle$ consists of terms in $e^{i\omega_0 t}$, $e^{-i\omega_0 t}$ and $e^{-3i\omega_0 t}$, whose integral over a period is zero. Hence on averaging the quadratic terms will disappear. The expression $\mathbf{N}_{\mathbf{uuu}}(0, \mathbf{0}|\mathbf{u}|\mathbf{u}|\mathbf{u})$ consists of terms in $a^3e^{3i\omega_0 t}$, $a^2\bar{a}e^{i\omega_0 t}$, $a\bar{a}^2e^{-i\omega_0 t}$ and $\bar{a}^3e^{-3i\omega_0 t}$, so that $\langle\mathbf{N}_{\mathbf{uuu}}(0, \mathbf{0}|\mathbf{u}|\mathbf{u}|\mathbf{u}), \boldsymbol{\phi}_0^*\rangle$ consists of terms in $a^3e^{2i\omega_0 t}$, $a^2\bar{a}$, $a\bar{a}^2e^{-2i\omega_0 t}$ and $\bar{a}^3e^{-4i\omega_0 t}$. On averaging, only the terms in $a^2\bar{a}$ will survive. Similarly, the constant term $\sigma'(0)a$ in (7.166) and the non-periodic diffusional term in (7.167) will remain, so that the equation becomes

$$a_t = \varepsilon^2\{\sigma'(0)a + ca^2\bar{a} + \delta\,\nabla^2 a\} \tag{7.168}$$

to leading order [cf. (7.114) and (7.144)], where $c = c_1 + ic_2$ is a complex constant and

$$\delta = (D_{11} - iD_{12} + iD_{21} + D_{22})/2 \tag{7.169}$$

We have assumed that the spatially uniform equations

$$a_t = \varepsilon^2\{\sigma'(0)a + ca^2\bar{a}\} \tag{7.170}$$

exhibit a supercritical Hopf bifurcation. But

$$(|a|^2)_t = a_t\bar{a} + a\bar{a}_t = 2\varepsilon^2\{\text{Re } \sigma'(0)|a|^2 + \text{Re } c|a|^4\} \tag{7.171}$$

Since $\text{Re } \sigma'(0) = \xi'(0) > 0$ (7.156), it follows from the supercriticality assumption that

$$\text{Re } c = c_1 < 0 \tag{7.172}$$

By defining $a = a_1 + ia_2$, $\delta = \delta_1 + i\delta_2$, so that

$$\delta_1 = (D_{11} + D_{22})/2, \qquad \delta_2 = (D_{21} - D_{12})/2 \tag{7.173}$$

the real and imaginary parts of (7.168) give

$$\begin{aligned} a_{1t} &= \varepsilon^2\{\Lambda(|a|^2)a_1 - \Omega(|a|^2)a_2\} + \delta_1 \nabla^2 a_1 - \delta_2 \nabla^2 a_2 \\ a_{2t} &= \varepsilon^2\{\Omega(|a|^2)a_1 + \Lambda(|a|^2)a_2\} + \delta_2 \nabla^2 a_1 + \delta_1 \nabla^2 a_2 \end{aligned} \tag{7.174}$$

where

$$\Lambda(|a|^2) = \xi'(0) + c_1|a|^2 \tag{7.175}$$

$$\Omega(|a|^2) = \eta'(0) + c_2|a|^2 \tag{7.176}$$

Note that $\Lambda(|a|^2)$ in Sections 7.6 to 7.10 corresponds to $\Lambda(|a|^2) + i\Omega(|a|^2)$ here.

This is an extension of the definition of a λ-ω system given by Kopell and Howard (1973), since they required $\delta_2 = 0$. This is rather unrealistic, since it requires that the original system (7.153), after the transformation that allows it to be written in canonical form (i.e. in the form (7.155) with $M(\lambda)$ given by (7.157)), has no cross-diffusional terms. This is a restrictive condition because such a transformation, which is in general necessary, is a rotation which normally introduces cross-diffusion. A particular case in which cross-diffusion is not introduced is that for which the original diffusion matrix is a scalar, which remains a scalar after rotation. However, this will not happen for reaction–diffusion systems in practice, since no two different substances have exactly equal diffusion coefficients.

Kopell and Howard introduced their λ-ω systems as simple systems with an explicit plane-wave solution. Our extended Λ-Ω system also has such a solution, given by

$$|a| = r_0, \qquad a_1 = r_0 \cos(p\varepsilon^2 t - \boldsymbol{\alpha} \cdot \mathbf{x}), \qquad a_2 = r_0 \sin(p\varepsilon^2 t - \boldsymbol{\alpha} \cdot \mathbf{x}) \quad (7.177)$$

where

$$\delta_1 \alpha^2 = \Lambda(r_0^2), \qquad p = \Omega(r_0^2) - \delta_2 \alpha^2 \quad (7.178)$$

and $\alpha = |\boldsymbol{\alpha}|$. Note that the waves may have any amplitude r_0 between 0 and $r_c = \sqrt{-\xi'(0)/c_1}$. At r_c we have $\Lambda(r_c^2) = 0$, so that the wave number $\alpha = 0$, and the plane waves reduce to a limit cycle solution. Since $\Lambda'(r_c^2) < 0$ this solution is stable and represents the Hopf bifurcation solution for the spatially uniform case. We have proved

THEOREM 7.179: Close to an ordinary supercritical Hopf bifurcation of the diffusionless system any reaction-diffusion system in which spatial variation is slow may be approximated by the Λ-Ω system (7.174), which has a plane-wave solution given by (7.177) with (7.178).

Spiral-wave solutions of (7.174) may be constructed by a formal asymptotic procedure (Hagan, 1982). These solutions are plausibly argued to model the spiral pattern formation which is observed in the development of the slime mould *dictyostelium discoideum* (Hagan and Cohen, 1981). However, as these authors point out, it seems likely that the spiral waves which arise in the Belousov–Zhabotinskii reaction (Winfree, 1972) generally occur in an excitable medium rather than close to a Hopf bifurcation point. Spiral waves in excitable media are discussed by Greenberg and Hastings (1978) and Mikhailov and Krinsky (1983), and computer simulations have been carried out by Madore and Freedman (1983). The stability of rotating chemical waves is discussed by Erneux (1981).

8. Singular Perturbations

8.1 Introduction

Singular perturbations were defined in the last chapter (Definition 7.5), and a class of perturbations were discussed whose singularity manifests itself as $t \to \infty$, although they are regular in any finite time interval. In this chapter we discuss perturbations which are more singular in the sense that the singularity manifests itself immediately at $t = 0$. Such problems occur when one of the variables in the system has the capacity to change much more quickly than another. Taking a second order system of ordinary differential equations for simplicity, as usual in non-dimensional variables, such a system has the form

$$\varepsilon \dot{u} = f(u, v), \qquad \dot{v} = g(u, v) \tag{8.1}$$

where f and g are $O(1)$ quantities, that is they are neither small nor large for general values of u and v, ε is a small parameter, and the dot represents differentiation with respect to t. The variable v changes on an $O(1)$ time scale, i.e., $\dot{v}$ is an $O(1)$ quantity, but u is a quicker variable. For if f is $O(1)$, then $\dot{u} = f/\varepsilon = O(1/\varepsilon)$, and u changes on a time scale of $O(\varepsilon)$. This is true unless $f(u, v)$ happens to be close to zero, and so in general u will adjust itself very quickly until this occurs. We are excluding here the possibility that u increases or decreases without limit, as is reasonable in all realistic models of biological systems.

The singular nature of such a problem is easily seen. We know from Picard's theorem (Theorem 2.1) that the system as it stands, given initial conditions $u(0) = u_0$, $v(0) = v_0$, has a solution in the neighbourhood of $t = 0$ as long as f and g satisfy Lipschitz conditions in u and v. However, the unperturbed ($\varepsilon = 0$) system clearly does not possess a solution unless the initial conditions satisfy $f(u_0, v_0) = 0$. This is to be expected since on setting $\varepsilon = 0$ we have reduced the order of the system from second to first, while still retaining two initial conditions.

In the next section we consider a problem in enzyme kinetics, in which u quickly adjusts itself so that $f(u, v)$ is small, and f then remains small indefinitely. The solution is obtained in two parts, the so-called inner solution representing the quick adjustment of u, and the outer solution, the part in which f is small. The inner and outer solution are matched together using an asymptotic matching procedure. We derive general conditions on f and g which ensure that this kind of solution is appropriate. We then discuss problems in which f does not remain small indefinitely after it has become so. The solution in this case is obtained by a rather more complicated matching procedure. Problems like this may exhibit threshold behaviour and solitary travelling wave solutions. They have applications to the cyclic AMP control system in the slime mould *dictyostelium discoideum*, trigger waves in the Belousov-Zhabotinskii reaction, and in nerve conduction. In the last case the threshold behaviour relates to the fact that a nerve impulse will not be triggered unless the stimulus reaches a certain level. The solitary travelling wave represents a wave of excitation travelling along the nerve axon followed by a refractory period and finally relaxation to the steady state, so that the nerve is again amenable to the passage of another impulse. Solitary travelling waves in reaction–diffusion systems may also be analysed by topological methods (Conley and Gardner, 1984; Gardner, 1982, 1984).

8.2 Michaelis–Menten Theory

As a practical example of such a system, let us consider a simple enzyme reaction scheme proposed by Michaelis and Menten (1913). The equations have been analysed from a singular perturbation point of view by Heineken *et al.* (1967); see also Lin and Segel (1974) and Murray (1977).

The basic assumptions of Michaelis and Menten are that an enzyme E catalyses the conversion of a substrate S to a product P by first forming a complex ES. This complex may break down to form E and S again, or to form E and P. Diagrammatically,

$$\mathrm{E}+\mathrm{S} \underset{k_{-1}}{\overset{k_1}{\rightleftharpoons}} \mathrm{ES} \xrightarrow{k_2} \mathrm{E}+\mathrm{P}$$

The k's are the rate constants of the reactions. By the law of mass action this gives

$$d[\mathrm{E}]/d\tau = -k_1[\mathrm{E}][\mathrm{S}] + k_{-1}[\mathrm{ES}] + k_2[\mathrm{ES}]$$

$$d[\mathrm{S}]/d\tau = -k_1[\mathrm{E}][\mathrm{S}] + k_{-1}[\mathrm{ES}]$$

$$d[\mathrm{ES}]/d\tau = k_1[\mathrm{E}][\mathrm{S}] - k_{-1}[\mathrm{ES}] - k_2[\mathrm{ES}]$$

$$d[\mathrm{P}]/d\tau = k_2[\mathrm{ES}]$$

where square brackets denote concentrations. The initial conditions we shall consider are applicable to the case in which substrate is added to free enzyme, i.e.

$$[\mathrm{E}]=\mathrm{E}_0, \qquad [\mathrm{S}]=\mathrm{S}_0, \qquad [\mathrm{ES}]=0, \qquad [\mathrm{P}]=0$$

at $\tau=0$. The equation for [P] is uncoupled from the system, that is if the first three equations are solved for [E], [S] and [ES], then [P] may be obtained by integrating the fourth. Adding the first and third equations gives

$$d([\mathrm{E}]+[\mathrm{ES}])/d\tau=0$$

which with the initial conditions gives

$$[\mathrm{E}]+[\mathrm{ES}]=\mathrm{E}_0 \tag{8.2}$$

This is essentially a law of conservation of the enzyme, i.e. the total (free and bound) enzyme concentration is constant. Using this equation we may eliminate [E] from the system. We also non-dimensionalise by defining

$$\begin{aligned} u&=[\mathrm{ES}]/\mathrm{E}_0, \qquad & v&=[\mathrm{S}]/\mathrm{S}_0, \qquad & t&=k_1\mathrm{E}_0\tau, \\ \kappa&=(k_{-1}+k_2)/k_1\mathrm{S}_0, \qquad & \lambda&=k_2/k_1\mathrm{S}_0, \qquad & \varepsilon&=\mathrm{E}_0/\mathrm{S}_0 \end{aligned} \tag{8.3}$$

to obtain

$$\varepsilon\dot{u}=v-(v+\kappa)u, \qquad \dot{v}=-v+(v+\kappa-\lambda)u \tag{8.4}$$

The initial conditions are

$$u(0)=0, \qquad v(0)=1 \tag{8.5}$$

In most biological situations there is very little enzyme present compared to the amount of substrate, that is $\varepsilon=\mathrm{E}_0/\mathrm{S}_0$ is a small number. This system is therefore of the form considered in the introduction, with the substrate the slow variable and the enzyme-substrate complex the fast one. Thus we expect the (non-dimensionalised) concentration u of this complex (the saturation of the enzyme) to change very rapidly until $f(u,v)\equiv v-(v+\kappa)u=0$. The pseudo-steady state hypothesis in biochemistry consists of assuming that

$$v-(v+\kappa)u=0 \tag{8.6}$$

everywhere, so that u can be written in terms of v and a single differential equation for v obtained, namely

$$\dot{v}=-v+(v+\kappa-\lambda)u=-v+\frac{(v+\kappa-\lambda)v}{v+\kappa}=-\frac{\lambda v}{v+\kappa} \tag{8.7}$$

with initial condition

$$v(0)=1 \tag{8.8}$$

This equation can be integrated to obtain an implicit equation for v from which v may be found by numerical methods,

$$v+\kappa \log v=1-\lambda t \tag{8.9}$$

and u is given by

$$u=v/(v+\kappa) \tag{8.10}$$

This solution is widely used and gives excellent agreement with experiment in very many cases. However, there is a difficulty with the initial conditions, since the solution is obtained by setting $\varepsilon=0$ and thus reducing the order of the system. It can be seen that $v(0)=1$ as required, but the initial value of u, which should be zero, is given by $u(0)=(1+\kappa)^{-1}$. To investigate the nature of the solution near $t=0$ we consider the phase plane (see Fig. 8.11). The effect of the $\varepsilon\dot{u}$ term is that all parts of trajectories

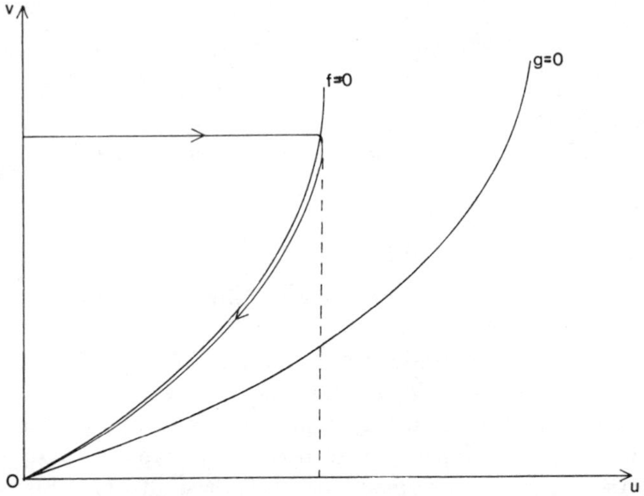

Fig. 8.11. The phase plane for the system (8.4) with the trajectory corresponding to initial conditions (8.5).

except those close to the line $f=0$ are almost flat; they are accomplished in times $O(\varepsilon)$, so that u may change appreciably but v cannot. Hence, with the initial conditions $u(0)=0$, $v(0)=1$, the trajectory shown in the figure is followed. Concentration u increases rapidly until $f=0$ is reached, and $u=(1+\kappa)^{-1}$. Then the line $f=0$ is followed to the origin, where all substrate has been converted to product ($v=0$) and there is no substrate–enzyme complex. Plotted against time, u and v are as shown in Fig. 8.12.

The pseudo-steady-state hypothesis gives the part of the solution after the dotted line, which is at some time $O(\varepsilon)$. It should therefore be expected

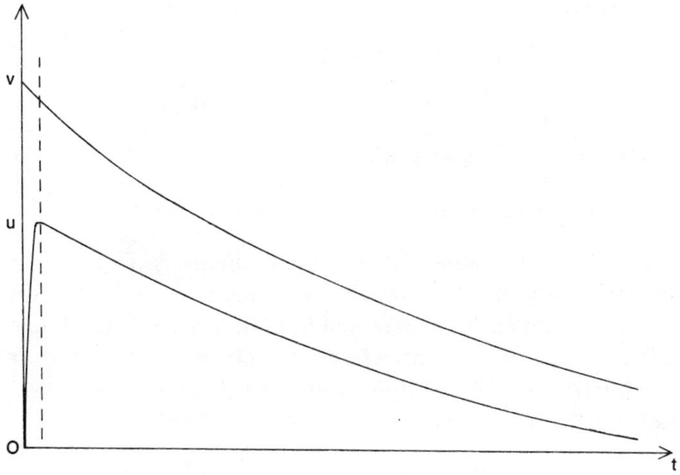

Fig. 8.12. Solutions of (8.4) with (8.5).

to give good agreement with experiment as long as no experimental measurements are made in the initial short period. In most practical applications this period is so short that no measurements can possibly be made there (it is normally about one second), and therefore the hypothesis adequately models the experimental results.

The mathematical technique used to obtain the solution for all times, including the initial short time period, to any required degree of accuracy is known as the method of matched asymptotic expansions. This method is discussed in the books on asymptotics by Nayfeh (1973, 1981), O'Malley (1974), Murray (1974a), van Dyke (1975) and Kevorkian and Cole (1981); see also Murray (1977, Appendix I) for a concise exposition and Lin and Segel (1974). Matched asymptotic expansions are used when the solution has a different character for different ranges of the independent variable. In Michaelis–Menten theory this variable is t and there are two distinct parts to the solution, the first when t is small, $O(\varepsilon)$, and the second when t is $O(1)$. The idea is to obtain asymptotic expansions in each of the regions, and then match them together to produce a smooth solution which is valid for all times. We shall obtain the leading order terms of such a representation of the solution.

The region close to $t = 0$ is known as the inner domain, and the solution there is the inner solution; the rest of the line $t > 0$ is known as the outer domain, with the outer solution. In the outer domain we have already stated (and this will be proved in the next section) that f is approximately zero. Here we may seek solutions for u and v in the form of asymptotic power series in ε, that is

$$\begin{aligned} u(t;\varepsilon) &\approx u_0(t)+\varepsilon u_1(t)+\cdots \\ v(t;\varepsilon) &\approx v_0(t)+\varepsilon v_1(t)+\cdots \end{aligned} \tag{8.13}$$

By substituting into Eqs. (8.4) and equating terms of $O(1)$, we obtain

$$0=v_0-(v_0+\kappa)u_0, \qquad \dot{v}_0=-v_0+(v_0+\kappa-\lambda)u_0 \tag{8.14}$$

These equations are the pseudo-steady state equations. In contrast to the simple pseudo-steady state hypothesis, this method provides correction terms u_1, v_1, etc., which may be obtained by equating terms of higher order. Also, no initial conditions are imposed on the solutions in this case. Instead, it will be required that the solutions found here match onto the solutions in the inner domain. The solutions are given (implicitly) by

$$v_0+\kappa \log v_0=A-\lambda t, \qquad u_0=v_0/(v_0+\kappa) \tag{8.15}$$

where A is a constant to be determined from the matching.

We must now obtain the solutions in the inner domain. Here a power series of the form given above is invalid, because of the problem of applying both initial conditions. It is clear from the previous discussion that f is not close to zero here, and u is changing on a fast $O(\varepsilon)$ time scale. To describe this mathematically, we must introduce a new fast time variable. Since t is $O(\varepsilon)$ in this domain, and we require a variable which is $O(1)$, we define $s=t/\varepsilon$. Then putting $U(s;\varepsilon)=u(t;\varepsilon)$, $V(s;\varepsilon)=v(t;\varepsilon)$, the equations become

$$\frac{dU}{ds}=V-(V+\kappa)U, \qquad \frac{1}{\varepsilon}\frac{dV}{ds}=-V+(V+\kappa-\lambda)U \tag{8.16}$$

Again we look for solutions as power series in ε, i.e.

$$\begin{aligned} U(s;\varepsilon) &\approx U_0(s)+\varepsilon U_1(s)+\cdots \\ V(s;\varepsilon) &\approx V_0(s)+\varepsilon V_1(s)+\cdots \end{aligned} \tag{8.17}$$

Since these solutions are valid in the initial time period, initial conditions are imposed. These are

$$U(0;\varepsilon)=0, \qquad V(0;\varepsilon)=1 \tag{8.18}$$

which gives, on equating powers of ε,

$$\begin{aligned} &U_n(0)=0 \qquad \text{for all } n\geq 0 \\ &V_0(0)=1, \qquad V_n(0)=0 \qquad \text{for all } n\geq 1 \end{aligned} \tag{8.19}$$

The leading order terms of the equations are

$$dU_0/ds = V_0 - (V_0 + \kappa)U_0, \qquad dV_0/ds = 0 \tag{8.20}$$

Hence $V_0 \equiv 1$ and $dU_0/ds = 1 - (1+\kappa)U_0$, so that

$$U_0 = \frac{1}{1+\kappa}\{1 - \exp(-(1+\kappa)s)\}, \qquad V_0 = 1 \tag{8.21}$$

Again the higher-order corrections may be obtained by equating terms in higher powers of ε in the equations.

We now have leading order terms of asymptotic expansions for u and v in both domains. It is necessary that these solutions match together, in other words there must be an overlap region where both asymptotic expansions are valid. In this region t will be between $O(\varepsilon)$ and $O(1)$, i.e. $\varepsilon \ll t \ll 1$, and therefore s will be between $O(1)$ and $O(1/\varepsilon)$, i.e. $1 \ll s \ll 1/\varepsilon$. But the outer solution holds when $t = O(1)$, so in the intermediate region the (first-order) outer solution is $\lim_{t\to 0}(u_0(t), v_0(t))$. Similarly, the (first-order) inner solution is $\lim_{s\to\infty}(U_0(s), V_0(s))$, and so for matching we require

$$\lim_{t\to 0}(u_0(t), v_0(t)) = \lim_{s\to\infty}(U_0(s), V_0(s)) \tag{8.22}$$

From (8.21),

$$\lim_{s\to\infty}(U_0(s), V_0(s)) = ((1+\kappa)^{-1}, 1) \tag{8.23}$$

This gives the first-order terms of the intermediate expansion. It can be seen from (8.15) that $\lim_{t\to 0}(u_0(t)v_0(t)) = ((1+\kappa)^{-1}, 1)$ if $A = 1$, so that matching is accomplished with this value of A.

We have thus obtained leading order terms of asymptotic expansions for u and v which are valid in different regions and match together between those regions. To obtain an expansion which is valid for all time, a uniformly valid expansion, the general procedure is to add the inner and outer expansions together and subtract the common part, i.e. the solution in the intermediate region. For then, taking u as an example,

$$u_{\text{uniform}} \approx u_{\text{inner}} + u_{\text{outer}} - u_{\text{intermediate}} \tag{8.24}$$

In the inner region $t \ll 1$ so that the outer expansion is replaced by its limit as $t \to 0$, and the second and third terms cancel out by definition of the intermediate expansion; thus $u_{\text{uniform}} \approx u_{\text{inner}}$, as required. Similarly, in the outer region $s \gg 1$ so that u_{inner} is replaced by its limit as $s \to \infty$, and cancels with $u_{\text{intermediate}}$, so that $u_{\text{uniform}} \approx u_{\text{outer}}$, as required. In the Michaelis–Menten case the first term in the uniformly valid expansion for v, $v_{0,\text{uniform}}$, is given implicitly by

$$v_{0,\text{uniform}} + \kappa \log v_{0,\text{uniform}} = 1 - \lambda t \tag{8.25}$$

and $u_{0,\text{uniform}}$ by

$$u_{0,\text{uniform}} = v_{0,\text{uniform}}/(v_{0,\text{uniform}} + \kappa) - (1+\kappa)^{-1} \exp(-(1+\kappa)t/\varepsilon) \tag{8.26}$$

Note as a check that the initial conditions are satisfied and that u and v tend to the pseudo-steady state solutions for $t = O(1)$.

8.3 Neglect of the Fast Variable Derivative Term

In the Michaelis–Menten system, neglect of the term $\varepsilon\dot{u}$ (the pseudo-steady state hypothesis) gave excellent agreement except in a small initial time period. With more complicated systems this is not always the case. Consider the phase planes of Fig. 8.27, which again refer to a system of the form

$$\varepsilon\dot{u} = f(u, v), \qquad \dot{v} = g(u, v) \tag{8.28}$$

Again the $\varepsilon\dot{u}$ term means that the parts of the trajectories away from the line, or nullcline, $f = 0$ are almost horizontal. From any initial point, u will change rapidly until the line $f = 0$ is reached, and the trajectory will then remain close to this nullcline for a time. However, in neither of these cases does it remain close for *all* time, as it did in the Michaelis–Menten system. For when the trajectories reach the point marked D on either diagram, g is negative and v must therefore continue to decrease. But this means that the trajectory can no longer remain on the nullcline $f = 0$. Hence, $\dot{u}$ becomes large (and positive in these cases), and so u changes quickly until the other branch of the nullcline is reached. Here g is positive, so v increases until either the steady state S is reached (in Fig. 8.27a) or until a similar process occurs at C (in Fig. 8.27b). In Fig. 8.27b the oscillation continues and in fact we have a limit cycle solution.

The question therefore arises of when the fast variable may be neglected except in the short initial period. This is equivalent to asking whether, if once $f(u, v)$ becomes small, it remains small for all time. It turns out that it does unless there is some autocatalysis of the fast variable.

THEOREM 8.29: If $(f(u, v))^2 < \varepsilon$ at some time $t = t_0$, f_v and g are $O(1)$ or smaller, and $f_u(u, v) < -k < 0$ for some $k = O(1)$ and for all values of u and v, then $(f(u, v))^2 < \varepsilon$ for all $t \geq t_0$, for ε sufficiently small.

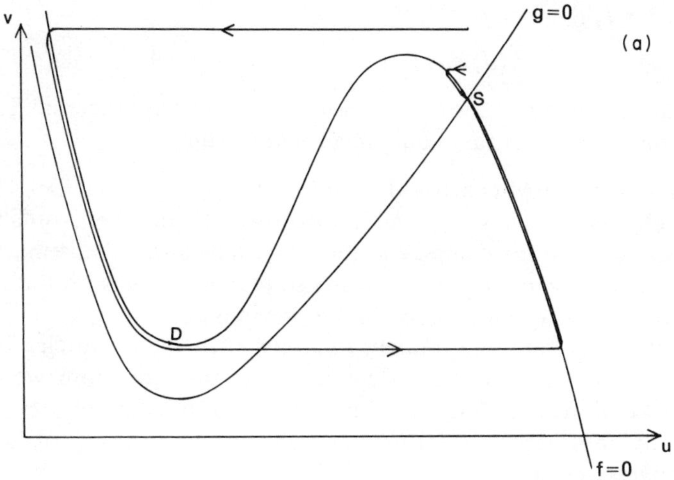

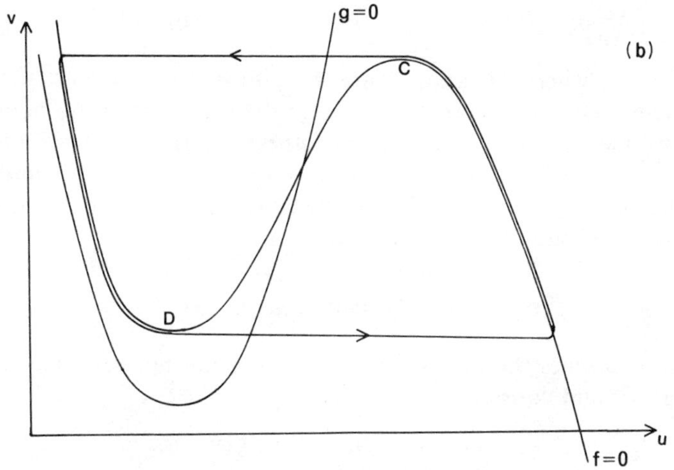

Fig. 8.27. Possible phase planes for the system (8.28).

REMARK 8.30: Autocatalysis of u is defined to occur when an increase in the concentration results in an increase in the rate of production of u. Thus the condition $f_u < 0$ ensures that there is no autocatalysis of the fast variable.

Proof: If the theorem does not hold, then there is a T such that $(f(u, v))^2 = \varepsilon$ and $(d/dt)(f(u, v))^2 \geqslant 0$ at $t = T$. But

$$\tfrac{1}{2}(d/dt)(f(u, v))^2 = f(f_u\dot{u} + f_v\dot{v}) = f(f_u f/\varepsilon + f_v g) = \varepsilon^{-1}f^2 f_u + ff_v g$$

Then at $t=T$, $f^2=\varepsilon$ and

$$\tfrac{1}{2}(d/dt)(f(u,v))^2=f_u\pm\sqrt{\varepsilon}f_v g<0$$

for ε sufficiently small, using the assumptions on the orders of f_u, f_v and g. But this contradicts the requirement that $(d/dt)(f^2)\geq 0$.

Note that in the Michaelis–Menten system $f_u(u,v)=-(v+\kappa)<-k<0$, k is an $O(1)$ quantity, and f_v and g are also $O(1)$. In other words there is no autocatalysis of the complex u; the conditions of the theorem hold and the only place where the pseudo-steady-state hypothesis is invalid is in the initial short period, as discussed in the previous section.

If a trajectory follows a nullcline closely, then there will clearly be no oscillatory solutions of the system of equations. Therefore we see the importance of autocatalysis in the production of oscillations.

We may consider the diffusional system in the same way. Addition of diffusional terms gives

$$\varepsilon u_t=f(u,v)+\varepsilon D_u\nabla^2 u,\qquad v_t=g(u,v)+D_v\nabla^2 v$$

in $\Omega\times(0,\infty)$, where D_u and D_v are the diffusion coefficients of u and v respectively. Let us assume that u, v and their derivatives are bounded in $\Omega\times(0,\infty)$ (see Chapter 5), and the hypotheses of Theorem 8.29 hold. Then the fast variable may again be neglected after a short $O(\varepsilon)$ initial time period as long as $f_u<-k<0$, if ε is sufficiently small. The proof is analogous to that for the spatially uniform case.

8.4 Threshold Phenomena

We now discuss the case when $f(u,v)$ does not remain small if once it becomes so, and consider the system

$$\varepsilon u_t=f(u,v),\qquad v_t=g(u,v) \tag{8.31}$$

where f and g are as shown in Fig. 8.32. The figure shows the case in which u and v are always positive (they may be concentrations or animal populations), but a similar argument holds if this restriction is not imposed. It can be seen that the steady state $S=(u_s, v_s)$ is stable, and in fact it can be shown to be asymptotically stable throughout the positive quadrant. In other words, as long as $u(0)>0$, $v(0)>0$, then $u(t)\to u_s$ and $v(t)\to v_s$ as $t\to\infty$. However, the nature of this approach to the steady state may be very different for initial conditions which are very close.

Consider the initial conditions represented by the point A in the figure. Here f is $O(1)$ and negative, so u decreases very quickly to attain the nullcline $f=0$. For these initial conditions this happens on the right-hand

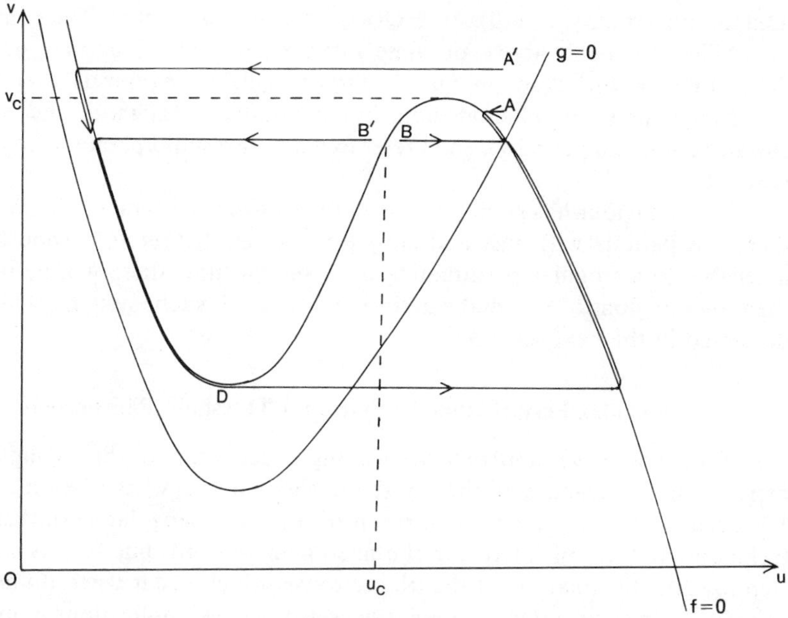

Fig. 8.32. Phase plane of a system with a threshold phenomenon.

branch of $f=0$ close to the steady state, which is then approached along the nullcline. Note that $f_u<0$ here so that there is no autocatalysis and the trajectory remains close to $f=0$. On the other hand, if the initial conditions are at A', then u again decreases quickly, but in this case the trajectory does not meet the nullcline $f=0$ close to the steady state but on the left-hand branch. This branch remains stable ($f_u<0$) until the point D is reached (where $f_u=0$) when u increases quickly to the right-hand branch of f. Since f is positive between the two branches, $f_u>0$ for the first part of this quick increase. The trajectory then remains close to this branch until the steady state is attained.

It can be seen that there is a threshold value of v, v_c, below which a quick return to the steady state occurs and above which there is a large excursion before an eventual return to the steady state. This is known as a threshold phenomenon. Similar behaviour occurs with perturbations in u, in which initial conditions at B, with $u>u_c$, result in a rapid return to the steady state, whereas starting at B', with $u<u_c$, the trajectory makes an excursion before returning.

The results stated above are true in the limit as $\varepsilon \to 0$, where the trajectories away from $f=0$ are perfectly horizontal. For the true situation with $\varepsilon>0$ this is only an approximation, and the trajectories have slopes $O(\varepsilon)$. This means that the threshold values of u and v are not sharp, and there is an area of uncertainty if $v(0)=v_c+O(\varepsilon)$, or $u(0)=u_c+O(\varepsilon)$. Here either possibility discussed above or some intermediate behaviour may occur, depending on the exact initial conditions. This behaviour is therefore not a true threshold phenomenon, but it approximates more and more closely to it as $\varepsilon \to 0$, and we shall refer to it as a threshold phenomenon if ε is small.

The behaviour has similarities with the Michaelis–Menten case in that there are periods with fast and slow time scales. It therefore should be amenable to a singular perturbation analysis, but now there will be more than two regions. The mathematical analysis of such systems will be discussed in the next section.

8.5 Singular Perturbation Analysis of a Threshold Phenomenon

In this section we construct the leading order terms in the asymptotic expansions for u and v in the threshold phenomenon of the last section. We do this not only as an application of a general singular perturbation technique and out of interest in the phenomenon itself, but also as a first step towards the analysis of the trigger waves which occur when diffusion is included in such systems. These trigger waves have applications in many fields, including physiology (nerve condution), chemistry, biochemistry and biology (e.g. slime mould aggregation). We shall follow the analysis of Britton (1982), but see also Keener (1980), Fife (1976b, 1982), Nishiura (1979, 1982), Mimura *et al.* (1980), Hosono and Mimura (1982), Ito (1984b), Ermentrout and Rinzel (1981) and Rinzel and Terman (1982).

The equations we consider are those of the last section, (8.31), with f and g as in Fig. 8.32. More rigorously, we require that the system satisfy the following conditions:

CONDITIONS 8.33: (i) f and g are C^1 functions of u and v.

(ii) $f \geq 0$ on $u=0$, $g \geq 0$ on $v=0$, which ensures that the positive quadrant is an invariant set for the system.

(iii) There is a unique steady state $S=(u_s, v_s)$, which is stable.

(iv) $f_v<0$.

(v) Let H be defined in Remark 8.34. Then there exist constants u_d, u_c such that $H'(u)<0$ for $u \in (0, u_d) \cup (u_c, \infty)$ and $H'(u)>0$ for $u \in (u_d, u_c)$, or equivalently, since $f_u+f_v H'(u)=0$ on $f=0$, $f_u<0$ for $u \in (0, u_d) \cup (u_c, \infty)$ and $f_u>0$ for $u \in (u_d, u_c)$ on $f=0$.

(vi) $v_c=H(u_c)>v_s>v_d=H(u_d)>0$.

(vii) Let h_1 and h_3 be as defined in Remark 8.35. Then $u_d'=h_3(v_d)$ and $u_c'=h_1(v_c)$ exist and are positive.

(viii) $\int_{h_1(v_s)}^{h_3(v_s)} f(u, v_s)\, du<0$ if $u_s>u_c$ and $\int_{h_1(v_s)}^{h_3(v_s)} f(u, v_s)\, ds>0$ if $u_s<u_d$.

REMARK 8.34: It follows from (iv) and the implicit function theorem (6.5) that the equation $f(u, v)=0$ may be solved for v in terms of u, $v=H(u)$, say. This function $H(u)$ is assumed to satisfy (v) and (vi).

REMARK 8.35: It follows from (v) and the implicit function theorem that the equation $f(u, v)=0$ may be solved for u in terms of v if u is restricted to one of the three intervals $(0, u_d]$, $[u_d, u_c]$, and $[u_c, \infty)$. Let the corresponding functions be $u=h_1(v)$, $u=h_2(v)$ and $u=h_3(v)$. Then $u_d=h_1(v_d)=h_2(v_d)$, $u_c=h_2(v_c)=h_3(v_c)$.

REMARK 8.36: Since the steady state is stable, the trace of the linearised matrix of the reaction terms is negative there, i.e. $f_u(u_s, v_s)+\varepsilon g_v(u_s, v_s)<0$. To highest order this implies that $f_u(u_s, v_s)<0$, which with (v) gives $u_s>u_c$ or $u_s<u_d$. It can be shown that if $u_d<u_s<u_c$, the system has an unstable steady state and admits a limit cycle solution.

REMARK 8.37: Since $f(u_c', v_c)=0$ and $f_u(u_c', v_c)<0$, we have $f(u, v_c)<0$ in (u_c', u_c). Thus, if $u_s=u_c$, then $v_s=v_c$ and

$$\int_{h_1(v_s)}^{h_3(v_s)} f(u, v_s)\, du=\int_{u_c'}^{u_c} f(u, v_c)\, du<0$$

Since the integral is a continuous function of v_s, then the first alternative of (viii) always holds if $u_s-u_c>0$ is small enough. Similarly, the second of (viii) holds if $u_d-u_s>0$ is small enough. We shall take $u_s>u_c$ here; the analysis for $u_s<u_d$ is similar.

REMARK 8.38: If (iv) is replaced by (iv′): $f_v>0$, the analysis still holds if we replace (vi) by $0<v_c=H(u_c)<v_s<v_d=H(u_d)$.

REMARK 8.39: The conditions have been stated for a reaction–diffusion system, in which u and v are necessarily positive. The analysis also holds for more general systems in which u and v are allowed to be negative if 0 is replaced by $-\infty$ in (v). The positivity requirements of (ii), (vi) and (vii) are then no longer necessary.

The phase plane for the system is shown in Fig. 8.40. We may determine the sign of g above and below the line $g=0$ by considering its derivative as we move along the nullcline $f=0$ from S. At S, $g=0$. On $f=0$ we have

$$dg/du=g_u+g_v\, dv/du=g_u-g_v f_u/f_v=-(f_u g_v-f_v g_u)/f_v$$

But stability of S requires that the determinant $f_u g_v-f_v g_u$ be positive at S, and f_v is known to be negative (Condition 8.33(iv)), so that $dg/du>0$. It

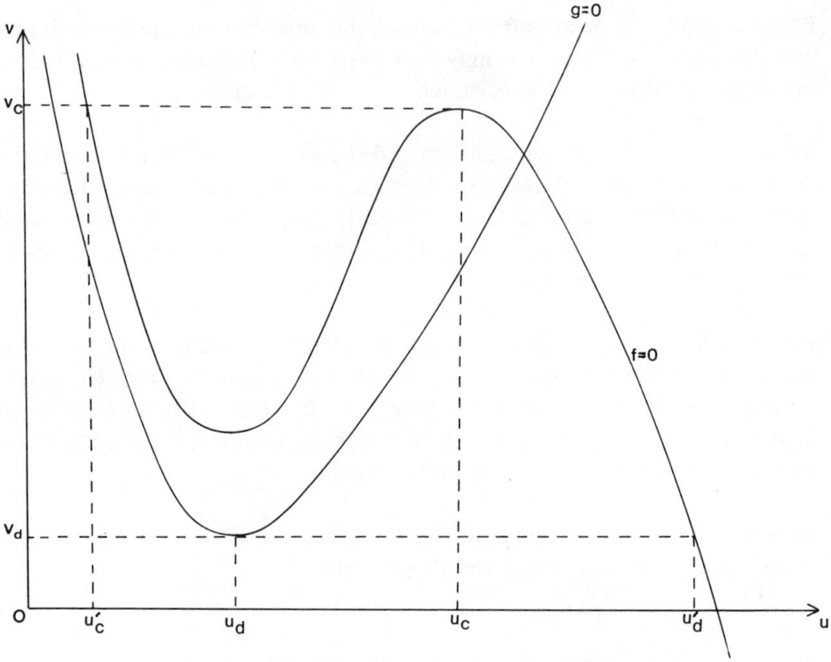

Fig. 8.40. Phase plane for the system (8.31).

follows that $g>0$ on $f=0$ for $u>u_s$, and by continuity that $g>0$ everywhere on that side of the line $g=0$. Similarly, $g<0$ on the opposite side. In particular,

REMARK 8.41: $g(h_1(v), v)<0$, and $g(h_3(v), v)>0$ for $v>v_s$ and <0 for $v_s<v<v_c$.

We shall need some preliminary results on the large time behaviour of solutions of certain differential equations.

LEMMA 8.42: Let u be the solution of $u_\tau = f(u, \bar{v})$, where $\bar{v}$ is a constant and f satisfies the conditions of this section. Then

$$u(\tau) \to \begin{cases} h_1(\bar{v}) & \text{as} \quad \tau\to\infty \quad \text{if} \quad 0<u(0)<h_2(\bar{v}) \\ h_3(\bar{v}) & \text{as} \quad \tau\to\infty \quad \text{if} \quad h_2(\bar{v})<u(0) \\ h_2(\bar{v}) & \text{as} \quad \tau\to-\infty \quad \text{if} \quad h_1(\bar{v})<u(0)<h_3(\bar{v}) \end{cases}$$

Proof: $f>0$ for $u\in(0, h_1(\bar{v}))\cup(h_2(\bar{v}), h_3(\bar{v}))$ and $f<0$ for $u\in(h_1(\bar{v}), h_2(\bar{v}))\cup(h_3(\bar{v}), \infty)$, and u correspondingly increases or decreases.

The solution of the differential equation is given implicitly by

$$\tau = \int_{u(0)}^{u} \frac{d\phi}{f(\phi, \bar{v})}$$

so that as $\tau \to \pm\infty$ u tends to a root of $f=0$. The result follows.

LEMMA 8.43: Let v be the solution of

$$v_t = g(h_3(v), v)$$

with g and h_3 as defined above, and let

$$v(0) = v_0 < v_c$$

Then $v(t) \to v_s$ as $t \to \infty$.

Proof: The result follows from Remark 8.41 and consideration of the implicit solution

$$t = \int_{v(0)}^{v} \frac{d\psi}{g(h_3(\psi), \psi)}$$

as $t \to \infty$.

We now construct an asymptotic expansion of the solution of the spatially uniform system

$$\varepsilon u_t = f(u, v), \qquad v_t = g(u, v) \tag{8.44}$$

We shall consider an initial perturbation in u from the steady state. Other perturbations may be treated similarly. We first take initial conditions

$$u(0) = u_a > h_2(v_s), \qquad v(0) = v_s \tag{8.45}$$

That is, u is perturbed from the steady state, but to a value which is still to the right of the part of the nullcline $f=0$ with positive gradient (see Fig. 8.46). It is obvious that we are dealing with a singular perturbation problem, and that we must define a short time scale by $\tau = t/\varepsilon$. Then

$$u_\tau = f(u, v), \qquad v_\tau = \varepsilon g(u, v) \tag{8.47}$$

We look for solutions in the form

$$\begin{aligned} u(\tau) &\approx u_0(\tau) + \varepsilon u_1(\tau) + \cdots \\ v(\tau) &\approx v_0(\tau) + \varepsilon v_1(\tau) + \cdots \end{aligned} \tag{8.48}$$

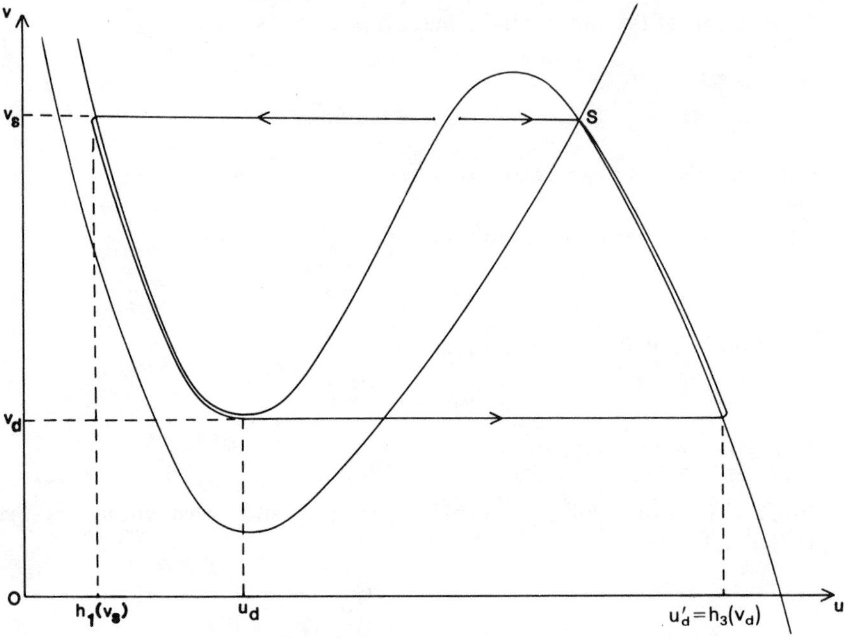

Fig. 8.46. Trajectories for (8.44) with initial conditions given by (8.45) and (8.50).

Substituting into (8.47), we have

$$\begin{aligned}(u_0+\varepsilon u_1+\cdots)_\tau &= f(u_0+\varepsilon u_1+\cdots, v_0+\varepsilon v_1+\cdots)\\ &= f(u_0, v_0)+O(\varepsilon)\\ (v_0+\varepsilon v_1+\cdots)_\tau &= \varepsilon g(u_0+\varepsilon u_1+\cdots, v_0+\varepsilon v_1+\cdots)\\ &= \varepsilon g(u_0, v_0)+O(\varepsilon^2)\end{aligned}$$

To $O(1)$ these give

$$u_0' = f(u_0, v_0), \qquad v_0' = 0$$

Thus $v \equiv v_s$, and u_0 satisfies

$$u_0' = f(u_0, v_s) \tag{8.49}$$

From Lemma 8.42 with $v = v_s$, u_0, given implicitly by

$$\tau = \int_{u_a}^{u_0} \frac{d\phi}{f(\phi, v_s)}$$

satisfies $u_0 \to h_3(v_s) = u_s$ as $\tau \to \infty$, and we therefore have an immediate return to the stable steady state.

If however the initial conditions are given by

$$u(0)=u_b, 0<u_b<h_2(v_s), \qquad v(0)=v_s \tag{8.50}$$

so that the perturbation is beyond the middle branch of $f=0$, then a similar analysis gives $v_0 \equiv v_s$ but from Lemma 8.42, u_0, given implicitly by

$$\tau=\int_{u_b}^{u_0} \frac{d\phi}{f(\phi, v_s)}$$

satisfies $u_0 \to h_1(v_s)$ as $\tau \to \infty$. The point $(h_1(v_s), v_s)$ is on the nullcline $f=0$, so that no more development takes place on the τ scale, but is not a steady state of the system, so that the trajectory cannot end here. Further development must take place on the original long time scale, so that we consider the Eqs. (8.44). We seek a solution in the form

$$\begin{aligned} u &= U(t) \approx U_0(t)+\varepsilon U_1(t)+\cdots \\ v &= V(t) \approx V_0(t)+\varepsilon V_1(t)+\cdots \end{aligned} \tag{8.51}$$

By substituting these into the system, we obtain

$$\begin{aligned} \varepsilon(U_0'+\varepsilon U_1'+\cdots) &= f(U_0+\varepsilon U_1+\cdots, V_0+\varepsilon V_1+\cdots) \\ V_0'+\varepsilon V_1'+\cdots &= g(U_0+\varepsilon U_1+\cdots, V_0+\varepsilon V_1+\cdots) \end{aligned} \tag{8.52}$$

Since the solution in the previous region tends to a point $(h_1(v_s), v_s)$, matching is straightforward. We require that the solution of the large time equations have initial conditions, or more strictly matching conditions as $t \to 0$, given by $u=h_1(v_s)$, $v=v_s$, that is

$$\begin{aligned} U_0(0) &= h_1(v_s), \qquad U_i(0)=0, \qquad i \geq 1 \\ V_0(0) &= v_s, \qquad V_i(0)=0, \qquad i \geq 1 \end{aligned} \tag{8.53}$$

Any intermediate time scale then has solution $u=h_1(v_s)$, $v=v_s$, and matching is accomplished.

The first-order terms in (8.52) are given by

$$0=f(U_0, V_0), \qquad V_0'=g(U_0, V_0) \tag{8.54}$$

The first of these has solution

$$U_0=h_1(V_0) \tag{8.55}$$

which satisfies the matching condition if V_0 does. Then V_0 satisfies

$$V_0'=g(h_1(V_0), V_0)$$

so that V_0 is given implicitly by

$$t = \int_{v_s}^{V_0} \frac{d\psi}{g(h_1(\psi), \psi)} \tag{8.56}$$

The solution here is on the left branch of the nullcline $f=0$, where $g<0$ (Remark 8.44) and in fact $g(h_1(V_0), V_0) < -k < 0$ for some constant k. Thus V_0 will decrease without limit or until $h_1(V_0)$ becomes undefined. But since $h_1(v)$ is defined for $v_d \leq v \leq v_s$ and not for $v < v_d$, $h_1(V_0)$ becomes undefined when V_0 drops below v_d. We can then no longer have $f=0$, so that the asymptotic expansion in terms of t breaks down. The condition obtained in Section 8.3 that a trajectory of (8.43), once in the vicinity of $f=0$, remains there for all time was that $f_u < 0$. Note that this condition breaks down at (u_d, v_d), since here we have $f_u = 0$. Inspection of the phase plane leads us to expect that u will increase quickly, and we define $\tilde{\tau} = (t - t_0)/\varepsilon$, where t_0 is some time in the period of rapid change of u. The equations become

$$u_{\tilde{\tau}} = f(u, v), \qquad v_{\tilde{\tau}} = \varepsilon g(u, v) \tag{8.57}$$

and we seek solutions in the form

$$\begin{aligned} u &= \tilde{u}(\tilde{\tau}) \approx \tilde{u}_0(\tilde{\tau}) + \varepsilon \tilde{u}_1(\tilde{\tau}) + \cdots \\ v &= \tilde{v}(\tilde{\tau}) \approx \tilde{v}_0(\tilde{\tau}) + \varepsilon \tilde{v}_1(\tilde{\tau}) + \cdots \end{aligned} \tag{8.58}$$

By substituting these expansions into Eqs. (8.57) and taking the first-order terms, we obtain

$$\tilde{u}_0' = f(\tilde{u}_0, \tilde{v}_0), \qquad \tilde{v}_0' = 0 \tag{8.59}$$

We require that the rapidly changing solution in this region match onto the slower one of the previous region. This means that matching must take place as $\tilde{\tau} \to -\infty$ and as $t \to t_d$, where t_d is the value of t at the end of the slowly varying period, where $u = u_d$ and $v = v_d$. Thus, to first order,

$$\lim_{t \to t_d} (U_0, V_0) = \lim_{\tilde{\tau} \to -\infty} (\tilde{u}_0, \tilde{v}_0) \tag{8.60}$$

But by definition of t_d,

$$\lim_{t \to t_d} (U_0, V_0) = (u_d, v_d)$$

so we require

$$\lim_{\tilde{\tau} \to -\infty} (\tilde{u}_0, \tilde{v}_0) = (u_d, v_d) \tag{8.61}$$

The solution for $\tilde{v}_0$ with this matching condition is

$$\tilde{v}_0 = v_d \tag{8.62}$$

so that $\tilde{u}_0$ satisfies

$$\tilde{u}_0' = f(\tilde{u}_0, v_d) \tag{8.63}$$

Since $\tilde{u}_0$ cannot remain at u_d for all time, as (u_d, v_d) is not a critical point, and $\tilde{u}_0' = f(\tilde{u}_0, v_d) > 0$ for $u \in (0, h_1(v_d)) \cup (h_2(v_d), h_3(v_d)) = (0, u_d) \cup (u_d, h_3(v_d))$, then $\tilde{u}_0$ must increase. Thus, since $\tilde{\tau} = 0$ is in the period of rapid change of u, we have $\tilde{u}_0(0) > u_d$, and the solution of the differential equation (8.63) for $\tilde{u}_0$ is given implicitly by

$$\tilde{\tau} = \int_{\tilde{u}_0(0)}^{\tilde{u}_0} \frac{d\phi}{f(\phi, v_d)} \tag{8.64}$$

Now Lemma 8.42 states that as $\tilde{\tau} \to -\infty$, $\tilde{u}_0 \to h_2(v_d) = u_d$, as required for matching, and that as $\tilde{\tau} \to \infty$, $\tilde{u}_0 \to h_3(v_d) = u_d'$.

The trajectory again tends to a stable branch of the nullcline $f = 0$ as $\tilde{\tau} \to \infty$, so it is again necessary to change to the slow time variable t. We seek solutions of (8.44) in the form

$$\begin{aligned} u &= \tilde{U}(t) \approx \tilde{U}_0(t) + \varepsilon \tilde{U}_1(t) + \cdots \\ v &= \tilde{V}(t) \approx \tilde{V}_0(t) + \varepsilon \tilde{V}_1(t) + \cdots \end{aligned} \tag{8.65}$$

and the equations become, to first order,

$$0 = f(\tilde{U}_0, \tilde{V}_0), \qquad \tilde{V}_0' = g(\tilde{U}_0, \tilde{V}_0) \tag{8.66}$$

On the t time scale, the time of entry to this region is $t_d + O(\varepsilon)$, or asymptotically just t_d. Thus for matching we require, to first order,

$$\lim_{\tilde{\tau} \to \infty} (\tilde{u}_0, \tilde{v}_0) = \lim_{t \to t_d^+} (\tilde{U}_0, \tilde{V}_0) \tag{8.67}$$

so that we must solve the equations with

$$\tilde{U}_0(t_d) = h_3(v_d) = u_d', \qquad \tilde{V}_0(t_d) = v_d \tag{8.68}$$

The solution $\tilde{U}_0 = h_3(\tilde{V}_0)$ satisfies the first equation of (8.66), so that

$$\tilde{V}_0' = g(h_3(\tilde{V}_0), \tilde{V}_0)$$

which has solution given implicitly by

$$t - t_d = \int_{v_d}^{\tilde{V}_0} \frac{d\psi}{g(h_3(\psi), \psi)}$$

These solutions also satisfy the matching conditions (8.68). By Lemma 8.43 $\tilde{V}_0 \to v_s$ as $t \to \infty$, so that $\tilde{U}_0 \to h_3(v_s) = u_s$, and the system returns to the stable steady state.

We have thus shown that the system exhibits a threshold phenomenon in its asymptotic behaviour as $\varepsilon \to 0$. That is, if the system is at the stable steady state and u is perturbed to a value $u_a > h_2(v_s)$, then there is an immediate return to the steady state. However, if the perturbation is to $u_b < h_2(v_s)$, then the system executes a large excursion before eventual return to the steady state. A similar statement is true for a perturbation in v, where the threshold is v_c.

8.6 Solitary Travelling Waves: The Leading Wave Front

In the next three sections we show how the threshold phenomenon just analysed may lead to solitary travelling wave solutions when diffusion is included in the system. For simplicity we consider the one-dimensional case, so that the equations become

$$\varepsilon u_t = f(u, v) + \varepsilon^2 u_{xx}, \qquad v_t = g(u, v) + \varepsilon\beta v_{xx} \tag{8.69}$$

where β is the ratio of the diffusion coefficients of v and u. We consider the Cauchy problem with initial conditions

$$u(x, 0) = \begin{cases} u_b < h_2(v_s), & x < 0 \\ u_s, & x > 0 \end{cases} \tag{8.70}$$

$$v(x, 0) = v_s$$

that is, we perturb u below its threshold value for x negative. Note that the steady state is still stable as long as β is an $O(1)$ quantity, which we assume to be the case, so that we expect eventual return to the steady state at any given point in space.

The perturbation is again singular in time, so that we transform to the fast time variable $\tau = t/\varepsilon$, and the equations become

$$u_\tau = f(u, v) + \varepsilon^2 u_{xx}, \qquad v_\tau = \varepsilon g(u, v) + \varepsilon^2 \beta v_{xx} \tag{8.71}$$

Initially, away from $x = 0$, diffusion is unimportant and the solution follows the spatially homogeneous trajectories. Close to $x = 0$, u changes by an $O(1)$ amount over a short distance so that diffusion is important. We therefore make the transformation

$$\xi = x/\varepsilon \tag{8.72}$$

and the equations become

$$u_\tau = f(u, v) + u_{\xi\xi}, \qquad v_\tau = \varepsilon g(u, v) + \beta v_{\xi\xi} \tag{8.73}$$

Essentially the problem in the initial short time period involves solving the

Cauchy problem for these equations with initial conditions

$$u(\xi, 0) = \begin{cases} u_b < h_2(v_s), & \xi < 0 \\ u_s, & \xi > 0 \end{cases} \tag{8.74}$$

$$v(\xi, 0) = v_s$$

But the solution for v is clearly

$$v(\xi, \tau) = v_s \tag{8.75}$$

so that

$$u_\tau = f(u, v_s) + u_{\xi\xi} \tag{8.76}$$

with the given initial conditions. Note that $f(u, v_s)$ satisfies

$$f(u, v_s)\begin{cases} >0 & \text{in} \quad (0, h_1(v_s)) \cup (h_2(v_s), h_3(v_s)) \\ <0 & \text{in} \quad (h_1(v_s), h_2(v_s)) \cup (h_3(v_s), \infty) \end{cases}$$

and that

$$\limsup_{\xi \to -\infty} u(\xi, 0) < h_2(v_s)$$

$$\liminf_{\xi \to \infty} u(\xi, 0) > h_2(v_s)$$

Equation (8.76) is thus a generalised Nagumo equation (Chapter 4), and it follows from Theorem 4.76 that there exists a travelling wave solution $u(\xi, \tau) = U(\xi - c\tau)$ (where c is the unique wave speed and is a positive constant), satisfying $U(\xi - c\tau) \to h_1(v_s)$ as $\xi \to -\infty$, and $U(\xi - c\tau) \to h_3(v_s) = u_s$ as $\xi \to \infty$. Moreover (Theorem 4.84),

$$u(\xi, \tau) \to U(\xi - \xi_0 - c\tau) \tag{8.77}$$

as $\tau \to \infty$ exponentially in time, where ξ_0 is a constant. The positivity of c is ensured by Condition 8.33(viii), on integrating the differential equation

$$-cU' = f(U, v_s) + U'' \tag{8.78}$$

satisfied by U with respect to U from $h_1(v_s)$ to $h_3(v_s)$. It follows that the wave front moves in the direction of $\xi > 0$, in other words, towards the region of steady state concentrations.

This completes the small time solution of the problem. To continue the analysis it is necessary to return to the longer time scale and seek a solution which matches onto the travelling wave front solution we have found. Since this travelling wave front is stable for $v = v_s$ and is moving into a region where $v = v_s$ it is natural to assume that it remains stable and propagates into the region with speed c. Numerical solutions for various systems seem to confirm this assumption [see for example Britton and Murray (1979)]. It therefore becomes the leading wave front of what we shall show to be a solitary travelling wave.

8.7 Solitary Travelling Waves: Behaviour before the Formation of the Trailing Wave Front

After the formation of the leading wave front the spatial domain may be divided into four distinct regions, as shown in Fig. 8.79. Region II is the leading wave front, which occurs around $\xi - c\tau = \xi_0$ (by 8.77). In terms of the x variable this is at $x = ct$, and is a shock layer between region I, where

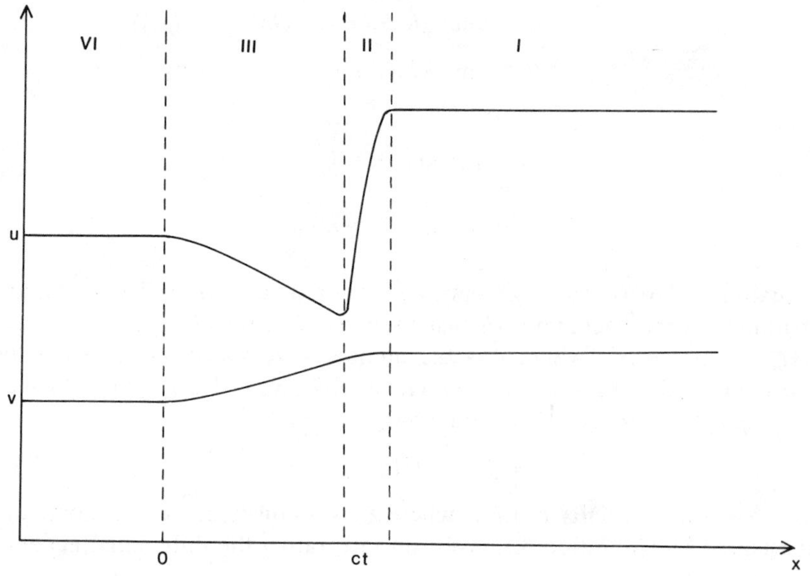

Fig. 8.79. Spatial variation in u and v before the formation of the trailing wave front.

the concentrations are at the steady state, and region III. Region III connects the tail of this shock to region VI, where the concentrations are unaffected by the wave front and follow the trajectories of the spatially uniform system. Regions IV and V, connected with the trailing wave front, have not yet been formed. We shall consider each of these regions in turn.

Region I, $x > ct$: Here the appropriate equations are

$$\varepsilon u_t = f(u, v) + \varepsilon^2 u_{xx}, \qquad v_t = g(u, v) + \varepsilon\beta v_{xx} \tag{8.80}$$

with initial conditions

$$u(x,0)=u_s, \qquad v(x,0)=v_s \tag{8.81}$$

Seeking a solution of the form

$$\begin{aligned} u &= u^{\mathrm{I}}(x,t) \approx u_0^{\mathrm{I}}(x,t)+\varepsilon u_1^{\mathrm{I}}(x,t)+\cdots \\ v &= v^{\mathrm{I}}(x,t) \approx v_0^{\mathrm{I}}(x,t)+\varepsilon v_1^{\mathrm{I}}(x,t)+\cdots \end{aligned} \tag{8.82}$$

we obtain

$$u_0^{\mathrm{I}}=u_s, \qquad v_0^{\mathrm{I}}=v_s, \qquad u_n^{\mathrm{I}}=v_n^{\mathrm{I}}=0, \qquad n\geqslant 1 \tag{8.83}$$

Region II, $x=ct$: Since this is a travelling wave front of thickness $O(\varepsilon)$, it is necessary to transform to the variables ξ and τ, and the equations become

$$u_\tau = f(u,v)+u_{\xi\xi}, \qquad v_\tau = \varepsilon g(u,v)+\beta v_{\xi\xi} \tag{8.84}$$

It is necessary to match this solution with the travelling wave front which develops in the initial time period. Thus we seek solutions of the form

$$\begin{aligned} u &= u^{\mathrm{II}}(\xi,\tau) \approx u_0^{\mathrm{II}}(\xi,\tau)+\varepsilon u_1^{\mathrm{II}}(\xi,\tau)+\cdots \\ v &= v^{\mathrm{II}}(\xi,\tau) \approx v_0^{\mathrm{II}}(\xi,\tau)+\varepsilon v_1^{\mathrm{II}}(\xi,\tau)+\cdots \end{aligned} \tag{8.85}$$

and obtain first-order solutions

$$u_0^{\mathrm{II}}(\xi,\tau)=U(\xi-\xi_0-c\tau), \qquad v_0^{\mathrm{II}}(\xi,\tau)=v_s \tag{8.86}$$

Note that the matching conditions with region I are automatically satisfied, and that $u_0^{\mathrm{II}}(\xi,\tau)\to h_1(v_s)$ as $\xi-c\tau\to-\infty$. This will provide a matching condition for region III.

Region III, $0<x<ct$: Here Eqs. (8.80) hold. We seek solutions in the form

$$\begin{aligned} u &= u^{\mathrm{III}}(x,t) \approx u_0^{\mathrm{III}}(x,t)+\varepsilon u_1^{\mathrm{III}}(x,t)+\cdots \\ v &= v^{\mathrm{III}}(x,t) \approx v_0^{\mathrm{III}}(x,t)+\varepsilon v_1^{\mathrm{III}}(x,t)+\cdots \end{aligned} \tag{8.87}$$

and the first-order terms become

$$f(u_0^{\mathrm{III}}, v_0^{\mathrm{III}})=0, \qquad \partial v_0^{\mathrm{III}}/\partial t = g(u_0^{\mathrm{III}}, v_0^{\mathrm{III}}) \tag{8.88}$$

Matching onto region II implies $u_0^{\mathrm{III}}\to h_1(v_s)$, $v_0^{\mathrm{III}}\to v_s$ as $x\to ct-$. The first of these with $f(u_0^{\mathrm{III}}, v_0^{\mathrm{III}})=0$ gives

$$u_0^{\mathrm{III}}=h_1(v_0^{\mathrm{III}}) \tag{8.89}$$

so that v_0^{III} satisfies

$$\partial v_0^{\mathrm{III}}/\partial t = g(h_1(v_0^{\mathrm{III}}), v_0^{\mathrm{III}}) \tag{8.90}$$

This has solution satisfying the matching condition $v_0^{\mathrm{III}} \to v_s$ as $x \to ct-$ given implicitly by

$$t - \frac{x}{c} = \int_{v_s}^{v_0^{\mathrm{III}}} \frac{d\psi}{g(h_1(\psi), \psi)} \tag{8.91}$$

This is well-defined for $x > 0$ as long as v_0^{III} remains greater than v_d, or $t < t_d$, where

$$t_d = \int_{v_s}^{v_d} \frac{d\psi}{g(h_1(\psi), \psi)}$$

For the moment we shall take this to be the case.

Region IV and V have not yet formed, as they are connected with the trailing wave front, so region III matches to region VI.

Region VI, $x < 0$: Here the equations are again given by (8.80), and the matching conditions from the small time solution are

$$u \to h_1(v_s), \qquad v \to v_s$$

as $t \to 0$. Seeking solutions of the form

$$\begin{aligned} u &= u^{\mathrm{VI}}(x, t) \approx u_0^{\mathrm{VI}}(x, t) + \varepsilon u_1^{\mathrm{VI}}(x, t) + \cdots \\ v &= v^{\mathrm{VI}}(x, t) \approx v_0^{\mathrm{VI}}(x, t) + \varepsilon v_1^{\mathrm{VI}}(x, t) + \cdots \end{aligned} \tag{8.92}$$

we have

$$u_0^{\mathrm{VI}} = h_1(v_0^{\mathrm{VI}}) \tag{8.93}$$

where v_0^{VI} is given implicity by

$$t = \int_{v_s}^{v_0^{\mathrm{VI}}} \frac{d\psi}{g(h_1(\psi), \psi)} \tag{8.94}$$

Note that this solution matches onto that of region III, and is defined as long as $t < t_d$. We have thus constructed the first terms of a matching asymptotic solution which is defined for all $t < t_d$.

8.8 Solitary Travelling Waves: Formation and Development of the Trailing Wave Front

At $t = t_d$ the concentration of u in region VI, which follows the spatially uniform solution of Section 8.5, jumps to a new value in a time $O(\varepsilon)$ on the t time scale. Thus a shock layer develops between region III and region VI; this is the trailing wave front. The development of this front is considered by Keener (1980). The concentration profiles after the formation of the shock are shown in Fig. 8.95. The position of the trailing wave front is not yet known; we shall take it to be at $x = y(t)$. The domain may now be divided into six regions. Regions I to III and VI are as before, except that III now occupies $y < x < ct$. Region IV is the trailing wave front and region V connects the tail of this to the spatially uniform region VI. We shall again consider the system asymptotically, taking each region in turn.

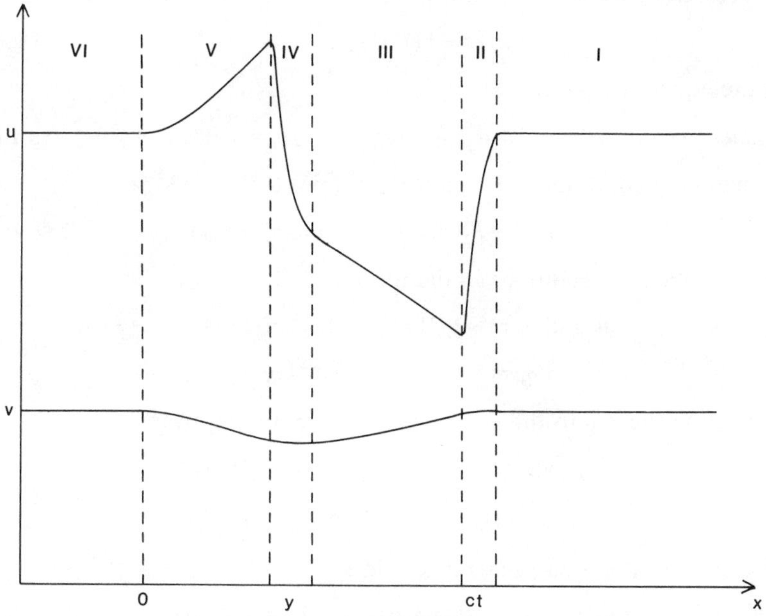

Fig. 8.95. Spatial variation in u and v after the formation of the trailing wave front.

In region I, $(x > ct)$, $u = u^{\mathrm{I}}(x, t)$ and $v = v^{\mathrm{I}}(x, t)$, where $u_0^{\mathrm{I}} = u_s$, $v_0^{\mathrm{I}} = v_s$. In region II $(x = ct)$, $u = u^{\mathrm{II}}(\xi, \tau)$ and $v = v^{\mathrm{II}}(\xi, \tau)$, where $u_0^{\mathrm{II}} = U(\xi - \xi_0 - c\tau)$, $v_0^{\mathrm{II}} = v_s$. In region III $(y(t) < x < ct)$, $u = u^{\mathrm{III}}(x, t)$ and $v = v^{\mathrm{III}}(x, t)$, where

$$u_0^{\mathrm{III}} = h_1(v_0^{\mathrm{III}}) \tag{8.96}$$

and v_0^{III} is given implicitly by

$$t - \frac{x}{c} = \int_{v_s}^{v_0^{\mathrm{III}}} \frac{d\psi}{g(h_1(\psi), \psi)} \tag{8.97}$$

Region III is defined to be the region where this representation is valid, so it follows that $v_0^{\mathrm{III}} \geqslant v_d$, or $t - x/c \leqslant t_d$ here. Thus $t - y(t)/c \leqslant t_d$, or $y(t) \geqslant$

$c(t-t_d)$. Let us define $\tilde{v}(t)$ to be the value of v_0^{III} at the end of this region, $\tilde{v}(t)=\lim_{x\to y^+} v_0^{\mathrm{III}}(x,t)$. Then $\tilde{v}$ is defined by

$$t-\frac{y}{c}=\int_{v_s}^{\tilde{v}} \frac{d\psi}{g(h_1(\psi),\psi)} \tag{8.98}$$

Region IV, $x=y$: This is the shock layer connecting values of u on one branch of the nullcline with values on the other. Since this region is of thickness $O(\varepsilon)$ we must transform the space variable. Hence we define

$$\eta=(x-y(t))/\varepsilon, \qquad \tau=t \tag{8.99}$$

and the equations become

$$\varepsilon u_\tau-\dot{y}(t)u_\eta=f(u,v)+u_{\eta\eta}, \qquad \varepsilon v_\tau-\dot{y}(t)v_\eta=\varepsilon g(u,v)+\beta v_{\eta\eta} \tag{8.100}$$

The matching conditions to region III require that

$$u(\eta,\tau)\to h_1(\tilde{v}(t)), \qquad v(\eta,\tau)\to\tilde{v}(t) \tag{8.101}$$

as $\eta\to\infty$. We seek solutions in the form

$$\begin{aligned} u=u(\eta,\tau)&\approx u_0^{\mathrm{IV}}(\eta,\tau)+\varepsilon u_1^{\mathrm{IV}}(\eta,\tau)+\cdots \\ v=v(\eta,\tau)&\approx v_0^{\mathrm{IV}}(\eta,\tau)+\varepsilon v_1^{\mathrm{IV}}(\eta,\tau)+\cdots \end{aligned} \tag{8.102}$$

Substituting these into the equations we have, to first order,

$$\begin{aligned} -\dot{y}(t)\,\partial u_0^{\mathrm{IV}}/\partial\eta&=f(u_0^{\mathrm{IV}},v_0^{\mathrm{IV}})+\partial^2 u_0^{\mathrm{IV}}/\partial\eta^2, \\ -\dot{y}(t)\,\partial v_0^{\mathrm{IV}}/\partial\eta&=\beta\,\partial^2 v_0^{\mathrm{IV}}/\partial\eta^2 \end{aligned} \tag{8.103}$$

The second of these has general solution

$$v_0^{\mathrm{IV}}(\eta,\tau)=A(\tau)+B(\tau)\exp(-\dot{y}(t)\eta/\beta) \tag{8.104}$$

which with the requirements that v_0^{IV} be bounded as $\eta\to\pm\infty$ and satisfy the matching condition as $\eta\to\infty$ gives

$$v_0^{\mathrm{IV}}(\eta,\tau)=\tilde{v}(t) \tag{8.105}$$

The equation for u_0^{IV} becomes

$$-\dot{y}(t)\,\partial u_0^{\mathrm{IV}}/\partial\eta=f(u_0^{\mathrm{IV}},\tilde{v}(t))+\partial^2 u_0^{\mathrm{IV}}/\partial\eta^2 \tag{8.106}$$

with $u_0^{\mathrm{IV}}(\eta,\tau)\to h_1(\tilde{v}(t))$ as $\eta\to\infty$, from (8.101). As $\eta\to-\infty$ then u_0^{IV} must tend to a solution of $f(u_0^{\mathrm{IV}},\tilde{v}(t))=0$, that is, to $h_i(v(t))$ for $i=1$, 2 or 3. If $i=1$, then there is no shock layer, and we know that $u=h_2(v)$ is unstable; thus

$$u_0^{\mathrm{IV}}(\eta,\tau)\to h_3(\tilde{v}(t)) \qquad \text{as} \quad \eta\to-\infty \tag{8.107}$$

This is a matching condition with region V. We shall complete the matching procedure before returning to a discussion of the unknown functions $\tilde{v}(t)$ (the value of v interior to the trailing shock) and $y(t)$ (the position of this shock). It will be shown that $\dot{y}(t)>0$, so that $x=y(t)$ may be inverted to obtain

$$t=y^{-1}(x) \tag{8.108}$$

the time at which the trailing shock passes position x.

Region V, $0<x<y(t)$: Here the equations are

$$\varepsilon u_t=f(u, v)+\varepsilon^2 u_{xx}, \qquad v_t=g(u, v)+\varepsilon\beta v_{xx} \tag{8.109}$$

We seek a solution of the form

$$\begin{aligned} u&=u^{\mathrm{V}}(x, t)\approx u_0^{\mathrm{V}}(x, t)+\varepsilon u_1^{\mathrm{V}}(x, t)+\cdots \\ v&=v^{\mathrm{V}}(x, t)\approx v_0^{\mathrm{V}}(x, t)+\varepsilon u_1^{\mathrm{V}}(x, t)+\cdots \end{aligned} \tag{8.110}$$

To first order, the equations become

$$f(u_0^{\mathrm{V}}, v_0^{\mathrm{V}})=0, \qquad \partial v_0^{\mathrm{V}}/\partial t=g(u_0^{\mathrm{V}}, v_0^{\mathrm{V}})$$

with matching conditions to region IV, from (8.105) and (8.107), given by $u_0^{\mathrm{V}}(x, t)\to h_3(\tilde{v}(t))$, $v_0^{\mathrm{V}}(x, t)\to \tilde{v}(t)$ as $x\to y(t)$ or as $t\to y^{-1}(x)$. Thus

$$u_0^{\mathrm{V}}(x, t)=h_3(v_0^{\mathrm{V}}(x, t)) \tag{8.112}$$

where $v_0^{\mathrm{V}}(x, t)$ is given implicitly by

$$t-y^{-1}(x)=\int_{\tilde{v}(t)}^{v_0^{\mathrm{V}}}\frac{d\psi}{g(h_3(\psi), \psi)} \tag{8.113}$$

Region VI, $x<0$: By using an argument similar to that for region V, we have

$$u_0^{\mathrm{VI}}(x, t)=h_3(v_0^{\mathrm{VI}}(x, t)) \tag{8.114}$$

where

$$t-t_d=\int_{v_d}^{v_0^{\mathrm{VI}}}\frac{d\psi}{g(h_3(\psi), \psi)} \tag{8.115}$$

Note that v_0^{VI} and hence u_0^{VI} are spatially independent, and follow the phase plane trajectory along the right-hand branch of the nullcline $f=0$ to the steady state.

It remains to consider the unknown functions y and $\tilde{v}$, which are connected by Eqs. (8.98) and (8.106). Writing $u=u_0^{\mathrm{IV}}$ for simplicity and prime for differentiation with respect to η, (8.106) becomes

$$-\dot{y}u'=f(u, \tilde{v})+u'' \tag{8.116}$$

with matching conditions, from (8.101) and (8.107),

$$u \to h_1(\tilde{v}) \quad \text{as } \eta \to \infty, \quad u \to h_3(\tilde{v}) \quad \text{as } \eta \to -\infty \tag{8.117}$$

Through this equation the value $\tilde{v}$ of v interior to the shock influences $\dot{y}$, the speed of the shock, which in turn determines $\tilde{v}$ through the Eq. (8.98), namely

$$t - \frac{y}{c} = \int_{v_s}^{\tilde{v}} \frac{d\psi}{g(h_1(\psi), \psi)}$$

There are two questions which need to be answered. First, does there exist a solitary travelling wave solution (implying the existence of a trailing wave front with speed equal to that of the leading wave front)? Second, does the solution of the initial value problem tend to such a solitary travelling wave? We now address ourselves to the first question, and consider the trailing wave front.

By differentiating (8.98) we have

LEMMA 8.118: $\tilde{v}$ is a constant if and only if $\dot{y} = c$.

We may now state a theorem on the dependence of $\dot{y}$ on $\tilde{v}$ through (8.116), in which c and $\tilde{v}$ are considered as parameters.

THEOREM 8.119: There exists a constant $c^* > 0$ such that

(i) if $\dot{y} = c$, where $c \geqslant c^*$, then Eq. (8.116) with the matching conditions (8.117) has a solution with $\tilde{v} = v_d$.

(ii) If $\tilde{v}$ is a given constant and satisfies $v_d < \tilde{v} < v_c$, then there is a unique constant c such that (8.116) with (8.117) has a solution with $\dot{y} = c$; moreover, $c < c^*$.

(iii) Conversely, if $\dot{y} = c$, where $0 < c < c^*$, then there exists a unique value of $\tilde{v}$ satisfying $v_d < \tilde{v} < v_c$ such that (8.116) with (8.117) has a solution.

Proof: The proof of (i) follows from Theorem 4.74, and the first statement of (ii) from Theorem 4.76. To show that if $v_d < \tilde{v} < v_c$, then $\dot{y} = c < c^*$, assume that the contrary holds, so that $\dot{y} = c \geqslant c^*$. Defining $u' = w$, (8.116) becomes

$$u' = w, \qquad w' = -cw - f(u, \tilde{v}) \tag{8.120}$$

and the matching conditions (8.117) become

$$(u, w) \to (h_1(\tilde{v}), 0) \quad \text{as } \eta \to \infty, \quad (u, w) \to (h_3(\tilde{v}), 0) \quad \text{as } \eta \to -\infty \tag{8.121}$$

Now the two phase planes for $\tilde{v} = v_d$, $c = c^*$, which has a solitary travelling wave by (i), and for $\tilde{v}$, $v_d < \tilde{v} < v_c$, $c \geqslant c^*$, which has a solitary travelling wave by assumption, are shown together in Fig. 8.122. It is clear that the trajectories $T(\tilde{v}, c)$ and $T(v_d, c^*)$ representing the travelling waves must cross at some point (u^+, w^+). But at this point the slopes of the two trajectories are given by

$$\left.\frac{dw}{du}\right|_{(\tilde{v},c)} = \frac{-cw - f(u, \tilde{v})}{w} = -\frac{cw^+ + f(u^+, \tilde{v})}{w^+}$$

for $T(\tilde{v}, c)$ and

$$\left.\frac{dw}{du}\right|_{(v_d,c^*)} = \frac{-cw - f(u, \tilde{v})}{w} = -\frac{c^* w^+ + f(u^+, v_d)}{w^+}$$

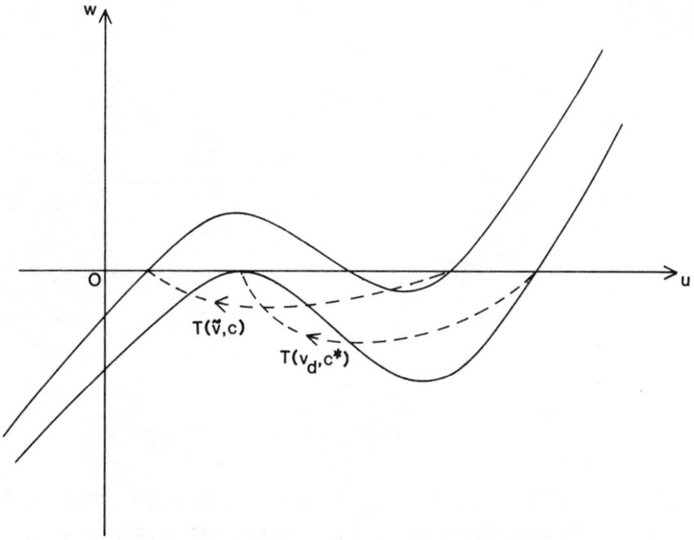

Fig. 8.122. Two phase planes for (8.120) superimposed, one with $v = v_d$, $c = c^*$, and the other with $v_d < v < v_c$, $c \geqslant c^*$.

for $T(v_d, c^*)$. It follows that

$$\left.\frac{dw}{du}\right|_{(\tilde{v},c)} - \left.\frac{dw}{du}\right|_{(v_d,c^*)} < 0$$

since, by Condition 8.33(iv), $f_v < 0$. But this contradicts the condition

$$\left.\frac{dw}{du}\right|_{(\tilde{v},c)} \geqslant \left.\frac{dw}{du}\right|_{(v_d,c^*)}$$

which must be fulfilled for crossing to occur, as can be seen from Fig. 8.122. Thus we have obtained a contradiction, so that $c < c^*$, as required.

To prove the existence statement of (iii), let $\dot{y} = c$, where $0 < c < c^*$. Then we are looking for a value $\tilde{v}$ such that (8.120) with (8.121) has a solution. Clearly if such a solution exists, then $v_d < \tilde{v} < v_c$. The phase plane for (8.120) with (8.121) for $v_d < \tilde{v} < v_c$ is shown in Fig. 8.123. The wave front required corresponds to a trajectory from $P_3 = (h_3(\tilde{v}), 0)$ to $P_1 = (h_1(\tilde{v}), 0)$. It is easy to show that P_3 is a saddle point, and we are concerned with the single trajectory leaving P_3 and entering the region $u < h_3(\tilde{v})$, $w < 0$. Let us call this trajectory $T(\tilde{v}, c)$. The idea is to prove the following three lemmas.

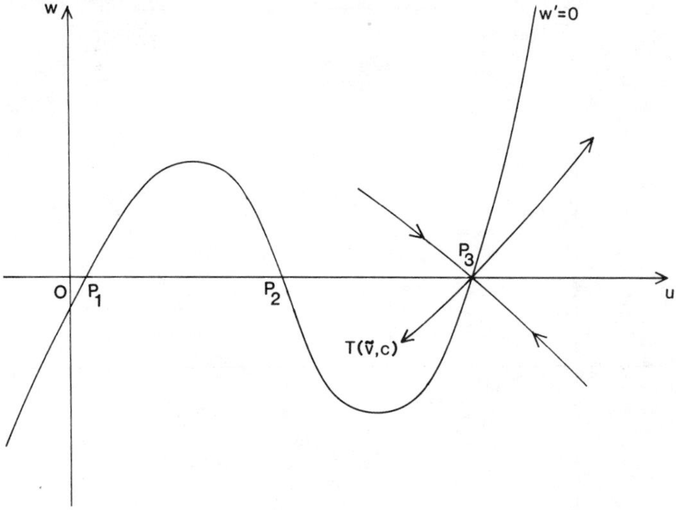

Fig. 8.123. The phase plane for (8.120) for $v_d < v < v_c$.

LEMMA 8.124: If $\tilde{v}$ is sufficiently close to v_c, then the trajectory $T(\tilde{v}, c)$ either tends to P_2 or crosses the u-axis between P_1 and P_2.

LEMMA 8.125: If $\tilde{v}$ is sufficiently close to v_d, then the trajectory $T(\tilde{v}, c)$ tends to infinity in the half-plane $w < 0$.

LEMMA 8.126: By continuity of the solution with respect to $\tilde{v}$, there is some intermediate value of $\tilde{v}$ such that $T(\tilde{v}, c) \to P_1$.

Proof of Lemma 8.124: If $\tilde{v}$ is close to v_c, P_2 and P_3 become close to each other (Fig. 8.127). An area, denoted by B in the figure, may then be constructed such that the vector field of (8.120) points into the area everywhere except at P_2 and P_3 and on the line segment P_2Q_3. It follows that the trajectory $T(\tilde{v}, \dot{y})$ either tends to P_2 or leaves B through the line segment. Since Q_3 is to the right of P_1 for $\tilde{v}$ sufficiently close to v_c, the result follows.

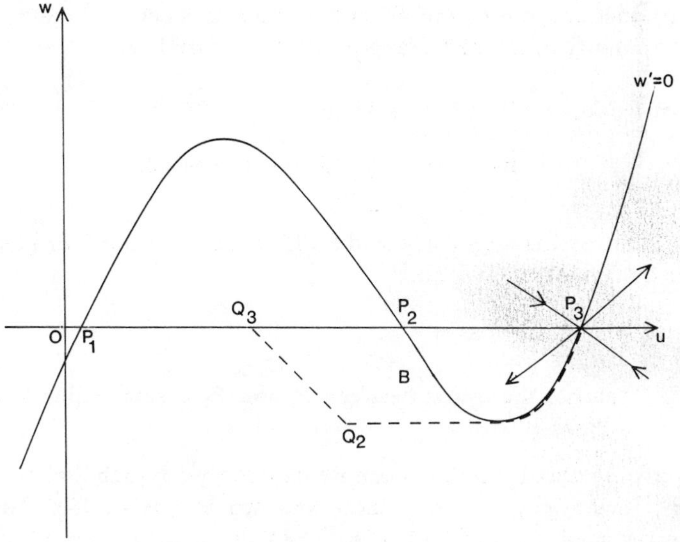

Fig. 8.127. The phase plane for (8.120) if v is close to v_c.

Proof of Lemma 8.125: If $\tilde{v} = v_d$, the phase planes for two different values of c are shown in Fig. 8.128. The first part of Theorem (8.119) ensures that $T(v_d, c^*)$ leaves P_3 and enters P_1. It follows from Lemma 4.72 that if $c < c^*$, then $T(v_d, c)$ is always strictly below the trajectory $T(v_d, c^*)$ joining P_3 and P_1, except at P_3 and possibly at P_1. But by the definition of c^*,

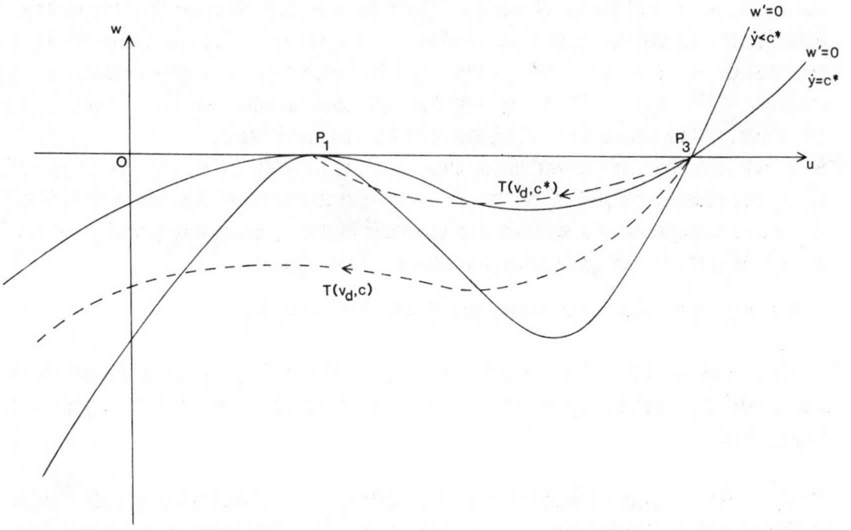

Fig. 8.128. Two phase planes for (8.120) with $v = v_d$ superimposed, one with $c = c^*$ and the other with $c < c^*$.

$T(v_d, c)$ does not join P_3 and P_1, and so must pass below P_1, and therefore tends to infinity in the half-plane $w<0$, as required.

Proof of Lemma 8.126: Let us define

$$\hat{v} = \sup_{v_d<\tilde{v}<v_c} \{\tilde{v} \mid T(\tilde{v}, c) \to \infty \text{ in } w<0\}$$

which exists by Lemmas 8.124 and 8.125. Then there are four possibilities for the behaviour of $T(\hat{v}, c)$:

(a) $T \to \infty$ in $w<0$ as $\eta \to \infty$.
(b) $T \to P_1$ as $\eta \to \infty$.
(c) T reaches the u-axis between P_1 and P_2 at some value of η.
(d) $T \to P_2$ as $\eta \to \infty$.

Let us assume that (a) holds. Then we may choose η such that one or both of $u(\eta; \hat{v})$ and $w(\eta; \hat{v})$ are as large and negative as we like. Then for ε sufficiently small $(u(\eta; \hat{v}+\varepsilon), w(\eta; \hat{v}+\varepsilon))$ is close to $(u(\eta; \hat{v}), w(\eta; \hat{v}))$, so that $T(\hat{v}+\varepsilon, \dot{y}) \to \infty$ in $w<0$ as well. But this contradicts the definition of $\hat{v}$, so that (a) cannot hold.

Let us assume that (c) holds. Then T enters the region $w>0$, so that we may choose an η such that $w(\eta; \hat{v})>0$. By continuity $w(\eta; \hat{v}-\varepsilon)>0$ for ε sufficiently small, so that $T(\hat{v}-\varepsilon, c)$ does not tend to infinity in $w<0$, contradicting the definition of $\hat{v}$. Thus (c) cannot hold.

Assume finally that (d) holds. Then we can choose η such that $(u(\eta; v), w(\eta; v))$ is as close to $P_2(\hat{v})$ as we like. Hence, by choosing ε sufficiently small, we can make $(u(\eta; \hat{v}-\varepsilon), w(\eta; \hat{v}-\varepsilon))$ as close to $P_2(\hat{v})$ and hence as close to $P_2(\hat{v}-\varepsilon)$ as we like. But P_2 is a stable critical point and so $T(\hat{v}-\varepsilon, \dot{y}) \to P_2$ as $\eta \to \infty$ for all sufficiently small ε. This again contradicts the definition of $\hat{v}$, so that (d) cannot hold.

It follows that (b) must hold, that is $T(\hat{v}, c) \to P_1$ as $\eta \to \infty$, as required. This completes the proof of the existence statement of Theorem 8.119(iii). To prove uniqueness a crossing argument is used, as in the proof of $c<c^*$ in (ii). We shall not give this argument explicitly.

We may now state the main result of this section.

COROLLARY 8.129: The kinetic system (8.31) satisfying conditions (8.33) has a solitary travelling wave solution, to leading order in the asymptotic expansions.

Proof: By the theorem there exists a trailing wave front with speed c, equal to the speed of the leading wave front. Let v_k be the value of v in this wave front, and define y_k by

$$-\frac{y_k}{c} = \int_{v_s}^{v_k} \frac{d\psi}{g(h_1(\psi), \psi)} \tag{8.130}$$

Then the leading order solitary travelling wave solution is described as follows:

Region I, $x > ct$: $u = u_s$, $v = v_s$.

Region II, $x = ct$: Leading wave front, $v = v_s$; $u \to h_1(v_s)$ as $(x - ct)/\varepsilon \to -\infty$, $u \to h_3(v_s)$ as $(x - ct)/\varepsilon \to \infty$.

Region III, $y_k < x - ct < 0$: $u = h_1(v)$, where

$$t - \frac{x}{c} = \int_{v_s}^{v} \frac{d\psi}{g(h_1(\psi), \psi)}$$

Note that $v = v_k$ at the tail of this region, by (8.130).

Region IV, $x - ct = y_k$: Trailing wave front, $v = v_k$; $u \to h_3(v_k)$ as $(x - y_k - ct)/\varepsilon \to -\infty$, $u \to h_1(v_k)$ as $(x - y_k - ct)/\varepsilon \to \infty$, guaranteed to exist by Theorem 8.119.

Region V, $x - ct < y_k$: $u = h_3(v)$, where

$$t - \frac{y_k}{c} = \int_{v_k}^{v} \frac{d\psi}{g(h_3(\psi), \psi)}$$

Note that as $t \to \infty$ in region V, $v \to v_s$, and $u \to h_3(v_s) = u_s$, so that the system returns to its stable steady state after the passage of the wave.

We now return to the initial value problem, and investigate whether its solution tends to the solitary travelling wave which we have shown to exist. There are two cases to consider.

First, let $c \geqslant c^*$. That is, the speed of the leading front is at least as great as the minimum possible speed for the trailing wave front with $\tilde{v} = v_d$. It is thus possible, by Theorem 8.119, that when the trailing wave front forms at $v = \tilde{v} = v_d$ it starts moving at speed $\dot{y} = c$. We shall show that it does so by a contradiction argument. First, assume that it starts moving at speed $\dot{y} < c$. But, from (8.98).

$$1 - (\dot{y}/c) = \dot{\tilde{v}}/g(h_1(\tilde{v}), \tilde{v}) \tag{8.131}$$

and $g(h_1(v_d), v_d) < 0$, so that $\dot{\tilde{v}} < 0$, and $\tilde{v}$ decreases from v_d. But in the region in front of the shock we have $u = h_1(v)$, and $h_1(v)$ does not exist for $v < v_d$, so that this is a contradiction. Assume therefore that the front moves with speed $\dot{y} > c$. In this case it follows that $\dot{\tilde{v}} > 0$, and $\tilde{v}$ increases from v_d. But then there is a unique wave speed $\dot{y} < c^* \leq c$, and this again leads to a contradiction. We conclude that $\dot{y} = c$, that is, that the solitary travelling wave solution obtains for $x > 0$.

The second possibility is that $c < c^*$, that is, there is no trailing wave front with speed $\dot{y} = c$ and $\tilde{v} = v_d$, but there is with $\tilde{v} = v_k > v_d$. Then on formation of the shock, with $v = v_d$, we have $\dot{y} \geq c^* > c$. We wish to investigate the behaviour of $\dot{y}$ and $\tilde{v}$ as $t \to \infty$. If they tend to c and v_k, respectively, then the trailing wave front tends to a front travelling with the same speed as the leading wave front, and thus the asymptotic solution which we have constructed tends to the solitary travelling wave solution of Corollary 8.129.

THEOREM 8.132: The solution of the Cauchy problem (8.69), satisfying Conditions (8.33), with initial conditions (8.70), tends to a solitary travelling wave as $t \to \infty$, to leading order in matched asymptotic expansions.

Proof: We must show that $\dot{y} \to c$, $\tilde{v} \to v_k$. But (8.131) gives

$$1 - (\dot{y}(\tilde{v})/c) = \dot{\tilde{v}}/g(h_1(\tilde{v}), \tilde{v})$$

Since $g < 0$ on $h_1(v)$, by Remark 8.40(a) we have

$$\dot{\tilde{v}} > 0 \quad \text{if} \quad \dot{y}(\tilde{v}) > c, \quad \text{i.e.} \quad \tilde{v} < v_k$$

$$\dot{\tilde{v}} < 0 \quad \text{if} \quad \dot{y}(\tilde{v}) < c, \quad \text{i.e.} \quad \tilde{v} > v_k$$

It follows that $\tilde{v} \to v_k$ as $t \to \infty$, and therefore that $\dot{y} \to c$, as required.

It therefore seems from an asymptotic analysis that reaction–diffusion systems which exhibit a threshold phenomenon in the spatially uniform case have solitary travelling wave solutions in the spatially non-uniform case, and that these waves are stable. They are triggered by a sufficiently large initial perturbation from the steady state, and after their passage the system again becomes excitable and amenable to the triggering of another such wave. Such behaviour is observed in nerve conduction, in the Belousov–Zhabotinskii reaction, in certain enzyme systems and in the emission of cyclic AMP by the slime mould *dictyostelium discoideum.* In the next section we shall consider mathematical models of such systems and show that the equations satisfy the conditions of Section 8.5.

Note that we have shown the existence of solitary travelling waves in the asymptotic limit as $\varepsilon \to 0$, and assumed that these waves exist for small
cyclic AMP by the slime mould *dictyostelium discoideum.* In the next section

positive value of ε. A rigorous proof of this fact may be obtained by using topological techniques; see Smoller (1983) for a review.

8.9 Systems with Solitary Travelling Waves

We have thus shown the existence of solitary travelling wave solutions of certain reaction–diffusion systems to leading order in an asymptotic expansion, and discussed their stability within the asymptotic framework. The analysis and numerical solutions of such systems suggest that the travelling waves are stable solutions of the equations. The waves are trigger waves, in that a finite perturbation from the steady state is necessary to initiate them. Trigger waves are observed physically in a number of contexts, and it is natural to consider whether the physical systems may be modelled by reaction–diffusion equations of the form analysed here. We shall discuss several systems arising in biology, chemistry and physiology.

8.9.1 The Cyclic AMP Control System in *Dictyostelium Discoideum*

Before aggregation the slime mould *dictyostelium discoideum* exists as separate uni-cellular organisms, or amoebae. As the food supply for these amoebae diminishes, certain of them begin to emit periodic wave trains of the enzyme cyclic AMP (cAMP), and the others move towards these pacemaker centres until they have collected in a dense mass, a process known as aggregation. These masses become fruiting bodies and release spores which are dispersed and become amoebae, thus completing the life cycle. The system has been discussed by Cohen (1977); see also previous references therein. He proposes a reaction–diffusion mechanism for the system, given by

$$u_t = f(u, v) + \varepsilon \nabla^2 u, \qquad v_t = \varepsilon g(u, v) + \varepsilon\beta \nabla^2 v$$

where f and g are shown in Fig. 8.133.

The nullcline $f = 0$ remains stationary while $g = 0$ moves from (a) to (b) to (c) to (d) as time passes. When it is in position (b), that is, sufficiently close to and to the left of (u_d, v_d), then the system will be in a stable but excitable state and Conditions 8.33 will hold. Hence the system admits a solitary travelling wave in this case. It therefore seems likely that the waves of cAMP are solitary travelling waves triggered by some mechanism, as yet unknown, at the pacemaker centre.

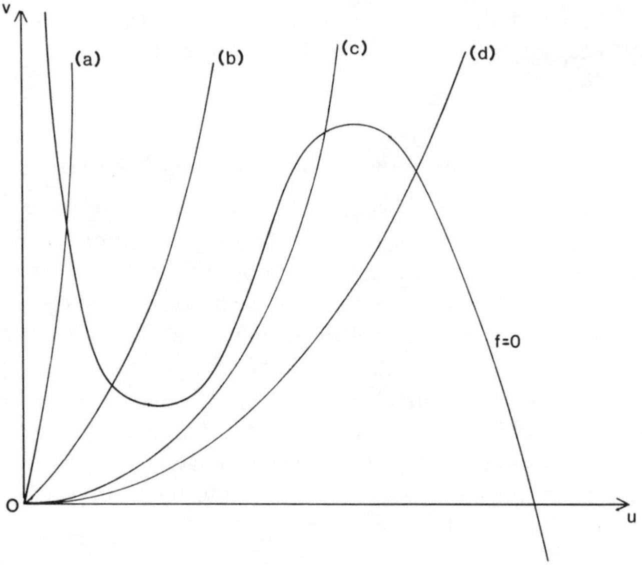

Fig. 8.133. Phase plane for the cAMP control system in *dictyostelium discoideum.*

8.9.2 The Belousov-Zhabotinskii Reaction

This chemical reaction provides a nice example of travelling waves which may be set up experimentally [see Winfree (1980)]. If the reaction is set up in a thin layer of liquid in a Petri dish, then circular waves typically propagate outwards from pacemaker centres, as in the aggregation process in *dictyostelium discoideum.* These waves again seem to be trigger waves, and have been studied experimentally by Winfree (1974) and theoretically by Murray (1976), Field and Troy (1979) and others. The associated threshold phenomenon was discussed by Troy (1977). The value of the wave speed c has been discussed theoretically by Murray (1976) and numerically by Manoranjan and Mitchell (1983). The mechanism of the pacemaker centres which trigger these solitary travelling waves is not known. It is sometimes assumed that the centre is a heterogeneity such as a particle of dust or a bubble of carbon dioxide. The other possibility is that the chemical composition of the mixture close to the centres is slightly different from that in the rest of the domain, so that the local kinetics are oscillatory instead of excitable (Tyson and Fife, 1980). In this case it is possible that the oscillations at the centre could trigger solitary waves in the excitable medium. The model most used to describe this system is the Oregonator of Field and Noyes (1974) (see Chapter 2), which when diffusion is included may be written, in a different notation from that of Chapter 2, as

$$\begin{aligned} \varepsilon' w_t &= w + u - wu - qw^2 + \varepsilon' d_1 \nabla^2 w \\ u_t &= 2fv - u - wu + d_2 \nabla^2 u \\ pv_t &= w - v + pd_3 \nabla^2 v \end{aligned} \tag{8.134}$$

where f is a stoichiometric parameter. Empirically, $\varepsilon' = O(10^{-4})$, $q = O(10^{-5})$ and, in many situations, $p = O(10^2)$. Simplifying the model still further by taking $\varepsilon' = 0$ (Tyson, 1976) and defining $\varepsilon = 1/p$, the system becomes

$$\begin{aligned} 0 &= w + u - wu - qw^2 \\ \varepsilon u_t &= -u + 2fv - wu + \varepsilon \nabla^2 u \\ v_t &= w - v + \varepsilon\beta \nabla^2 v \end{aligned} \tag{8.135}$$

where $\beta = d_3/d_2$ and we have re-scaled the space and time variables. The first of these gives an equation for w in terms of u, so that $w = w(u)$. By substituting this expression into the second and third we obtain a system of two reaction–diffusion equations

$$\begin{aligned} \varepsilon u_t &= -u + 2fv - uw(u) + \varepsilon \nabla^2 u = f(u, v) + \varepsilon \nabla^2 u \\ v_t &= w(u) - v + \varepsilon \nabla^2 v = g(u, v) + \varepsilon\beta \nabla^2 v \end{aligned} \tag{8.136}$$

which is in the form considered in this chapter. It may be shown that this system satisfies Conditions 8.33 (but with $f_v > 0$; see Remark 8.38) as long as the stoichiometric parameter f satisfies $f < \frac{1}{4}$ or $f > (1+\sqrt{2})/2$ (Britton, 1982), and that therefore the system has solitary wave solutions for these values of f. With regard to f these results are the best possible, since we know that for $\frac{1}{4} < f < (1+\sqrt{2})/2$ the system has no stable steady state and therefore cannot exhibit the required threshold behaviour.

8.9.3 The Fitzhugh–Nagumo Equations

In 1952 Hodgkin and Huxley published their classic model for conduction in nerve. Fitzhugh (1961, 1969) and Nagumo *et al.* (1962) proposed simplifications which retained the essential features of the model, and which may be written in the form

$$\varepsilon u_t = -F(u) - v + \varepsilon u_{xx}, \qquad v_t = u - bv \tag{8.137}$$

where $b \geq 0$ and ε is a small parameter (see Section 4.4). A perturbation analysis for this system was carried out by Casten *et al.* (1975). Solitary travelling waves were considered using topological techniques by Conley (1975), Hastings (1976b), who showed the existence of a fast and a slow pulse, and Carpenter (1977). Early work on stability was done by Evans

(1972a,b,c, 1975) and the slow pulse was shown to be unstable by Evans and Feroe (1977) and Sleeman (1980). The stability of the fast pulse is more difficult to prove: recent work has been carried out by Jones (1983c, 1984) and Yanagida (1985). Multiple pulse solutions were analysed by Hastings (1982). Periodic travelling waves were considered by Maginu (1980). The system is of the form described in this chapter with $f(u, v) = -F(u) - v$, $g(u, v) = u - bv$, and the ratio of the diffusion coefficients $\beta = 0$; $F(u)$ is as shown in Fig. 8.138, and $\int_0^1 F(u)\, du < 0$.

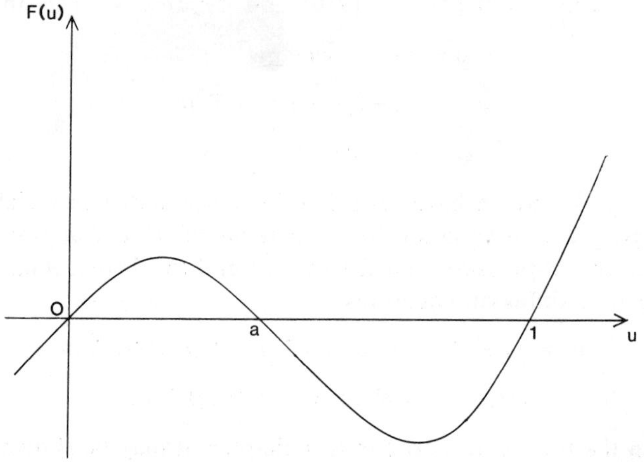

Fig. 8.138. The function $F(u)$ in the nerve conduction model (8.137).

It is easy to show that the required conditions hold for this system, and that there is therefore a solitary travelling wave solution of the equations for ε small, corresponding to the triggered response of a nerve (the fast pulse solution). The threshold behaviour ensures that only stimuli above a certain level will trigger a nerve impulse. This impulse consists of a wave of excitation, the leading wave front, followed by a refractory period, when the nerve is not amenable to the triggering of another impulse, followed by relaxation to the steady state. Another nerve impulse may then be triggered. Other systems based on Hodgkin and Huxley's (1952) model have also been widely studied; see for example the review of Rinzel (1981), Evans *et al.* (1982) and references therein, Feroe (1982), Terman (1983), and McKean (1970, 1983, 1984). Travelling wave solutions of the Hodgkin–Huxley model itself have been considered by Hastings (1976c).

Existence of travelling wave solutions in predator–prey equations has been proved by Dunbar (1983, 1984). Stationary waves for systems have been considered by Ermentrout *et al.* (1984) and Klaasen and Troy (1984). The method of Hofer and Toland (1984) is sometimes applicable to such problems.

9. Macromolecular Carriers: Asymptotic Techniques

9.1 Introduction

This chapter is again concerned with the technique of matched asymptotic expansions. In the last chapter expansions in domains where a short time scale was appropriate were matched to those in domains with a long time scale. Here we shall be concerned with matching in spatial domains. This leads to different problems, since the spatial derivatives are of course second order whereas the time derivatives are first order. The subject will be discussed using examples from physiology and medicine on the role of myoglobin in muscle and on carbon monoxide poisoning.

9.2 Macromolecular Carriers

Many physical processes require the movement of essential substances from one part of the body to another, and often diffusion plays a large part in these transport processes. For example, respiration requires that oxygen in the alveoli of the lung diffuse into the blood in the pulmonary capillaries. In many cases, however, diffusion by itself cannot transport enough of the required substance, and facilitated diffusion takes place. Here the flux of the substance is enhanced by the presence of carrier molecules which combine reversibly with it. This effect was investigated independently by Wittenberg (1959) and Scholander (1960), who conducted experiments on the facilitated diffusion of oxygen through solutions containing haemoglobin. The increase in oxygen flux in the presence of the haemoglobin was measured. These experiments were directed towards an understanding of the role of myoglobin, a protein similar to haemoglobin, in muscle. Since myoglobin combines reversibly with oxygen, releasing it when needed, it had been thought that it was simply an oxygen reservoir in the muscle. However, simple calculations show that even in the rest state in man the amount of oxygen bound to myoglobin is only sufficient for the requirements of the muscle for about five seconds. It was therefore proposed that the myoglobin facilitated the diffusion of oxygen in the muscle in some way. This has come to be widely accepted and the subject has been definitively reviewed by Wittenberg (1970).

In general facilitation of transport of a substrate by a carrier may take place in two ways. Either the carrier is stationary and the substrate molecule moves from one carrier molecule to the next (the "bucket-brigade" mechanism) or the movement of the carrier itself with the attached substrate is important. The bucket-brigade mechanism applies to ion-exchange mechanisms, but there is no experimental evidence that it occurs in haemoglobin- or myoglobin-facilitated oxygen diffusion. In fact Wittenberg (1966) reported that the flux of oxygen decreased at high protein concentrations which argues against such a mechanism. One motivation for the mathematical modelling of such systems was to discover whether simple diffusion of the carrier–substrate complex could be sufficient to account for the increased flux of oxygen which was observed. A reaction–diffusion model for the haemoglobin–oxygen case was proposed by Wyman (1966) and modified by Murray (1971), who obtained analytic solutions agreeing well with Wittenberg's experiments. Numerical work on Wyman's model has been carried out by Kreuzer and Hoofd (1970), Kutchai *et al.* (1970) and Jacquez *et al.* (1972), whose results are also in general agreement with experiment. Further asymptotic analysis in a more general setting has been carried out by Kolkka and Salathé (1984). Murray's analysis is presented very readably in his book (1977), and we shall therefore give only a brief summary of the main ideas.

9.3 Facilitation of Oxygen Diffusion into Muscle by Myoglobin

We concentrate our attention on the case of myoglobin, which has been investigated experimentally by Wittenberg *et al.* (1975). A mathematical model for a muscle fibre containing myoglobin was set up by Wyman (1966) and analysed by Murray (1974b). Physiologically a muscle fibre is approximately cylindrical in cross-section and its length is much greater than its thickness. We therefore take it to be cylindrical, of radius a, and neglect any variations in concentration along its length. We also assume radial symmetry, so that the concentrations of oxygen, myoglobin and the complex oxymyoglobin depend only on the distance from the centre of the fibre. We assume that the concentration of oxygen at the surface of the muscle is constant, and that no myoglobin is transported out of the muscle.

The reaction between myoglobin and oxygen may be adequately modelled by

$$\mathrm{Mb} + \mathrm{O}_2 \underset{k_{-1}}{\overset{k_1}{\rightleftharpoons}} \mathrm{MbO}_2 \tag{9.1}$$

where k_1 and k_{-1} are the forward and backward rate constants. We assume that oxygen is consumed in the muscle at a constant rate q. A more realistic Michaelis–Menten consumption law was discussed by Taylor and Murray (1977), who showed that for normal conditions in man the assumption of constant consumption is a good approximation. The differential equations are therefore

$$\begin{aligned}\frac{\partial[\mathrm{Mb}]}{\partial t} &= -k_1[\mathrm{Mb}][\mathrm{O_2}]+k_{-1}[\mathrm{MbO_2}]+D_2\,\nabla^2[\mathrm{Mb}]\\ \frac{\partial[\mathrm{MbO_2}]}{\partial t} &= k_1[\mathrm{Mb}][\mathrm{O_2}]-k_{-1}[\mathrm{MbO_2}]+D_2\,\nabla^2[\mathrm{MbO_2}]\\ \frac{\partial[\mathrm{O_2}]}{\partial t} &= -k_1[\mathrm{Mb}][\mathrm{O_2}]+k_{-1}[\mathrm{MbO_2}]-q+D_1\,\nabla^2[\mathrm{O_2}]\end{aligned} \tag{9.2}$$

where we have taken the diffusion coefficients of myoglobin and oxymyoglobin to be equal, since the molecular weight of oxygen is very much smaller than that of myoglobin. The boundary conditions are

$$[\mathrm{O_2}]=c_0, \qquad \frac{\partial[\mathrm{Mb}]}{\partial R}=\frac{\partial[\mathrm{MbO_2}]}{\partial R}=0 \quad \text{on} \quad R=a \tag{9.3}$$

where R is the radial coordinate, and

$$\frac{\partial[\mathrm{O_2}]}{\partial R}=\frac{\partial[\mathrm{Mb}]}{\partial R}=\frac{\partial[\mathrm{MbO_2}]}{\partial R}=0 \quad \text{on} \quad R=0 \tag{9.4}$$

by symmetry. These conditions will be modified in the case of oxygen debt, as we shall see. We look for a steady-state solution, so that $\partial/\partial t\equiv 0$, and as discussed above we consider only variations with respect to R. Then it can easily be seen that

$$[\mathrm{Mb}]+[\mathrm{MbO_2}]=c_t=\text{const} \tag{9.5}$$

Defining non-dimensional variables by

$$r=R/a, \qquad u=u(r)=[\mathrm{O_2}]/c_0, \qquad v=v(r)=[\mathrm{MbO_2}]/c_t \tag{9.6}$$

we obtain

$$\begin{aligned}0&=c_tk_1c_0(1-v)u-c_tk_{-1}v+\frac{D_2c_t}{a^2}\frac{1}{r}\frac{d}{dr}\left(r\frac{dv}{dr}\right)\\ 0&=-c_tk_1c_0(1-v)u+c_tk_{-1}v-q+\frac{D_1c_0}{a^2}\frac{1}{r}\frac{d}{dr}\left(r\frac{du}{dr}\right)\end{aligned} \tag{9.7}$$

The boundary conditions become

$$u=1,\quad \frac{dv}{dr}=0 \quad \text{at} \quad r=1, \qquad \frac{du}{dr}=\frac{dv}{dr}=0 \quad \text{at} \quad r=0 \tag{9.8}$$

Adding the two equations in (9.7),

$$\frac{D_1c_0}{a^2}\frac{1}{r}\frac{d}{dr}\left(r\frac{du}{dr}\right)+\frac{D_2c_t}{a^2}\frac{1}{r}\frac{d}{dr}\left(r\frac{dv}{dr}\right)=q \tag{9.9}$$

and integrating twice,

$$D_1c_0u+D_2c_tv=qa^2r^2/4+A\log r+B \tag{9.10}$$

Physically, $D_1c_0u+D_2c_tv$ cannot be infinite, so that A must be equal to zero. If we denote the value of v at $r=1$ by v^* (at present unknown), then since $u=1$ at $r=a$,

$$D_1c_0u+D_2c_tv=D_1c_0+D_2c_tv^*-qa^2(1-r^2)/4 \tag{9.11}$$

Eliminating v from (9.9) and (9.11) we obtain, after some algebra,

$$\frac{1}{r}\frac{d}{dr}\left(r\frac{du}{dr}\right)=-(\beta+\gamma r^2)-(\kappa+\lambda r^2)u+\delta u^2 \tag{9.12}$$

where

$$\beta=-\frac{qa^2}{c_0D_1}+\frac{k_{-1}a^2c_tv^*}{c_0D_1}-\frac{k_{-1}qa^4}{4c_0D_1D_2}+\frac{k_{-1}a^2}{D_2}$$

$$\kappa=-\frac{c_tk_1a^2}{D_1}\left\{1-v^*+\frac{qa^2}{4c_tD_2}-\frac{D_1c_0}{D_2c_t}\right\}-\frac{k_{-1}a^2}{D_2} \tag{9.13}$$

$$\gamma=\frac{k_{-1}qa^4}{4D_1D_2c_0},\qquad \delta=\frac{k_1a^2c_0}{D_2},\qquad \lambda=\frac{k_1qa^4}{4D_1D_2}$$

Note that v^* is unknown, but order of magnitude estimates for β and κ can be found, since we do know that $0\leq v^*\leq 1$. Typical values for the parameters (Wittenberg, 1970; Riveros-Moreno and Wittenberg, 1972, or Murray, 1977) give

$$\beta=\beta(v^*)=O(10^3),\qquad \kappa=\kappa(v^*)=O(10^4),$$
$$\gamma=O(10^2),\qquad \delta=O(10^4),\qquad \lambda=O(10^3)$$

It can be seen that these are all large parameters, and this suggests an asymptotic approach to the problem. Defining

$$\varepsilon=D_1/k_1c_ta^2 10^{-4},\qquad b=\varepsilon\beta,\qquad k=\varepsilon\kappa,$$
$$g=\varepsilon\gamma,\qquad d=\varepsilon\delta,\qquad l=\varepsilon\lambda \tag{9.14}$$

we obtain

$$\varepsilon \frac{1}{r}\frac{d}{dr}\left(r\frac{du}{dr}\right) = -(b+gr^2)-(k+lr^2)u+du^2 \tag{9.15}$$

where

$$b = b(v^*) = O(10^{-1}), \qquad k = k(v^*) = O(1),$$
$$g = O(10^{-2}), \qquad d = O(1), \qquad l = O(10^{-1}) \tag{9.16}$$

This therefore becomes a singular perturbation problem, since there is a small parameter multiplying the highest derivative, and putting this equal to zero, in general, means that we will not be able to satisfy the boundary conditions. We therefore might expect boundary layers close to $r=0$ or $r=1$, where the concentration of the oxymyoglobin and the oxygen change by a large amount over a short distance. The normal method of solution is the method of matched asymptotic expansions discussed in Chapter 8. Away from the boundary layers the derivative term is $O(\varepsilon)$. It will become clear that for matching with the solution in the boundary layer we must look for an asymptotic expansion for u in powers of $\sqrt{\varepsilon}$, that is,

$$u(r;\varepsilon) \approx u_0(r)+\sqrt{\varepsilon}\,u_1(r)+\cdots \tag{9.17}$$

This solution, valid away from the boundary layers, is known as the outer solution. By substituting into the equation and equating terms of $O(1)$, we obtain

$$0 = -(b+gr^2)-(k+lr^2)u_0+du_0^2 \tag{9.18}$$

The higher-order corrections u_n for $n \geq 1$ may be obtained by equating terms of $O(\varepsilon^{n/2})$, but we shall not do this here. Since u_0 is a concentration it is positive, and hence is given by the positive solution of the quadratic equation,

$$2d\,u_0(r) = k+lr^2+\sqrt{(k+lr^2)^2+4d(b+gr^2)} \tag{9.19}$$

It can be seen that this satisfies the boundary condition $du_0/dr = 0$ at $r=0$, and therefore we do not expect a boundary layer here. We may however have a boundary layer at $r=1$. If we do the derivative term in the equation becomes important, and it can be seen that changes must be taking place on a space scale of $O(\sqrt{\varepsilon})$. We therefore define

$$\rho = (1-r)/\sqrt{\varepsilon} \tag{9.20}$$

and let

$$U(\rho;\varepsilon) = u(r;\varepsilon) \tag{9.21}$$

The equation becomes

$$\frac{1}{1-\sqrt{\varepsilon}\,\rho}\frac{d}{d\rho}\left\{(1-\sqrt{\varepsilon}\,\rho)\frac{dU}{d\rho}\right\}$$
$$=-\{b+g(1-\sqrt{\varepsilon}\,\rho)^2\}-\{k+l(1-\sqrt{\varepsilon}\,\rho)^2\}U+dU^2 \qquad (9.22)$$

Since powers of $\sqrt{\varepsilon}$ occur in this equation, we must look for a solution as a power series in $\sqrt{\varepsilon}$, that is,

$$U(\rho;\,\varepsilon)\approx U_0(\rho)+\sqrt{\varepsilon}\,U_1(\rho)+\cdots \qquad (9.23)$$

This is the inner solution, valid near the boundary. Substituting into the equation and equating terms of $O(1)$ (again we shall not consider higher-order terms), we obtain

$$\frac{d^2U_0}{d\rho^2}=-(b+g)-(k+l)U_0+dU_0^2 \qquad (9.24)$$

The conditions to be applied are boundary conditions at $r=1$, or $\rho=0$, and matching conditions with the outer solution. Since $u=1$ at $r=1$, (9.8), one boundary condition is

$$U_0=1 \qquad \text{at} \quad \rho=0 \qquad (9.25)$$

However, we also have a second boundary condition which must be satisfied at $\rho=0$. Differentiating (9.11) and noting that $dv/dr=0$ at $r=1$, we have

$$\frac{du}{dr}=\frac{ga^2}{2D_1c_0}=0(1) \quad \text{at } r=1.$$

Thus

$$\frac{dU_0}{d\rho}=0 \quad \text{at } \rho=0 \qquad (9.26)$$

The matching condition may be expressed as

$$\lim_{\rho\to\infty}U_0(\rho)=\lim_{r\to 1}u_0(r)=u_0(1) \qquad (9.27)$$

In other words, the two solutions must match together smoothly at the edge of the boundary layer. The three conditions (9.25), (9.26) and (9.27) are sufficient to determine the required solution of (9.24) *and* the unknown quantity v^*.

Let us consider the problem in the phase plane. Introducing $V_0=dU_0/d\rho$

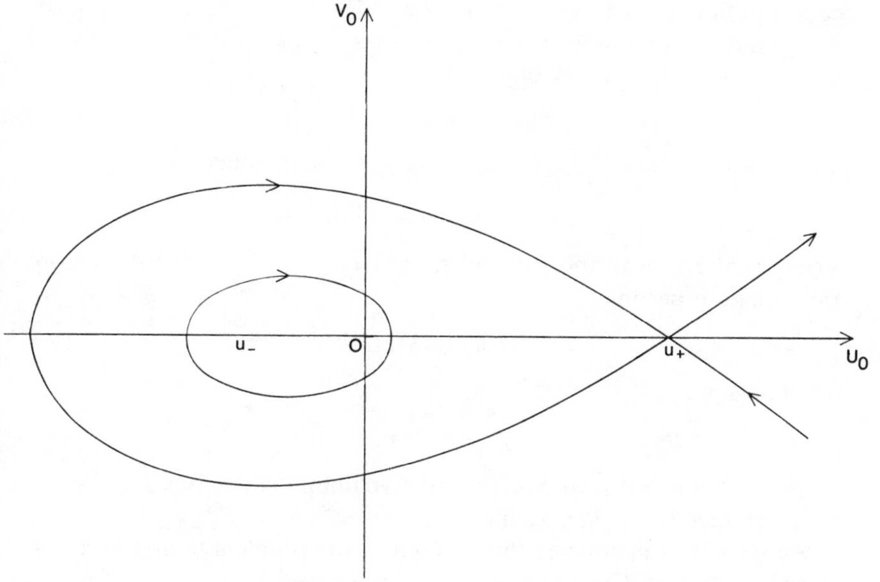

Fig. 9.29. The phase plane for the system (9.28).

we have

$$\frac{dU_0}{d\rho} = V_0, \qquad \frac{dV_0}{d\rho} = -(b+g) - (k+l)U_0 + dU_0^2 \tag{9.28}$$

The phase plane for the system is as shown in Fig. 9.29. The two critical points are at $(u_-, 0)$ and $(u_+, 0)$, where u_- and u_+ are the two roots of the quadratic $-(b+g)-(k+l)u+du^2=0$. It follows that $u_+=u_0(1)$, since from (9.18) $u_0(1)$ is the positive solution of the same quadratic. We require that $U_0(0)=1$, $V_0(0)=0$ and that $U_0(\rho)\to u_0(1)$, $V_0(0)\to 0$ as $\rho\to\infty$. In other words, the trajectory must start on the U_0 axis at $(1, 0)$ and tend towards the saddle point at $(u_0(1), 0)$ as $\rho\to\infty$. The only possibility is that the trajectory remains at the saddle point for all ρ, and that $u_+=u_0(1)=1$. Thus u does not change, to first order, in the boundary layer at $r=1$. This result was originally proved by Mitchell and Murray (1973) in the similar case of haemoglobin and oxygen [see also Murray 1977)] using a longer method but one which does not require any argument about derivatives. It is reasonable biologically, since the existence of a boundary layer would mean that the concentrations were changing in a length of $O(\sqrt{\varepsilon})$, which in dimensional terms is about 10^{-5} cm. It seems very unlikely that this could occur, and this is confirmed by the analysis.

We now have the first-order solution for u, $u=u_0(r)$, throughout the muscle fibre. However, it still depends on v^*, the value of the myoglobin concentration at $r=1$, which is unknown. The requirement that u_0 satisfies the boundary conditions at $r=1$ gives us an expression for v^*. From the quadratic (9.18), $u_0(1)=1$ implies that

$$0=-b-g-k-l+d \tag{9.30}$$

Substituting for these parameters from (9.13) and (9.14), we obtain

$$v^*=[k_1c_0+(q/c_t)]/(k_1c_0+k_{-1}) \tag{9.31}$$

For typical values of the parameters, $q/k_1c_0c_t=0(\varepsilon)$, so that to first order this equation becomes

$$v^*=k_1c_0/(k_1c_0+k_{-1}) \tag{9.32}$$

or, equivalently,

$$k_1c_0(1-v^*)-k_{-1}v^*=0 \tag{9.33}$$

i.e. the reaction between oxygen and myoglobin is near equilibrium close to the surface of the muscle fibre.

We have now completed the solution of the problem to first order, that is, with an error of $O(\sqrt{\varepsilon})$ or about 1%. Summarising, we have the following result.

THEOREM 9.34: The equations (9.7) with boundary conditions (9.8) have a uniformly valid asymptotic solution for u whose first term is given by

$$2\delta u_0(r)=\kappa+\lambda r^2+\{(\kappa+\lambda r^2)^2+4\delta(\beta+\gamma r^2)\}^{1/2} \tag{9.35}$$

where the parameters are given by (9.13), and v^* is given by (9.32). From (9.11), v_0 is then given by

$$v_0(r)=\frac{D_1c_0(1-u_0)+D_2c_tv^*-\frac{1}{4}qa^2(1-r^2)}{D_2c_t} \tag{9.36}$$

We have assumed that there is no region in oxygen debt. A region of muscle goes into oxygen debt when the oxygen concentration according to the differential equation becomes negative. It can be seen from (9.11) that this will happen if q is sufficiently large; for example the centre of the muscle will be in oxygen debt if $qa^2/4>D_1c_0+D_2c_t$ (since v^* is at most one). Of course it is physically unrealistic to have negative concentrations, and in the region of oxygen debt we must replace the differential equations by the condition that the concentrations of oxygen and oxymyoglobin are zero [see Murray (1977)]. A region of muscle may go into oxygen debt temporarily, but a measure of its efficiency is its maximum rate of working without this happening. At this rate $\bar{q}$, say, the oxygen concentration at the centre of the muscle, where the concentration is a minimum, is zero, i.e.

$u(0) = 0$. From the quadratic (9.18) this occurs when b is zero, that is when

$$\frac{1}{4}\bar{q}a^2 = \frac{D_1 c_0 + D_2 c_t v^*}{1 + 4D_2/(k_{-1}a^2)} = D_1 c_0 + D_2 c_t v^* \tag{9.37}$$

to first order, since $4D_2/k_{-1}a^2 = 0(\varepsilon)$. If there were no myoglobin, then c_t would be zero, so the maximum rate of working without oxygen debt would be given by $\bar{q}a^2/4 = D_1 c_0$. Thus the myoglobin increases this maximum rate of working by $D_2 c_t v^*/D_1 c_0$, or about 40%. It can be seen that the increase is due to the diffusive flux of the oxymyoglobin, which is given by $D_2 c_t v^*$. The oxymyoglobin is therefore facilitating the flux of oxygen into the muscle by combining with it and diffusing itself. This is the basis of facilitated diffusion.

9.4 Carbon Monoxide Poisoning

An analysis similar to that of the previous section shows that haemoglobin plays a comparable role in facilitating the diffusion of oxygen into the blood of the pulmonary capillaries from the alveoli of the lung. In carbon monoxide poisoning this role is greatly reduced. Carbon monoxide binds to the same sites on the haemoglobin molecule as oxygen does, and the complex formed, carboxyhaemoglobin, is very stable. Hence even a small concentration of carbon monoxide in the air leads to a large proportion of the haemoglobin molecules of the blood being bound to carbon monoxide. They then cannot form oxyhaemoglobin, and the facilitation due to the diffusion of this complex is lost. In effect the diffusion constant of the lung, or the effectiveness of the lung in absorbing oxygen through diffusion, is reduced. A second more well-known effect of the high concentration of carboxyhaemoglobin is that there is less bound oxygen to be transferred to the tissues.

For these reasons it is essential in the treatment of carbon monoxide poisoning to reduce the concentration of carboxyhaemoglobin as quickly as possible. This may be done by blood transfusion, but a more frequently used technique is the transfer of the patient to an environment of high-pressure oxygen. The increased oxygen pressure in the lungs results in the formation of more oxyhaemoglobin and carbon monoxide is displaced from the haemoglobin molecules. It also provides the tissues of the body with more oxygen directly by increasing the amount dissolved in the blood.

Unfortunately this technique carries a risk of oxygen poisoning. It is therefore important to know how long it will take for the blood carboxyhaemoglobin to fall to a safe level at a given oxygen pressure in order that this risk may be minimised.

To investigate this problem we set up a model for oxygen and carbon monoxide as competing ligands of haemoglobin, based on that of Wyman

(1966). The treatment follows that of Britton (1979). Wyman assumed that the kinetics of the haemoglobin–oxygen reaction could be adequately described by

$$\mathrm{Hb} + \mathrm{O_2} \underset{k_{-1}}{\overset{k_1}{\rightleftharpoons}} \mathrm{HbO_2} \tag{9.38}$$

Similarly we take the kinetics of the haemoglobin–carbon monoxide reaction to be

$$\mathrm{Hb} + \mathrm{CO} \underset{k_{-2}}{\overset{k_2}{\rightleftharpoons}} \mathrm{HbCO} \tag{9.39}$$

These assumptions are not strictly true, as each haemoglobin molecule has four sites at which oxygen or haemoglobin may be bound, so that k_1 and k_2 in fact depend on the saturation of the haemoglobin molecules, but they are an acceptable approximation and have given good agreement with experiment in several studies (Wyman, 1966; Murray, 1971, 1977; Murray and Wyman, 1971; Britton and Murray, 1977).

These reactions take place in the pulmonary capillaries which we take to be rectangular in cross-section (for geometrical convenience only) and to run between two alveoli. We assume that the blood is homogeneous and is steadily convected through the pulmonary capillaries, or that any effects of inhomogeneity or non-steady flow are negligible. We assume that the blood is well-mixed on arrival at the pulmonary capillaries, as is reasonable, and that there is no sudden change in concentrations at the exit, since these have attained a steady state long before the exit is reached.

We consider the case of a man recovering from carbon monoxide poisoning. In this case the carbon monoxide concentrations in the blood and the rest of the body will have reached an equilibrium, and we therefore neglect any loss of carbon monoxide from the blood to the body. As the carboxyhaemoglobin concentration is reduced, the body will give up its carbon monoxide to the blood, but this is a very slow process [see Root (1965)] and we therefore neglect this as well. This simplification may produce errors in the last stages of recovery from carbon monoxide poisoning when the concentration of carboxyhaemoglobin is very low, but is a good approximation when this concentration is higher and changing quite quickly.

The boundary conditions at the walls of the capillaries are that there is no flux of haemoglobin and its compounds through the walls, and that the concentrations of oxygen and carbon monoxide are equal to their alveolar concentrations. In other words, the capillary wall is permeable to the gases but impermeable to the haemoglobin.

The alveolar oxygen concentration is taken to be linearly dependent on its value in the atmosphere, following data from Brink (1944). This use of an empirical law departs from normal practice in models of gas exchange in the lung, where the alveolar oxygen pressure is calculated from theoretical considerations as follows [see for example Coburn *et al.* (1965)]. The rate of uptake of oxygen into the pulmonary capillaries is equal to $\dot{V}_a(P_{aO_2} - P_{iO_2})/(P_b - P_{H_2O})$, where P is the pressure (alveolar, inspired, barometric, etc.), and $\dot{V}_a$ is the alveolar ventilation. It is also said to be equal to $D(\bar{P}_{cO_2} - P_{aO_2})$, where D is the diffusion constant of the lung and $\bar{P}_{cO_2}$ is the average oxygen pressure in the capillaries. From this the alveolar oxygen pressure, P_{aO_2}, may be calculated. This is subject to inaccuracy on two counts. First, the use of an average neglects variations in uptake along the length of the capillaries. Second, the diffusion of oxygen which occurs in the lung is not purely molecular diffusion but also includes facilitated diffusion due to the presence of molecules of free haemoglobin, as already discussed. This facilitated diffusion, which accounts for a large part of the oxygen diffusing into the blood in normal circumstances, is drastically reduced if there is an appreciable amount of COHb in the blood. This therefore reduces the effective diffusion constant of the lung. We take the alveolar carbon monoxide concentration to be zero, since it is certainly very small.

The coordinate system is as shown in Fig. 9.40. We neglect any variations in the z-direction. If the blood flows at speed s, then the equations are

$$\begin{aligned}
\frac{\partial[O_2]}{\partial\tau} + s\frac{\partial[O_2]}{\partial Y} &= -k_1[O_2][Hb] + k_{-1}[HbO_2] + D_1\nabla^2[O_2] \\
\frac{\partial[CO]}{\partial\tau} + s\frac{\partial[CO]}{\partial Y} &= -k_2[CO][Hb] + k_{-2}[HbCO] + D_2\nabla^2[CO] \\
\frac{\partial[Hb]}{\partial\tau} + s\frac{\partial[Hb]}{\partial Y} &= -k_1[O_2][Hb] + k_{-1}[HbO_2] - k_2[CO][Hb] \\
&\quad + k_{-2}[HbCO] + D\nabla^2[Hb] \\
\frac{\partial[HbO_2]}{\partial\tau} + s\frac{\partial[HbO_2]}{\partial Y} &= k_1[O_2][Hb] - k_{-1}[HbO_2] + D\nabla^2[HbO_2] \\
\frac{\partial HbCO]}{\partial\tau} + s\frac{\partial[HbCO]}{\partial Y} &= k_2[CO][Hb] - k_{-2}[HbCO] + D\nabla^2[HbCO]
\end{aligned} \tag{9.41}$$

where D_1 and D_2 are the diffusion coefficients of oxygen and carbon monoxide, respectively, and we have taken the diffusion coefficients of haemoglobin and its compounds to be equal to D. We seek steady-state

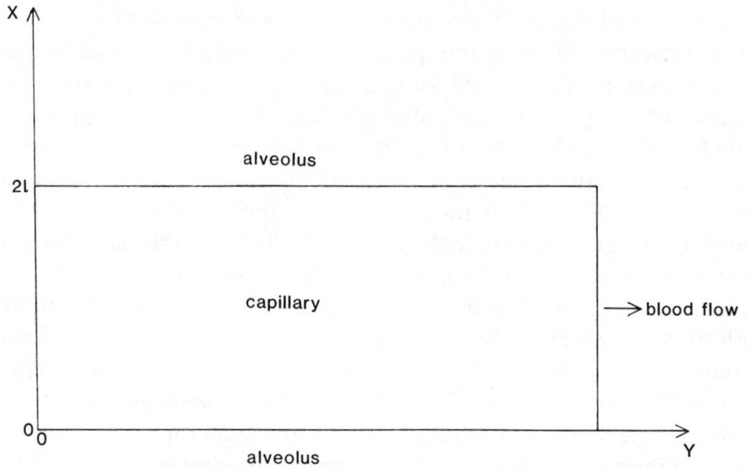

Fig. 9.40. Diagrammatic representation of a pulmonary capillary.

solutions so that $\partial/\partial\tau \equiv 0$. Then adding the last three equations we obtain

$$\left(s\frac{\partial}{\partial Y} - D\,\nabla^2\right)([\mathrm{Hb}]+[\mathrm{HbO_2}]+[\mathrm{HbCO}])=0$$

which can be integrated (using appropriate boundary conditions which we shall derive later) to give

$$[\mathrm{Hb}]+[\mathrm{HbO_2}]+[\mathrm{HbCO}]=c_t \tag{9.42}$$

a constant, in other words, we have a conservation equation for the haemoglobin.

Let us non-dimensionalise the equations by defining

$$\begin{aligned} &u_1=[\mathrm{O_2}]/\gamma_1, \quad u_2=[\mathrm{CO}]/\gamma_2 \\ &v=[\mathrm{Hb}]/c_t, \quad v_1=[\mathrm{HbO_2}]/c_t, \quad v_2=[\mathrm{HbCO}]/c_t \\ &y=Y/l, \quad x=X/l \end{aligned} \tag{9.43}$$

where γ_1 and γ_2 are typical concentrations of oxygen and carbon monoxide. Then the steady-state equations become

$$\delta_1\frac{\partial u_1}{\partial y}=-u_1v+K_{-1}v_1+\varepsilon_1\,\nabla^2 u_1 \tag{9.44}$$

$$\delta_2\frac{\partial u_2}{\partial y}=-u_2v+K_{-2}v_2+\varepsilon_2\,\nabla^2 u_2 \tag{9.45}$$

$$\delta_1'\frac{\partial v_1}{\partial y}=u_1v+K_{-1}v_1+\varepsilon_1'\,\nabla^2 v_1 \tag{9.46}$$

$$\delta_2' \frac{\partial v_2}{\partial y} = u_2 v - K_{-2} v_2 + \varepsilon_2' \nabla^2 v_2 \tag{9.47}$$

$$v + v_1 + v_2 = 1 \tag{9.48}$$

where

$$\delta_1 = s/k_1 c_t l = O(10^{-4}), \qquad \delta_2 = s/k_2 c_t l = O(10^{-3})$$

$$\delta_1' = s/k_1 \gamma_1 l = O(10^{-2}), \qquad \delta_2' = s/k_2 \gamma_2 l = O(10^{-1}\lambda^{-1})$$

$$\varepsilon_1' = D_1/k_1 c_t l^2 = O(10^{-3}), \qquad \varepsilon_2' = D_2/k_2 c_t l^2 = O(10^{-2})$$

$$\varepsilon_1' = D/k_1 \gamma_1 l^2 = O(10^{-2}), \qquad \varepsilon_2' = D/k_2 \gamma_2 l^2 = O(10^{-2}\lambda^{-1})$$

$$K_{-1} = k_{-1}/k_1 \gamma_1 \simeq 1/14{\cdot}25, \qquad K_{-2} = k_{-2}/k_2 \gamma_2 \simeq 2 \cdot 10^{-4}\lambda^{-1}$$

and the values of the parameters are given in Table 9.50. Here we have taken the values of the parameters in Table 9.50 based on data from Wyman (1966), Fishman (1963) and Roughton (1945) (or see Britton and Murray, 1977). The typical value for the carbon monoxide concentration is based on a blood carboxyhaemoglobin level of about 75% of the total haemoglobin, or a reasonably severe case of carbon monoxide poisoning.

Table 9.50
Parameter Values

Parameter	Value	Parameter	Value
γ_1	2×10^{-7} mole cm^{-3}	k_1	2.85×10^{9} (mole cm^{-3})$^{-1}$ s^{-1}
γ_2	2×10^{-9} mole cm^{-3}	k_{-1}	40 s^{-1}
D_1	1.2×10^{-5} cm^2 s^{-1}	k_2	2×10^{8} (mole cm^{-3})$^{-1}$ s^{-1}
D_2	1.3×10^{-5} cm^2 s^{-1}	k_{-2}	8×10^{-3} s^{-1}
D	2.45×10^{-7} cm^2 s^{-1}	s	10^{-3} cm s^{-1}
l	4×10^{-4} cm		

The boundary conditions are as follows:

(i) The blood is well mixed on arrival at the capillary, and is in equilibrium, i.e.

$$\begin{aligned} u_i(x,0) &= u_i^0, \qquad i=1,2 \\ v(x,0) &= v^0 \\ v_i(x,0) &= v_i^0, \qquad i=1,2 \end{aligned} \tag{9.51}$$

where u_i^0, v^0 and v_i^0 are constants and

$$k_i u_i^0 v^0 - k_{-i} v_i^0 = 0, \qquad i=1,2 \tag{9.52}$$

(ii) There are no sudden changes in any concentration at the exit from the capillary.

(iii) There is no flux of haemoglobin or its compounds through the walls of the capillaries, i.e.

$$\frac{\partial v}{\partial x}(0, y) = 0, \qquad \frac{\partial v_i}{\partial x}(0, y) = 0, \qquad i = 1, 2 \tag{9.53}$$

(iv) The concentrations of oxygen and carbon monoxide at the walls are equal to the alveolar concentrations, i.e.

$$u_1(0, y) = u_1^a, \qquad u_2(0, y) = u_2^a = 0 \tag{9.54}$$

where u_1^a is a constant.

(v) By symmetry, at the midpoint of the capillary,

$$\frac{\partial u_i}{\partial x}(1, y) = \frac{\partial v}{\partial x}(1, y) = \frac{\partial v_i}{\partial x}(1, y) = 0, \qquad i = 1, 2 \tag{9.55}$$

This is again a singular perturbation problem, since we have the small parameters ε_1, ε_2 and ε_1' multiplying the highest (second) order derivatives, and δ_1, δ_2 and δ_1' multiplying the first-order derivatives in the first three equations, (9.44)–(9.46). We therefore expect an outer solution to hold away from the boundaries of the domain, where diffusion is unimportant, and an inner solution to hold in boundary layers near the walls and close to the entrance to the capillary where the derivative terms do become important. The boundary condition (9.52) states that there is no boundary layer at the exit to the capillary; this is physically reasonable, since the blood reaches equilibrium long before the exit is reached.

In the outer region all derivatives are taken to be $O(1)$. From (9.44) this implies

$$u_1 v - K_{-1} v_1 = O(10^{-3}) \tag{9.56}$$

and from (9.45)

$$u_2 v - K_{-2} v_2 = O(10^{-2}) \tag{9.57}$$

With (9.47),

$$\delta_2' \frac{\partial v_2}{\partial y} - \varepsilon_2' \nabla^2 v_2 = O(10^{-2}) \tag{9.58}$$

We therefore wish to find, to first order, solutions satisfying these three conditions and (9.48). We shall take the view that if we can construct solutions which satisfy all the conditions of the problem, then these are the solutions required. In fact, we assume that the problem has a unique solution.

Obvious outer solutions are

$$\begin{aligned} u_i &= u_i^{\mathrm{f}}, \qquad i=1,2 \\ v &= v^{\mathrm{f}} \\ v_i &= v_i^{\mathrm{f}}, \qquad i=1,2 \end{aligned} \tag{9.59}$$

where u_i^{f}, v^{f} and v_i^{f} are constants and the superscript f, for final, denotes the values of the various concentrations at the exit to the capillary. These final concentrations must satisfy

$$u_i^{\mathrm{f}} v^{\mathrm{f}} - K_{-i} v_i^{\mathrm{f}} = 0, \qquad i=1,2, \qquad v^{\mathrm{f}} + v_1^{\mathrm{f}} + v_2^{\mathrm{f}} = 1 \tag{9.60}$$

These solutions then satisfy the required differential equations (to leading order) and the requirement that there be no boundary layer at the exit to the capillary. They also satisfy the symmetry boundary conditions (9.55) at the centre of the capillary, that the gradients of the concentrations there be zero. There is therefore no boundary layer here, as we might expect from physical considerations.

Now consider the boundary conditions (9.53) and (9.54) at the wall of the capillary. Those for haemoglobin and its compounds, the zero-flux conditions, are again satisfied by the outer solutions, and there are no boundary layers for these. In particular $\partial v_1/\partial y$ and $\nabla^2 v_1$ are $O(1)$ so that from (9.46) we have

$$u_1 v - K_{-1} v_1 = 0 \tag{9.61}$$

in the boundary layer. Hence

$$u_1 = K_{-1} v_1^{\mathrm{f}} / v^{\mathrm{f}} = u_1^{\mathrm{f}} \tag{9.62}$$

in the boundary layer. But the boundary condition for oxygen is that $u_1(0, y) = u_1^{\mathrm{a}}$, the alveolar concentration. To satisfy this we must have

$$u_1^{\mathrm{f}} = u_1^{\mathrm{a}} \tag{9.63}$$

The boundary condition for carbon monoxide is $u_2(0, y) = u_2^{\mathrm{a}} = 0$. This is not satisfied by the outer solution, so there must be a boundary layer in this case. In the boundary layer the x-derivative term $\varepsilon_2\, \partial^2 u_2/\partial x^2$ in Eq. (9.45) must be important. The boundary layer equations are obtained by re-scaling the x variable so that this becomes $O(1)$. To do this we define $\xi = x/\sqrt{\varepsilon_2}$, and the equation becomes

$$\delta_2 \frac{\partial u_2}{\partial y} = -u_2 v^{\mathrm{f}} + K_{-2} v_2^{\mathrm{f}} + \frac{\partial^2 u_2}{\partial \xi^2} + \varepsilon_2 \frac{\partial^2 u_2}{\partial y^2} \tag{9.64}$$

The leading order terms are

$$0 = -u_2 v^{\mathrm{f}} + K_{-2} v_2^{\mathrm{f}} + \frac{\partial^2 u_2}{\partial \xi^2} \tag{9.65}$$

which has solution

$$u_2 = (K_{-2} v_2^{\mathrm{f}} / v^{\mathrm{f}}) + A \exp\{-\xi\sqrt{v^{\mathrm{f}}}\} + B \exp\{\xi\sqrt{v^{\mathrm{f}}}\} \tag{9.66}$$

Since we require $u_2 \to u_2^{\mathrm{f}} = K_{-2} v_2^{\mathrm{f}} / v^{\mathrm{f}}$ as $\xi \to \infty$, we must have $B = 0$, and since $u_2 = u_2^{\mathrm{a}} = 0$ at $\xi = 0$, then $A = -u_2^{\mathrm{f}}$, and

$$u_2 = u_2^{\mathrm{f}}(1 - \exp(-\xi\sqrt{v^{\mathrm{f}}})) = K_2 v_2^{\mathrm{f}}\{1 - \exp(-\xi\sqrt{v^{\mathrm{f}}})\}/v^{\mathrm{f}} \tag{9.67}$$

This completes the description of the boundary layers at the wall of the capillary.

It remains to consider the boundary layer close to the entrance to the capillary. This is a boundary layer in y, the coordinate across the capillary, and we find the boundary layer equations by considering short y scales. Let us assume that there is a boundary layer in the carboxyhaemoglobin concentration v_2, and that the appropriate y scale is $\eta = y/\varepsilon$, where ε is some small parameter. Then the v_2 equation (9.47) becomes

$$\frac{\delta_2'}{\varepsilon}\frac{\partial v_2}{\partial \eta} = u_2 v - K_{-2} v_2 + \frac{\varepsilon_2'}{\varepsilon^2}\frac{\partial^2 v_2}{\partial \eta^2} + \varepsilon_2' \frac{\partial^2 v_2}{\partial x^2} \tag{9.68}$$

The dominant terms in this equation are

$$\frac{\delta_2'}{\varepsilon}\frac{\partial v_2}{\partial \eta} = \frac{\varepsilon_2'}{\varepsilon^2}\frac{\partial^2 v_2}{\partial \eta^2} \tag{9.69}$$

since $\delta_2'/\varepsilon \gg 1$ and $\varepsilon_2'/\varepsilon \gg 1$ whereas the remaining terms are $O(1)$. This has solution

$$v_2 = A + B \exp(\varepsilon \delta_2' \eta / \varepsilon_2') \tag{9.70}$$

Since we require $v_2 \to v_2^{\mathrm{f}}$ as $\eta \to \infty$, we must have $B = 0$ and $A = v_2^{\mathrm{f}}$. Hence

$$v_2 = v_2^{\mathrm{f}} \tag{9.71}$$

throughout the boundary layer. To satisfy the boundary condition at $\eta = 0$, or $y = 0$, we must have $v_2^{\mathrm{f}} = \mathrm{v}_2^0$.

We could go through and solve the boundary layer equations for the remaining concentrations u_1, u_2, v and v_1, but we are really interested in the carboxyhaemoglobin concentration alone. Our analysis so far has told us that this concentration does not change, to first order, throughout the pulmonary capillary. To obtain the small correction to this it would be possible to go to second order in the perturbation scheme, in other words, to consider the effects of the terms we have so far neglected. However, it turns out that this is not necessary. This is because we already have an expression which will allow us to calculate the amount of carbon monoxide which diffused out of the capillary as the blood passed through it.

Consider an element of blood from the time t_0 when once it leaves the pulmonary capillary to the time t_1 when it leaves one circuit of the body later. The concentration of carboxyhaemoglobin is the same, to first order, at t_0 and t_1. The concentration of oxygen is equal to the alveolar concentration and hence is the same at t_0 and t_1. Since the oxygen reaction is in equilibrium at t_0 and t_1, it follows that the concentrations of haemoglobin and oxyhaemoglobin are the same, to first order, at both times. Also the carbon monoxide reaction is in equilibrium so that this concentration is the same at both times, again to first order. But we know that there has been a loss of carbon monoxide by diffusion into the lungs. Neglecting any exchange of carbon monoxide between the blood and other parts of the body, this carbon monoxide must either have come from the free carbon monoxide in the blood or from carbon monoxide bound to carboxyhaemoglobin. The flux is so large that it cannot have come from the free carbon monoxide, since this would have changed the carbon monoxide concentration considerably, contradicting the fact that it remains unchanged to first order. It must therefore have come from the carboxyhaemoglobin. This does not change the carboxyhaemoglobin concentration appreciably because the amount of carbon monoxide which is bound to haemoglobin is orders of magnitude greater than the amount of free carbon monoxide, and hence the change is negligible.

We therefore wish to calculate the rate of loss of carbon monoxide to the lungs. The (dimensional) flux of carbon monoxide per second per unit area of capilllary wall is given by

$$F = -\frac{D_2\gamma_2}{l}\frac{\partial u_2}{\partial x}(0, y) \tag{9.72}$$

so that the rate of loss per unit volume of blood is

$$\frac{D_2\gamma_2}{l^2}\frac{\partial u_2}{\partial x}(0, y) \tag{9.73}$$

and the average rate of loss over one circuit is

$$\frac{D_2\gamma_2}{l^2}\frac{\partial u_2}{\partial x}(0, y)\frac{t_L}{t_C} \tag{9.74}$$

taking account of the fact that carbon monoxide is only lost when the blood is in the pulmonary capillary. Here t_L is the time spent in the pulmonary capillary and t_C the time taken for a full circuit of the body. Hence

$$\frac{d\bar{v}_2}{dt} = -\frac{D_2\gamma_2}{l^2 c_t}\frac{t_L}{t_C}\frac{\partial u_2}{\partial x}(0, y) \tag{9.75}$$

where $\bar{v}_2$ is the average carboxyhaemoglobin concentration in the body. From (9.67)

$$\frac{\partial u_2}{\partial x}(0, y) = \frac{K_{-2}v_2^f}{\sqrt{\varepsilon_2}\sqrt{v^f}} \tag{9.76}$$

Since $v_2^f = v_2^0$, the carboxyhaemoglobin concentration does not change markedly throughout the body, and we may drop the superfixes and the bar on v_2. From (9.48) and (9.60)

$$v^f + u_1^f v^f / K_{-1} + v_2 = 1 \tag{9.77}$$

so that, using the fact that $u_1^f = u_1^a$,

$$v^f = K_{-1}(1 - v_2)/(K_{-1} + u_1^a) \tag{9.78}$$

Since $K_{-1} \simeq 0.07$ and $u_1^a = 1$ in the case of a man breathing air at atmospheric pressure and greater than this in hyperbaric oxygen, K_{-1} may be neglected compared to u_1^a, so that

$$v^f \simeq K_{-1}(1 - v_2)/u_1^a \tag{9.79}$$

and if t is now the time in minutes

$$\frac{dv_2}{dt} = -b\sqrt{u_1^a}\frac{v_2}{\sqrt{1 - v_2}} \tag{9.80}$$

say, where

$$b = \frac{60 D_2\gamma_2 K_{-2} t_L}{\sqrt{\varepsilon_2}\, l^2 c_t \sqrt{K_{-1}}\, t_C} \simeq 6 \times 10^{-3} \tag{9.81}$$

Since there are no data for t_L/t_C in a state of carbon monoxide poisoning this has been estimated to be 2×10^{-2}. This is greater than its value of 10^{-2} at rest [see Daly (1979) and Roughton (1945)], since it is known that the presence of carboxyhaemoglobin in the blood leads to an increased pulse rate and deeper breathing (Haldane, 1895).

The differential equation (9.80) for v_2 can be integrated to obtain the time required for a drop in v_2 from a value v_2^1 to a value v_2^2. This is given by

$$\begin{aligned} b\sqrt{u_1^a}\, t &= \int_{v_2^2}^{v_2^1} \frac{\sqrt{1 - v_2}}{v_2}\, dv_2 \\ &= \left[2\sqrt{1 - v_2} + \log\frac{1 - \sqrt{1 - v_2}}{1 + \sqrt{1 - v_2}}\right]_{v_2^2}^{v_2^1} \end{aligned} \tag{9.82}$$

Data from Brink (1944) gives u_1^a in terms of the pressure of oxygen in the atmosphere. If this pressure is n atmospheres, then

$$u_1^a = 4.6n - 0.25 \tag{9.83}$$

Thus the time for such a reduction can be found. If the initial carboxyhaemoglobin concentration is not too high (less than 30%, say), then the formula may be approximated by

$$bt\sqrt{u_1^a} = \log(v_2^1/v_2^2) \tag{9.84}$$

The time required for the carboxyhaemoglobin to attain a level of half its initial value is therefore fixed for a given pressure of oxygen and given by $t_{1/2}$, where

$$bt_{1/2}\sqrt{u_1^a} = \log 2 \tag{9.85}$$

$t_{1/2}$ is known as the half-clearance time. A graph of $t_{1/2}$ against external oxygen pressure is given in Fig. 9.86. Experimental measurements of half-clearance times have been made by Pace *et al.* (1950). Their values are compared with ours in Table 9.87.

It can be seen that the calculated values for 1 and 2.5 atmospheres of oxygen agree well with the experimental results. The reason for the disagree-

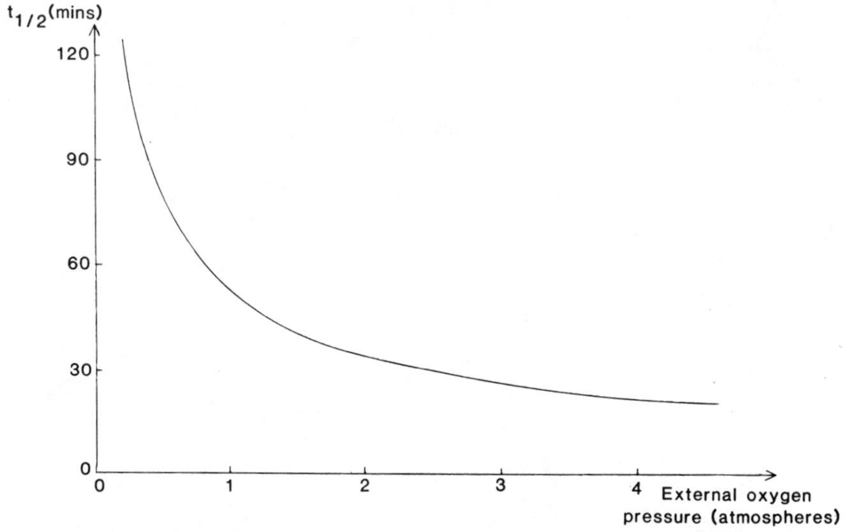

Fig. 9.86. The half-clearance time as a function of external oxygen pressure.

Table 9.87
Half-Clearance Times for Carboxyhaemoglobin in the Blood at Various Oxygen Pressures

Atmospheres of oxygen	Experimental half-clearance time (min)	Calculated half-clearance time (min)
0.21	249	104
1.0	47	41
2.5	22	25

ment in the case of air at atmospheric pressure is probably that the process in this case is much slower, so that the approximations we have made in neglecting transfer of carbon monoxide between the blood and body tissues are no longer valid. Since the medically interesting values are for high pressures of oxygen, this is not too much of a drawback.

References

Alikakos, N. D. (1981). Remarks on invariance in reaction-diffusion equations. *Nonlinear Anal. TMA* **5**, 593-614.

Amann, H. (1978a). Invariant sets and existence theorems for semilinear parabolic and elliptic systems. *J. Math. Anal. Appl.* **65**, 432-467.

Amann, H. (1978b). Existence and stability of solutions for semilinear parabolic systems, and applications to some diffusion-reaction equations. *Proc. R. Soc. Edinburgh, Sect. A: Math.* **81**, 35-47.

Aris, R. (1975). "The Mathematical Theory of Diffusion and Reaction in Permeable Catalysts", Vols. I and II. Oxford Univ. Press (Clarendon), London and New York.

Aronson, D. G. (1976). Topics in nonlinear diffusion. *CBMS/NSF Lecture Notes, Reg. Conf. Nonlinear Diffusion, Houston, 1976.*

Aronson, D. G. (1978). Comparison method for stability analysis of nonlinear parabolic problems. *SIAM Rev.* **20**, 245-264.

Aronson, D. G. and Weinberger, H. F. (1975). Nonlinear diffusion in population genetics, combustion and nerve propagation. In Proceedings of the Tulane Program in Partial Differential Equations and Related Topics, pp. 5-49. "Lecture Notes in Mathematics", Vol. 446, Springer-Verlag, Berlin.

Aronson, D. G. and Weinberger, H. F. (1978). Multidimensional nonlinear diffusions arising in population genetics. *Adv. Math.* **30**, 33-76.

Arrowsmith, D. K. and Place, C. M. (1982). "Ordinary Differential Equations". Chapman and Hall, London.

Auchmuty, J. F. G. and Nicolis, G. (1976). Bifurcation analysis of reaction-diffusion equations. III. Chemical oscillations. *Bull. Math. Biol.* **38**, 325-350.

Bailey, N. T. J. (1975). "The Mathematical Theory of Infectious Diseases and its Applications". Griffin, London.

Bard, J. (1981). A model for generating aspects of zebra and other mammalian coat patterns. *J. Theor. Biol.* **93**, 363-385.

Bates, P. W. and Brown, K. J. (1984). Convergence to equilibrium in a reaction-diffusion system. *Nonlinear Anal. TMA* **8**, 227-235.

Belousov, B. P. (1959). An oscillating reaction and its mechanism. *Sborn. referat. med.*, Medgiz, Moscow, 145.

Beloussov, L. V., Badenko, L. A., Katchurin, A. L. and Kurilo, L. F. (1972). Cell movements in morphogenesis of hydroid polyps. *J. Embryol. Exp. Morphol.* **27**, 317.

Blat, J. and Brown, K. J. (1984). Bifurcation of steady-state solutions in predator-prey and competition systems. *Proc. R. Soc. Edinburgh, Sect. A: Math.* **97**, 21-34.

Bogoliubov, N. N. and Mitropolsky, Y. A. (1961). Asymptotic methods in the theory of nonlinear oscillations. Hindustan.

Bramson, M. (1983). The convergence of solutions of the Kolmogorov nonlinear differential equation to travelling waves. *Mem. Amer. Math. Soc.* **44**, 285.

Bridge, J. F. and Angrist, S. E. (1962). An extended table of roots of $J'_n(x)Y'_n(\beta x) - J'_n(\beta x)Y'_n(x) = 0$. *Math. Comp.* **16**, 198–204.

Brink, F. (1944). *In* "Handbook of Respiratory Data in Aviation". OSRD committee, Washington, D.C.

Britton, N. F. (1979). A mathematical model for carbon monoxide poisoning in man. *Nonlinear Anal. TMA* **3**, 361–377.

Britton, N. F. (1982). Threshold phenomena and solitary travelling waves in a class of reaction–diffusion systems. *SIAM J. Appl. Math.* **42**, 188–217.

Britton, N. F. and Murray, J. D. (1977). The effect of carbon monoxide on haem-facilitated oxygen diffusion. *Biophys. Chem.* **7**, 159–167.

Britton, N. F. and Murray, J. D. (1979). Threshold, wave and cell–cell avalanche behaviour in a class of substrate inhibition oscillators. *J. Theor. Biol.* **77**, 317–322.

Britton, N. F., Joly, G. and Duban, M.-C. (1983). A model for hydranth regeneration in *Tubularia*. *Bull. Math. Biol.* **45**, 311–321.

Brown, K. J. (1983). Spatially inhomogeneous steady state solutions for systems of equations describing interacting populations. *Math. Anal. Appl.* **95**, 251–264.

Brown, P. N. (1980). Decay to uniform states in ecological interactions. *SIAM J. Appl. Math.* **38**, 22–37.

Brown, P. N. (1983). Decay to uniform states in competitive systems. *SIAM J. Math. Anal.* **14**, 659–673.

Brown, P. N. (1985). Decay to uniform states in predator–prey chains. *SIAM J. Appl. Math.* **45**, 465–478.

Capasso, V. and Fortunato, D. (1980). Stability results for semilinear evolution equations and their application to some reaction–diffusion problems. *SIAM J. Appl. Math.* **39**, 37–47.

Capasso, V. and Maddalena, L. (1981). Convergence to equilibrium states for a reaction-diffusion system modelling the spatial spread of a class of bacterial and viral diseases. *J. Math. Biol.* **13**, 173–184.

Capasso, V. and Maddalena, L. (1982a). Saddle point behaviour for a reaction–diffusion system: application to a class of epidemic models. *Math. Comp. Simulation* **24**, 540–547.

Capasso, V. and Maddalena, L. (1982b). Asymptotic behaviour for a system of nonlinear diffusion equations modelling the spread of oro-faecal diseases. *Rend. Accad. Sci. Fis. Mat. Napoli 4*, **48**, 475–495.

Capasso, V. and Maddalena, L. (1983). Periodic solutions for a reaction–diffusion system modelling the spread of a class of epidemics. *SIAM J. Appl. Math.* **43**, 417–427.

Carpenter, G. (1977). A geometric approach to singular perturbation problems with applications to nerve impulse equations. *J. Differ. Equations* **23**, 335–367.

Carr, J. and Muncaster, R. G. (1983a). The application of centre manifolds to amplitude expansions. I. Ordinary differential equations. *J. Differ. Equations* **50**, 260–279.

Carr, J. and Muncaster, R. G. (1983b). The application of centre manifolds to amplitude expansions. II. Infinite dimensional problems. *J. Differ. Equations* **50**, 280–288.

Casten, R. G. and Holland, C. J. (1977). Stability properties of solutions to systems of reaction–diffusion equations. *SIAM J. Appl. Math.* **33**, 353–364.

Casten, R. G. and Holland, C. J. (1978). Instability results for reaction–diffusion equations with Neumann boundary conditions. *J. Differ. Equations* **27**, 266–273.

Casten, R. G., Cohen, H. and Lagerstrom, P. (1975). Perturbation analysis of an approximation to Hodgkin–Huxley theory. *Q. Appl. Math.* **32**, 365–402.

Cerrai, P. (1983). Studiodi una sistema direazione-diffusione par la dinamica di due specie biologiche in competizione. *Boll. Un. Mat. Ital. C(6)*, **21(1)**, 253–276.

Cesari, L. (1963). "Asymptotic Behaviour and Stability Problems in Ordinary Differential Equations", 2nd Ed. Springer-Verlag, Berlin.

Chafee, N. (1975). Asymptotic behaviour for solutions of a one-dimensional parabolic equation with homogeneous Neumann boundary conditions. *J. Differ. Equations* **18**, 111–135.

Chandra, J. and Davis, P. W. (1979). Comparison theorems for systems of reaction–diffusion equations. *In* "Applied Nonlinear Analysis" (V. Lakshmikantham, ed.). Academic Press, New York.

Chow, S.-N. and Hale, J. K. (1982). "Methods of Bifurcation Theory". Springer-Verlag, New York.

Chueh, K. N., Conley, C. C. and Smoller, J. A. (1977). Positively invariant regions for systems of nonlinear diffusion equations. *Indiana Univ. Math. J.* **26**, 373–392.

Churchill, R. (1972). Isolated invariant sets in compact metric spaces. *J. Differ. Equations* **12**, 330–352.

Coburn, R. F., Forster, R. E. and Kane, P. B. (1965). Consideration of physiological variables that determine the blood carboxyhaemoglobin concentration in man. *J. Clin. Invest.* **44**, 1899–1910.

Coddington, E. A. and Levinson, N. (1955). "Theory of Ordinary Differential Equations". McGraw-Hill, New York.

Cohen, D. S. and Murray, J. D. (1981). A generalised diffusion model for growth and dispersion in a population. *J. Math. Biol.* **12**, 237–249.

Cohen, D. S., Hoppensteadt, F. C. and Miura, R. M. (1977). Slowly modulated oscillations in nonlinear diffusion processes. *SIAM J. Appl. Math.* **33**, 217–229.

Cohen, D. S., Neu, J. C. and Rosales, R. R. (1978). Rotating spiral wave solutions of reaction–diffusion equations. *SIAM J. Appl. Math.* **35**, 536–547.

Cohen, D. S., Contsias, E. and Neu, J. C. (1979a). Stable oscillations in single species growth models with hereditary effects. *Math. Biosci.* **44**, 255–268.

Cohen, D. S., Hagan, P. S. and Simpson, H. C. (1979b). Spatial structures in predator–prey communities with hereditary effects and diffusion. *Math. Biosci.* **44**, 167–178.

Cohen, M. S. (1977). The cyclic AMP control system in the development of *Dictyostelium Discoideum*. I. Cellular dynamics. *J. Theor. Biol.* **69**, 57–85.

Conley, C. C. (1975). On travelling wave solutions of nonlinear diffusion equations. Math. Research Center Technical Report 1492, Univ. of Wisconsin, Madison.

Conley, C. C. and Easton, R. (1971). Isolated invariant sets and isolating blocks. *Trans. Am. Math. Soc.* **158**, 35–61.

Conley, C. C. and Gardner, R. A. (1984). An application of the generalised Morse index to travelling wave solutions of a competitive reaction–diffusion model. *Indiana Univ. Math. J.* **33**, 319–343.

Conley, C. C. and Smoller, J. A. (1980a). Topological techniques in reaction–diffusion *In* "Biological Growth and Spread" (W. Jäger, H. Rost and P. Tautu, eds), Lecture notes in Biomathematics 38. Springer-Verlag, Berlin.

Conley, C. C. and Smoller, J. A. (1980b). Remarks on the stability of steady-state solutions of reaction–diffusion equations. *In* "Bifurcation Phenomena in Mathematical Physics and Related Topics" (C. Bardos and D. Bessis, eds) pp. 47–56. Reidel.

Conway, E. D. (1984). Diffusion and predator–prey interaction: pattern in closed systems. *In* "Partial Differential Equations and Dynamical Systems" (W. E. Fitzgibbon, ed.) pp. 85–133. Pitman, Boston.

Conway, E. D. and Smoller, J. A. (1977a). A comparison technique for systems of reaction–diffusion equations. *Comm. PDEs* **2**, 679–697.

Conway, E. D. and Smoller, J. A. (1977b). Diffusion and the predator–prey interaction. *SIAM J. Appl. Math.* **33**, 673–686.

Conway, E. D., Hoff, D. and Smoller, J. A. (1978). Large time behaviour of solutions of systems of nonlinear reaction–diffusion equations. *SIAM J. Appl. Math.* **35**, 1–16.

Conway, E. D., Gardner, R. A. and Smoller, J. A. (1982). Stability and bifurcation of steady-state solutions for predator-prey equations. *Adv. Appl. Math.* **3**, 288-334.

Cosner, C. and Lazer, A. C. (1984). Stable co-existence states in the Volterra-Lotka competition model with diffusion, *SIAM J. Appl. Math.* **44**, 1112-1132.

Courant, R. and Hilbert, D. (1962). "Methods of Mathematical Physics". Vol. 2. Wiley (Interscience), New York.

Crandall, M. and Rabinowitz, P. A. (1971). Bifurcation from simple eigenvalues. *J. Funct. Anal.* **8**, 321-340.

Crandall, M. and Rabinowitz, P. A. (1973). Bifurcation, perturbation of simple eigenvalues, and linearised stability. *Arch. Ration. Mech. Anal.* **52**, 161-220.

Crandall, M. and Rabinowitz, P. A. (1977). The Hopf bifurcation in infinite dimensions. *Arch. Ration. Mech. Anal.* **67**, 53-72.

Crank, J. (1975). "The Mathematics of Diffusion". Clarendon, Oxford.

Cronin, J. (1980). "Differential Equations: Introduction and Qualitative Theory". Dekker, New York.

Daly, M.,de B. (1979). *In* "Mechanics of Breathing in Human Physiology" (O. C. J. Lippold and F. R. Winton, eds.), 7th ed., p. 272. Edinburgh.

Deimling, K. and Lakshmikantham, V. (1980). Quasi-solutions and their role in the qualitative theory of differential equations. *Nonlinear Anal. TMA* **4**, 657-663.

de Mottoni, P. (1984). Space structures of some migrating populations. *In* "Mathematical Ecology" (S. A. Levin and T. G. Hallam, eds), Lecture Notes in Biomathematics 54, 502-513. Springer, Berlin.

de Mottoni, P. and Rothe, F. (1979). Convergence to homogeneous equilibrium state for generalised Volterra-Lotka systems with diffusion. *SIAM J. Appl. Math.* **37**, 648-663.

Diekmann, O. and Temme, N. M. (1976). "Nonlinear Diffusion Problems". Mathematisch Centrum, Amsterdam.

Duffy, M. R., Murray, J. D. and Britton, N. F. (1980). Spiral wave solutions of practical reaction-diffusion equations. *SIAM J. Appl. Math.* **39**, 8-13.

Dunbar, S. R. (1983). Travelling wave solutions of diffusive Lotka-Volterra equations. *J. Math. Biol.* **17**, 11-32.

Dunbar, S. R. (1984). Travelling wave solutions of diffusive Lotka-Volterra equations: a heteroclinic connection in $\mathbb{R}^4$. *Trans. Amer. Math. Soc.* **286**, 557-594.

Ermentrout, G. B. and Rinzel, J. (1981). Waves in a simple excitable or oscillatory reaction-diffusion model. *J. Math. Biol.* **11**, 269-294.

Ermentrout, G. B., Hastings, S. P. and Troy, W. C., (1984). Large amplitude stationary waves in an excitable lateral-inhibitory medium. *SIAM J. Appl. Math.* **44**, 1133-1149.

Erneux, T. (1981). Stability of rotating chemical waves. *J. Math. Biol.* **12**, 199-214.

Erneux, T. and Herschkowitz-Kaufman, M. (1979). Bifurcation diagram of a model chemical reaction—I. Stability changes of time-periodic solutions. *Bull. Math. Biol.* **41**, 21-38.

Evans, J. W. (1972a). Nerve axon equations: I. linear approximations. *Indiana Univ. Math. J.* **21**, 877-885.

Evans, J. W. (1972b). Nerve axon equations: II. stability at rest. *Indiana Univ. Math. J.* **22**, 75-90.

Evans, J. W. (1972c). Nerve axon equations: III. stability of the nerve impulse. *Indiana Univ. Math. J.* **22**, 577-593.

Evans, J. W. (1975). Nerve axon equations: IV. the stable and unstable impulse. *Indiana Univ. Math. J.* **24**, 1169-1190.

Evans, J. W. and Feroe, (1977). Local stability theory of the nerve impulse. *Math. Biosci.* **37**, 23.

Evans, J. W., Fenichel, N. and Feroe, J. A. (1982). Double impulse solutions in nerve axon equations. *SIAM J. Appl. Math.* **42**, 219-234.

Field, R. J. and Noyes, R. M. (1974). Oscillations in chemical systems. IV. Limit cycle behaviour in a model of a real chemical reaction. *J. Chem. Phys.* **60**, 1877–1884.

Field, R. J. and Troy, W. C. (1979). Existence of solitary travelling wave solutions of a model of the Belousov-Zhabotinskii reaction. *SIAM J. Appl. Math.* **37**, 561–587.

Field, R. J., Körös, E. and Noyes, R. M. (1972). Oscillations in chemical systems. II. Thorough analysis of temporal oscillations in the bromate-cerium-malonic acid system. *J. Am. Chem. Soc.* **94**, 8649–8664.

Fife, P. C. (1976a). Pattern formation in reacting and diffusing systems. *J. Chem. Phys.* **64**, 854–864.

Fife, P. C. (1976b). Boundary and interior transition layer phenomena for pairs of second-order differential equations. *J. Math. Anal. Appl.* **54**, 497–521.

Fife, P. C. (1979a). Mathematical aspects of reacting and diffusing systems. "Lecture Notes in Biomathematics", Vol. 28. Springer-Verlag, New York.

Fife, P. C. (1979b). The bistable nonlinear diffusion equation: basic theory and some applications, *in* "Applied Nonlinear Analysis" (V. Lakshmikantham, ed.). Academic Press, New York.

Fife, P. C. (1982). Sigmoidal Systems and Layer Analysis. *In* "Competition and Cooperation in Neural Nets" (S. Amari and M. A. Arbib, eds). Lecture Notes in Biomathematics 45, pp. 29–56, Springer, Berlin.

Fife, P. C. and McLeod, J. B. (1975). The approach of solutions of nonlinear diffusion equations to travelling wave solutions. *Bull. Am. Math. Soc.* **89**, 1075–1078.

Fife, P. C. and McLeod, J. B. (1977). The approach of solutions of nonlinear diffusion equations to travelling front solutions. *Arch. Ration. Mech. Anal.* **65**, 335–361.

Fife, P. C. and Tang, M. M. (1981). Comparison principles for reaction-diffusion systems: irregular comparison functions and applications to questions of stability and speed of propagation of disturbances. *J. Differ. Equations* **340**, 168–185.

Fisher, R. A. (1937). The wave of advance of advantageous genes. *Ann. Eugenics* **7**, 355–369.

Fishman, A. P. (1963). Dynamics of pulmonary circulation. *In* "Handbook of Physiology, 2", Vol. II, Chapter 48. Washington.

Fitzgibbon, W. E. and Walker, H. F., eds. (1977). "Nonlinear Diffusion". Research Notes in Mathematics 14. Pitman, London.

Fitzhugh, R. (1961). Impulses and physiological states in theoretical models of nerve membrane. *Biophys. J.* **1**, 445–466.

Fitzhugh, R. (1969). Mathematical models of excitation and propagation nerve. *In* "Biological Engineering" (H. Schwan, ed.). McGraw-Hill, New York.

Freedman, H. I. and Waltman, P. (1975a). Perturbation of two-dimensional predator-prey equations. *SIAM J. Appl. Math.* **28**, 1–10.

Freedman, H. I. and Waltman, P. (1975b). Perturbation of two-dimensional predator-prey equations with an unperturbed critical point. *SIAM J. Appl. Math.* **29**, 719–733.

Friedman, A. (1964). "Partial Differential Equations of Parabolic Type". Prentice-Hall, Englewood Cliffs, New Jersey.

Fujii, H., Mimura, M. and Nishiura, Y. (1982). A picture of the global bifurcation diagram in ecological interacting and diffusing systems. *Physica D* **5**, 1–42.

Gardner, R. A. (1981). Comparison and stability theorems for reaction-diffusion systems. *SIAM J. Math. Anal.* **12**, 603–616.

Gardner, R. A. (1982). Existence and stability of travelling wave solutions of competition models: a degree theoretic approach. *J. Differ. Equations* **44**, 343–364.

Gardner, R. A. (1984). Existence of travelling wave solutions of predator-prey systems via the connection index. *SIAM J. Appl. Math.* **44**, 56–79.

Gidas, B., Ni, W. M. and Nirenberg, L. (1979). Symmetry and related principles via the maximum principle. *Comm. Math. Phys.* **68**, 207–243.

Gierer, A. (1981). Generation of biological patterns and form: some physical, mathematical and logical aspects. *Prog. Biophys. Molec. Biol.* **27**, 1–47.

Gierer, A. and Meinhardt, H. (1972). A theory of biological pattern formation. *Kybernetika* **12**, 20–39.

Goodwin, B. C. (1976). "Analytical Physiology of Cells and Developing Organisms". Academic Press, London.

Greenberg, J. M. (1978). Axisymmetric time-periodic solutions to $\lambda-\omega$ systems. *SIAM J. Appl. Math.* **34**, 391–397.

Greenberg, J. M. and Hastings, S. P. (1978). Spatial patterns for discrete models of diffusion in excitable media. *SIAM J. Appl. Math.* **34**, 515–523.

Grindrod, P. and Sleeman, B. D. (1984). Comparison principles in the analysis of reaction-diffusion systems modelling unmyelinated nerve fibres. University of Dundee Report AA843.

Hadeler, K. P. and Rothe, F. (1975). Travelling fronts in nonlinear diffusion equations. *J. Math. Biol.* **2**, 251–263.

Hadeler, K. P., Rothe, F. and Vogt, H. (1979). Stationary solutions of reaction–diffusion equations. *Math. Methods Appl. Sci.* **1**, 545.

Hagan, P. S. (1981). Target patterns in reaction–diffusion systems. *Adv. Appl. Math.* **2**, 400–416.

Hagan, P. S. (1982). Spiral waves in reaction–diffusion equations. *SIAM J. Appl. Math.* **42**, 762–786.

Hagan, P. S. and Cohen, M. S. (1981). Diffusion-induced morphogenesis in the development of *dictyostelium*. *J. Theor. Biol.* **93**, 881–908.

Haldane, J. S. (1895). The action of carbonic oxide on man. *J. Physiol.* (*London*) **18**, 430.

Hanusse, P. (1972). On the existence of a limit cycle in evolution of open chemical systems. *Acad. Sci.* (*Paris*), *Ser. C* **274**, 1245–1247.

Hassard, B., Kazarinoff, N. and Wan, Y. (1981). "Theory and Applications of Hopf Bifurcation". Cambridge Univ. Press, London and New York.

Hastings, S. P. (1976a). Periodic plane waves for the Oregonator. *Stud. Appl. Math.* **55**, 293–299.

Hastings, S. P. (1976b). On the existence of homoclinic and periodic orbits for the Fitzhugh–Nagumo equations. *Q. J. Math.* (*Oxford*) **27**, 123–134.

Hastings, S. P. (1976c). On travelling wave solutions of the Hodgkin–Huxley equations. *Arch. Ration. Mech. Anal.* **60**, 229–257.

Hastings, S. P. (1982). Single and multiple pulse waves for the Fitzhugh–Nagumo equations. *SIAM J. Appl. Math.* **42**, 247–260.

Hastings, S. P. and Murray, J. D. (1975). On the existence of oscillatory solutions in the Field–Noyes model for the Belousoc–Zhabotinskii reaction. *SIAM J. Appl. Math.* **28**, 678–688.

Heineken, F. G., Tsuchiya, H. M. and Aris, R. (1967). On the mathematical status of the pseudo-steady state hypothesis of biochemical kinetics. *Math. Biosci.* **1**, 95–113.

Henry, D. (1981). Geometric Theory of Semilinear Parabolic Equations. "Lecture Notes in Mathematics", Vol. 840, 2nd ed. Springer-Verlag, Berlin.

Herschkowitz-Kaufman, M. (1975). Bifurcation analysis of nonlinear reaction–diffusion equations. II. Steady state solutions and comparison with numerical simulations. *Bull. Math. Biol.* **37**, 589–635.

Hethcote, H. W. (1976). Qualitative analysis of communicable disease models. *Math. Biosci.* **28**, 335–336.

Hiernaux, T. and Erneux, J. (1979). Chemical patterns in circular morphogenetic fields. *Bull. Math. Biol.* **41**, 461–468.

Hodgkin, A. L. and Huxley, A. F. (1952). A quantitative description of membrane current and its application to conduction and excitation in nerve. *J. Physiol.* **117**, 500–544.

Hofer, H. W. H. and Toland, J. F. (1984). Homoclinic, heteroclinic and periodic orbits for a class of indefinite Hamiltonian systems. *Math. Ann.* **268**, 387–403.

Hopf, E. (1942). Abzweigung einer periodischen Lösung von einer stationaren Lösung eines Differential systems. *Ber. Math.-Phys. Kl. Sachs Acad. Wiss. Leipzig* **94**, 3–22; (translation by N. Kopell and L. N. Howard *in* Marsden and McCracken (1976). "The Hopf Bifurcation and its Application". Springer-Verlag, New York.

Hoppensteadt, F. C. (1975). Mathematical Theories of Populations: Demographics, Genetics, and Epidemics. *Reg. Conf. Ser. Appl. Math. 20.* SIAM, Philadelphia.

Hosono, Y. and Mimura, M. (1982), Singular perturbation approach to travelling waves in competing and diffusing species models. *J. Math. Kyoto. Univ.* **22**, 435–461.

Hunding, A. (1983). Bifurcation of nonlinear reaction–diffusion systems in prolate spheroids. *J. Math. Biol.* **17**, 223–239.

Hunding, A. (1984). Bifurcation of nonlinear reaction–diffusion systems in oblate spheroids. *J. Math. Biol.* **19**, 249–263.

Iooss, G. and Joseph, D. D. (1980). "Elementary Stability and Bifurcation Theory". Springer-Verlag, Berlin.

Ito, M. (1984a). Global aspects of steady states for competitive–diffusive systems with homogeneous Dirichlet conditions. *Physica D* **14**, 1–28.

Ito, M. (1984b). A remark on singular perturbation methods. *Hiroshima Math. J.* **14**, 619–626.

Jacquez, J. A., Kutchai, H. and Daniels, E. (1972). Haemoglobin-facilitated diffusion of oxygen: interfacial and thickness effects. *Respir. Physiol.* **15**, 166–181.

Jones, C. K. R. T. (1983a). Spherically symmetric solutions of a reaction-diffusion equation. *J. Differ. Equations* **49**, 142–169.

Jones, C. K. R. T. (1983b). Asymptotic behaviour of a reaction–diffusion equation in higher space dimensions. *Rocky Mtn. J. Math.* **13**, 355–364.

Jones, C. K. R. T. (1983c). Some ideas in the proof that the Fitzhugh-Nagumo pulse is stable. In "Nonlinear Partial Differential Equations" (J. A. Smoller, ed.), Contemporary Mathematics Vol. 17, pp. 287–292, AMS, Providence, RI.

Jones, C. K. R. T. (1984). Stability of the travelling wave solution of the Fitzhugh-Nagumo system. *Trans. Am. Math. Soc.* **286**, 431–469.

Jones, D. S. and Sleeman, B. D. (1983). "Differential Equations and Mathematical Biology." George Allen and Unwin, London.

Jordan, D. W. and Smith, P. (1979). "Nonlinear ordinary differential equations". Oxford Univ. Press (Clarendon), London.

Joseph, D. D. and Sattinger, D. H. (1972). Bifurcating time-periodic solutions and their stability. *Arch. Ration. Mech. Anal.* **45**, 75–109.

Källén, A. (1984). Thresholds and travelling waves in an epidemic model for rabies. *Nonlinear Anal. TMA* **8**, 851–856.

Källén, A., Arcuri, P. and Murray, J. D. (1985). A simple model for the spatial spread and control of rabies, *J. Theor. Biol.* **116**, 377–394.

Kametaka, Y. (1976). On the nonlinear diffusion equation of Kolmogorov–Petrovskii–Piskunov type. *Osaka J. Math.* **13**, 11–66.

Kanel', Ya. I. (1962). On the stabilisation of solutions of the Cauchy problem for the equations arising in the theory of combustion. *Math. Sb.* **59**, 245–288.

Kaplan, C. (1977). "Rabies, The Facts". Oxford Univ. Press, London.

Kareiva, P. (1983). Local movements in herbivorous inserts: applying a passive diffusion model to mark-recapture experiments. *Oecologia* **57**, 322–324.

Kauffman, S. A., Shymko, R. M. and Trabert, K. (1978). Control of sequential compartment formation in *Drosphila Science* **199**, 259–270.

Keener, J. P. (1980). Waves in excitable media. *SIAM J. Appl. Math.* **32**, 528–548.

Kemmner, W. (1984). A model of head regeneration in hydra. *Differentiation* **26**, 83–90.

Kermack, W. O. and McKendrick, A. G. (1927). Contributions to the mathematical theory of epidemics. *Proc. R. Soc. Edinburgh, Sect. A: Math.* **115**, 700–721, reprinted in Oliveira-Pinto and Conolly (1982).

Kernevez, J. P. (1980). "Enzyme Mathematics". North-Holland Publ., Amsterdam.

Kernevez, J. P., Joly, G., Duban, M.-C., Bunow, B. and Thomas, D. (1979). Hysteresis, oscillations and pattern formation in realistic immobilised enzyme systems. *J. Math. Biol.* **7**, 41–56.

Kevorkian, J. and Cole, J. D. (1981) "Perturbation Methods in Applied Mathematics", Springer-Verlag, New York.

Keyfitz, B. L. and Kuiper, H. J. (1983). Bifurcation resulting from changes in domain in a reaction–diffusion equation. *J. Differ. Equations* **47**, 378–405.

Kielhöfer, H. (1976). On the Lyapunov stability of stationary solutions of semilinear parabolic equations. *J. Differ. Equations* **22**, 193–208.

Kishimoto, K. (1982). The diffusive Lotka–Volterra system with three species can have a stable non-constant equilibrium solution. *J. Math. Biol.* **16**, 103–122.

Kishimoto, K., Mimura, M. and Yoshida, K. (1983). Stable spatio-temporal oscillations of diffusive Lotka–Volterra systems with three or more species. *J. Math. Biol.* **18**, 213–221.

Klaasen, G. A. and Troy, W. C. (1981). The stability of travelling wave front solutions of a reaction–diffusion system. *SIAM J. Appl. Math.* **41**, 145–167.

Kolkka, R. W. and Salathé, E. P. (1984). A mathematical analysis of carrier-facilitated diffusion. *Math. Biosci.* **71**, 147–179.

Kolmogorov, A. N., Petrovskii, I. G. and Piskunov, N. S. (1937). A study of the equation of diffusion with increase in the quantity of matter, and its application to a biological problem. *Bjull. Moskovskovo Gos. Univ.* **17**, 1–72, partially reprinted in Oliveira-Pinto and Conolly (1982).

Kopell, N. and Howard, L. N. (1973). Plane wave solutions to reaction–diffusion equations. *Stud. Appl. Math.* **52**, 291–328.

Krasnoselskii, M. (1964). "Topological Methods in the Theory of Nonlinear Integral Equations". Macmillan, New York.

Kreuzer, F. and Hoofd, L. J. C. (1970). Facilitated diffusion of oxygen in the presence of haemoglobin. *Respir. Physiol.* **8**, 280–302.

Kuiper, H. J. (1977). Existence and comparison theorems for nonlinear diffusion systems. *J. Math. Anal. Appl.* **60**, 166–181.

Kuiper, H. J. (1980). Invariant sets for nonlinear elliptic and parabolic systems. *SIAM J. Math. Anal.* **11**, 1075–1103.

Kutchai, H., Jacquez, J. A. and Mather, F. J. (1970). Non-equilibrium facilitated oxygen transport in haemoglobin solutions. *Biophys. J.* **10**, 38–54.

Lacalli, T. (1980). Morphogenetic models for hydranth development. *In* "Development and Cellular Biology of Coelenterates" (P. Tardent and R. Tardent, eds.) Elsevier/North-Holland, Amsterdam.

Lacalli, T. and Harrison, L. G. (1979). Turing's conditions and the analysis of morphogenetic models. *J. Theor. Biol.* **76**, 419–436.

Ladde, G. S., Lakshmikantham, V. and Vatsala, A. S. (1985). Existence of coupled quasi-solutions of systems of nonlinear reaction–diffusion equations. *J. Math. Anal. Appl.* **108**, 249–266.

Lakshmikantham, V. (1979). Some problems of reaction-diffusion equations. *In* "Applied Nonlinear Analysis", Academic Press, New York.

Lakshmikantham, V. (1984). The present state of the method of upper and lower solutions. *In* "Trends in theory and practice of nonlinear differential equations" (V. Lakshmikantham, ed.), pp. 285-299. Dekker, New York and Basel.

Larson, D. A. (1978). Transient bounds and time asymptotic behaviour of solutions to nonlinear equations of Fisher type. *SIAM J. Appl. Math.* **34**, 93-103.

LaSalle, J. P. (1968). Stability theory for ordinary differential equations. *J. Differ Equations* **4**, 57-65.

Lau, K.-S. (1985). On the nonlinear diffusion equation of Kolmogorov, Petrovskii and Piskunov. *J. Differ Equations* **59**, 44-70.

Lazer, A. C. and McKenna, J. P. (1982). On steady state solutions of a system of reaction-diffusion equations from biology. *Nonlinear Anal.* **6**, 523-530.

Lefschetz, S. (1957). "Differential Equations: Geometric Theory". Wiley (Interscience), New York.

Leung, A. W. (1983). Reaction-diffusion equations for competing species singularly perturbed by a small diffusion rate *Rock Mtn Math.* **13**, 177-190.

Leung, A. W. (1984). A study of 3-species prey-predator reaction-diffusions by monotone schemes. *J. Math. Anal. Appl.* **100**, 583-604.

Levin, S. A. and Segel, L. A. (1985). Pattern generation in space and aspect. *SIAM Rev.* **27**, 45-67.

Lin, C. C. and Segel, L. A. (1974). "Deterministic Problems in the Natural Sciences". Macmillan, New York.

Lotka, A. (1920). Undamped oscillations derived from the law of mass action. *J. Am. Chem. Soc.* **42**, 1595-1599.

Ludwig, D., Jones, D. D. and Holling, C. S. (1978). Qualitative analysis of insect outbreak systems: the spruce budworm and the forest. *J. Anim. Ecol.* **47**, 315-332.

Ludwig, D., Aronson, D. G. and Weinberger, H. F. (1979). Spatial patterning of the spruce budworm. *J. Math. Biol.* **8**, 217-258.

MacDonald, D. W. and Bacon, P. J. (1982). Fox society, contact rate and rabies epizootiology. *Comp. Immun. Microbiol. Infect. Dis.* **5**, 247-256.

Madore, B. F. and Freedman, W. L. (1983). Computer simulations of the Belousov-Zhabotinskii reaction. *Science* **222**, 615-616.

Maginu, K. (1975). Reaction-diffusion equations describing morphogenesis. I. *Math. Biosci.* **27**, 17-98.

Maginu, K. (1980). Existence and stability of periodic travelling wave solutions to Nagumo's nerve equation. *J. Math. Biol.* **10**, 133-153.

Maginu, K. (1981). Stability of periodic travelling wave solutions with large spatial periods in reaction-diffusion systems. *J. Differ Equations* **39**, 73-99.

Manoranjan, V. S. and Mitchell, A. R. (1983). A numerical study of the Belousov-Zhabotinskii reaction using Galerkin finite element methods. *J. Math. Biol.* **16**, 251-260.

Markus, L. (1956). Asymptotically autonomous differential systems. *In* "Contributions to the Theory of Nonlinear Oscillations", Vol. 3, pp. 17-29. Annals of Mathematical Studies 36, Princeton Univ. Press, Princeton, New Jersey.

Marsden, J. E. and McCracken, M. (1976). "The Hopf Bifurcation and its Applications". Springer-Verlag, New York.

Matano, H. (1979). Asymptotic behaviour and stability of semilinear diffusion equations. *Publ. Res. Inst. Math., Kyoto* **15**, 401-451.

McKean, H. P. (1970). Nagumo's equation. *Adv. Math.* **4**, 209-223.

McKean, H. P. (1975). Application of Brownian motion to the equation of Kolmogorov-Petrovskii-Piskunov. *Comm. Pure Appl. Math.* **28**, 323-331.

McKean, H. P. (1983). Stabilisation of solutions of a caricature of the Fitzhugh-Nagumo equation 1. *Comm. Pure Appl. Math.* **36**, 291-324.

McKean, H. P. (1984). Stabilisation of solutions of a caricature of the Fitzhugh-Nagumo equation 2. *Comm. Pure Appl. Math.* **37**, 299-301.

McNabb, A. (1961). Comparison and existence theorems for multicomponent diffusion systems. *J. Math. Anal. Appl.* **3**, 133-144.

Michaelis, L. and Menten, M. I. (1913). Die Kinetik der Invertinwirkung. *Biochem. Z.* **49**, 333-369.

Meinhardt, H. (1982). "Models of Biological Pattern Formation". Academic Press, New York.

Mikhailov, A. S. and Krinsky, V. I. (1983). Rotating spiral waves in excitable media: the analytical results. *Physica D* **9**, 346-371.

Mimura, M. (1979). Asymptotic behaviours of a parabolic system related to a planktonic prey and predator model. *SIAM J. Appl. Math.* **37**, 499-512.

Mimura, M. and Murray, J. D. (1978a). On a diffusive prey-predator model which exhibits patchiness. *J. Theor. Biol.* **75**, 249-262.

Mimura, M. and Murray, J. D. (1978b). Spatial structures in a model substrate inhibition reaction-diffusion system. *Z. Naturforsch.* **33**, 580-586.

Mimura, M. and Nishida, T. (1978). On a certain semilinear parabolic system related to the Lotka-Volterra ecological model. *Publ. RIMS Kyoto Univ.* **14**, 269-282.

Mimura, M., Nishiura, Y. and Yamaguti, M. (1979). Some diffusive prey and predator systems and their bifurcation problems. *Ann. NY Acad. Sci.* **316**, 490-510.

Mimura, M., Tabata, M. and Hosono, Y. (1980). Multiple solutions of two-point boundary value problems of Neumann type with a small parameter. *SIAM J. Math. Anal.* **11**, 613-631.

Minorsky, N. (1974). "Nonlinear Oscillations", 2nd Ed. Krieger, New York (1st Ed. 1962, Van Nostrand. Reinhold, Princeton, New Jersey).

Mitchell, P. J. and Murray, J. D. (1973). Facilitated diffusion: the problem of boundary conditions. *Biophysik (Berlin)* **9**, 177-190.

Miura, R. M. (1977). A nonlinear WKB method and slowly-modulated oscillations in nonlinear diffusion processes. *In* "Nonlinear Diffusion". Research Notes in Mathematics 14, pp. 155-170. Pitman, London.

Mollison, D. (1977). Spatial contact models for ecological and epidemiological spread. *J. R. Stat. Soc. B* **39**, 283-326.

Mollison, D. (1983). Simplifying simple epidemic models (preprint).

Mollison, D. and Kuulasmaa, K. (1984). Spatial epidemic models: theory and simulations. *In* "Mathematical Aspects of Rabies Epizootics" (P. J. Bacon, ed.). Academic Press, London.

Montgomery, J. (1973). Cohomology of isolated invariant sets under perturbation. *J. Differ. Equations* **13**, 257-299.

Morrison, J. A. (1966). Comparison of the modified method of averaging and the two-variable expansion procedure. *SIAM Rev.* **8**, 66-85.

Murray, J. D. (1971). On the molecular mechanism of facilitated oxygen diffusion by haemoglobin and myoglobin. *Proc. R. Soc. B* **178**, 95.

Murray, J. D. (1974a). "Asymptotic Analysis". Oxford Univ. Press, London and New York.

Murray, J. D. (1974b). On the role of myoglobin in muscle respiration. *J. Theor. Biol.* **47**, 115-126.

Murray, J. D. (1975). Nonexistence of wave solutions for the class of reaction-diffusion equations given by the Volterra interacting population equations with diffusion. *J. Theor. Biol.* **52**, 459-469.

Murray, J. D. (1976). On travelling wave solutions in a model for the Belousov-Zhabotinskii reaction. *J. Theor. Biol.* **56**, 329-353.

Murray, J. D. (1977). "Lectures on Nonlinear Differential-Equation Models in Biology". Oxford Univ. Press (Clarendon), London and New York.

Murray, J. D. (1982). Parameter space for Turing instability in reaction-diffusion mechanisms: a comparison of models. *J. Theor. Biol.* **98**, 143-163.

Murray, J. D. and Wyman, J. (1971). Facilitated diffusion: the case of carbon monoxide. *J. Biol. Chem.* **246**, 5903-5906.

Murray, J. D. and Sperb, R. P. (1983). Minimum domains for spatial patterns in a class of reaction-diffusion equations. *J. Math. Biol.* **18**, 169-184.

Murray, J. D. and Oster, G. F. (1984). Generation of biological pattern and form. *IMAJ. Math. Appl. Med. Biol.* **1**, 51-76.

Murray, J. D., Oster, G. F. and Harris, A. K. (1983). A mechanical model for mesenchymal morphogenesis. *J. Math. Biol.* **17**, 125-129.

Nagumo, J., Arimoto, S. and Yoshizawa, S. (1962). An active pulse transmission line simulating nerve axon. *Proc. Inst. Radio Eng.* **50**, 2061-2070.

Nallaswamy, R. (1983). Global stability of a system of two interacting species with convective and dispersive migration.*Math. Biosci.* **67**, 101-111.

Nayfeh, A. H. (1973). "Perturbation Methods". Wiley (Interscience), New York.

Nayfeh, A. H. (1981). "Introduction to Perturbation Techniques". Wiley, New York.

Neu, J. C. (1979). Chemical waves and the diffusive coupling of limit cycle oscillators. *SIAM J. Appl. Math.* **36**, 509-515.

Nicolis, G. and Prigogine, I. (1977). "Self-Organisation in Nonequilibrium Systems". Wiley (Interscience), New York.

Nishiura, Y. (1979). Global branching theorem for spatial patterns of reaction-diffusion systems. *Proc. Japan. Acad. A* **55**, 201-204.

Nishiura, Y. (1982). Global structure of bifurcating solutions of some reaction-diffusion systems. *SIAM J. Math. Anal.* **13**, 555-593.

Noyes, R. M. and Jwo, J.-J. (1975). Oscillations in chemical systems X. Implications of cerium oxidation mechanisms for the Belousov-Zhabotinskii reaction. *J. Am. Chem. Soc.* **97**, 5431-5433.

Odell, G. M. (1980). Qualitative theory of systems of ordinary differential equations, including phase plane and the use of the Hopf bifurcation theorem. *In* "Mathematical Models in Molecular and Cellular Biology" (L. A. Segel, ed.) Cambridge Univ. Press, London and New York.

Odell, G. M., Oster, G. F., Alberch, P. and Burnside, B. (1981). The mechanical basis of morphogenesis. I. Epithelial folding and invagination. *Dev. Biol.* **85**, 446-462.

Okubo, A. (1980). "Diffusion and Ecological Problems: Mathematical Models". Springer-Verlag, Berlin.

Oliveira-Pinto, F. and Conolly, B. W. (1982). "Applicable Mathematics of Non-physical Phenomena". Ellis Horwood, Chichester.

O'Malley, R. E. (1974). "Introduction to Singular Perturbations". Academic Press, New York.

Oster, G. F. and Odell, G. M. (1984). The mechanochemistry of cytogels. *In* "Fronts, Interfaces and Patterns" (A. Bishop, ed.). Elsevier, Amsterdam.

Othmer, H. G. (1977). Current problems in pattern formation. *In* "Lectures on Mathematics in the Life Sciences", Vol. 9, 57-85. American Mathematical Society, Providence, Rhode Island.

Pace, N., Strajman, E. and Walker, E. (1950). Acceleration of carbon monoxide elimination in man by high pressure oxygen. *Science* **111**, 652.

Pao, C. V. (1980). Mathematical analysis of enzyme-substrate reaction-diffusion in some biochemical systems. *Nonlinear Anal. TMA* **4**, 369-392.

Pao, C. V. (1981). Co-existence and stability of a competition-diffusion system in population dynamics. *J. Math. Anal. Appl.* **83**, 54-96.

Pao, C. V. (1982). On nonlinear reaction-diffusion systems. *J. Math. Anal. Appl.* **87**, 165-198.

Pao, C. V. (1984). Monotone convergence of time-dependent solutions for coupled reaction-diffusion systems. *In* "Trends in Theory and Practice of Nonlinear Differential Equations" (V. Lakshmikantham, ed.), pp. 455-466. Dekker, New York and Basel.

Pao, C. V. (1985). Asymptotic behaviour of a reaction-diffusion system in bacteriology. *J. Math. Anal. Appl.* **108**, 1-14.

Pauwelussen, J. P. and Peletier, L. A. (1981). Clines in the presence of asymmetric migration. *J. Math. Biol.* **11**, 207-233.

Poincaré, H. (1892-1899). "Les Méthodes Nouvelles de la Méchanique Celeste" (3 volumes). Gauthiers-Villars, Paris.

Protter, M. H. and Weinberger, H. F. (1967). "Maximum Principles in Differential Equations". Prentice-Hall, Englewood Cliffs, New Jersey.

Rabinowitz, P. (1971). Some global results for nonlinear eigenvalue problems. *J. Funct. Anal.* **1**, 487-513.

Rauch, J. and Smoller, J. A. (1978). Qualitative theory of the Fitzhugh-Nagumo equations. *Adv. Math.* **27**, 12-44.

Redheffer, R. and Walter, W. (1978). Invariant sets for systems of partial differential equations. I. Parabolic equations. *Arch. Ration. Mech. Anal.* **67**, 41-52.

Redheffer, R. and Walter, W. (1983). On parabolic systems of the Volterra predator-prey type. *Nonlinear Anal. TMA* **7**, 333-347.

Rinzel, J. (1981). Models in neurobiology. *In* "Nonlinear Phenomena in Physics and Biology" (R. H. Enns *et al.*, eds), pp. 347-367. Plenum, New York and London.

Rinzel, J. and Keller, J. B. (1973). Travelling wave solutions of a nerve conduction equation. *Biophys. J.* **13**, 1313-1337.

Rinzel, J. and Terman, D. (1982). Propagation phenomena in a bistable reaction-diffusion system. *SIAM J. Appl. Math.* **42**, 1111-1137.

Rinzel, J. and Keener, J. P. (1983). Hopf bifurcation to repetitive activity in nerve. *SIAM J. Appl. Math.* **43**, 907-922.

Riveros-Moreno, V. and Wittenberg, J. B. (1972). The self-diffusion coefficients of myoglobin and haemoglobin in concentrated solutions. *J. Biol. Chem.* **247**, 895-901.

Robinson, C. (1985). Phase plane analysis using the Poincaré map. *Nonlinear Anal. TMA* **9**, 1159-1164.

Root, W. (1965). Carbon monoxide, *in* "Handbook of Physiology, 3", vol. II, p. 1037. Washington.

Rothe, F. (1981). Convergence to pushed fronts. *Rocky M. J. Math.* **11**, 617-633.

Rothe, F. (1982). Uniform bounds from bounded L^p-functionals in reaction-diffusion equations. *J. Differ. Equations* **45**, 207-233.

Rothe, F. (1984). "Global Solutions of Reaction-Diffusion Equations". Lecture Notes in Mathematics, vol. 1072. Springer-Verlag, Berlin.

Rothe, F. and de Mottoni, P. (1979). A simple system of reaction-diffusion equations describing morphogenesis: asymptotic behaviour. *Ann. Mat. Pure Appl.* **122**, 141-157.

Roughton, F. J. W. (1945). The average time spent by the blood in the lung and its relation to the rates of carbon monoxide uptake and elimination in man. *Am. J. Physiol.* **143**, 621.

Sansone, G. and Conti, R. (1964). "Nonlinear Differential Equations". Pergamon, Oxford.

Sattinger, D. H. (1971). Stability of bifurcation solutions by Leray-Schauder degree. *Arch. Ration. Mech. Anal.* **43**, 154-166.

Sattinger, D. H. (1973). Topics in stability and bifurcation theory. "Lecture Notes in Mathematics", Vol. 309, Springer-Verlag, Berlin.

Sattinger, D. H. (1976). On the stability of waves of nonlinear parabolic systems. *Adv. Math.* **22**, 312–355.

Schneider, K. R. (1983). A note on the existence of periodic travelling wave solutions with large periods in generalised reaction–diffusion systems. *Z. Angew. Math. Phys.* **34**, 236–240.

Scholander, P. F. (1960). Oxygen transport through haemoglobin solutions. *Science* **131**, 585–590.

Seelig, F. F. (1976). Chemical oscillations by substrate inhibition—a parametrically universal oscillator type in homogeneous catalysis by metal complex formation. *Ber. Bunsenges. Phys. Chem.* **80**, 1126–1131.

Segel, L. A. (1984). Taxes in ecology and cell biology. *In* "Mathematical Ecology" (S. A. Levin and T. Hallam, eds.). Springer-Verlag, Berlin.

Shukla, V. P. and Das, P. C. (1982). Effects of dispersion on stability of multi-species prey-predator systems. *Bull. Math. Biol.* **44**, 571–578.

Sleeman, B. D. (1980). Instability of certain travelling wave solutions to the Fitzhugh–Nagumo equations. *J. Math. Anal. Appl.* **74**, 106–119.

Sleeman, B. D. (1981). Analysis of diffusion equations in biology. *Bull. IMA* **17**, 7–13.

Sleeman, B. D. and Tuma, E. (1984). On exact solutions of a class of reaction–diffusion equations. *IMA J. Appl. Math.* **33**, 153–168.

Smoller, J. A. (1983). "Shock Waves and Reaction–Diffusion Equations". Springer-Verlag, Berlin.

Sperb, R. (1981). "Maximum Principles and Their Applications". Academic Press, New York and London.

Stokes, A. N. (1976). On two types of moving front in quasilinear diffusion. *Math. Biosci.* **31**, 307–315.

Taylor, B. A. and Murray, J. D. (1977). Effect of the rate of oxygen consumption on muscle respiration. *J. Math. Biol.* **4**, 1–20.

Terman, D. (1983). Threshold phenomena for a reaction–diffusion system. *J. Differ. Equations* **47**, 406–443.

Thomas, D. (1976). Artificial enzyme membranes, kinetic and transport phenomena. *Phys. Veget.* **14**, 843–848.

Troy, W. C. (1977). A threshold phenomenon in the Field–Noyes model of the Belousov–Zhabotinskii reaction. *J. Math. Anal. Appl.* **58**, 233–248.

Turing, A. M. (1952). The chemical basis of morphogenesis. *Philos. Trans. R. Soc. London, Ser. B* **237**, 37–72.

Tyson, J. J. (1976). The Belousov–Zhabotinskii reaction. "Lecture Notes in Biomathematics", Vol. 10. Springer-Verlag, New York.

Tyson, J. J. (1981). On scaling the Oregonator equations. *In* "Synergetics: Nonlinear Phenomena in Chemical Dynamics" (A. Pacault and C. Vidal, eds.), 222–227. Springer-Verlag, Berlin.

Tyson, J. J. (1982). Scaling and reducing the Field–Körös–Noyes mechanism of the Belousov–Zhabotinskii reaction. *J. Phys. Chem.* **86**, 3006–3012.

Tyson, J. J. (1984). A quantitative account of oscillations, bistability and travelling waves in the Belousov–Zhabotinskii reaction. *In* "Oscillations and Travelling Waves in Chemical Systems" (R. J. Field and M. Burger, eds.). Wiley, New York.

Tyson, J. J. and Fife, P. C. (1980). Target patterns in a realistic model of the Belousov–Zhabotinskii reaction. *J. Chem. Phys.* **73**, 2224–2237.

van Dyke, M. (1975). "Perturbation Methods in Fluid Dynamics". Parabolic Press, Stanford.

Volterra, V. (1926). Variazionie fluttuazioni del numero d'individui in specie animali conviventi. *Mem. Acad. Lincei.* **2**, 31–113; Translation in R. N. Chapman (1931). "Animal Ecology" 409–448. McGraw-Hill, New York and in Oliveira-Pinto and Conolly (1982).

Voorhees, B. H. (1982). Dissipative structures associated to certain reaction–diffusion equations in mathematical ecology. *Bull. Math. Biol.* **44**, 339–348.

Webster, G. C. (1971). Morphogenesis and pattern formation in hydroids. *Biol. Rev.* **46**, 1.

Weinberger, H. F. (1975). Invariant sets for weakly coupled parabolic and elliptic systems. *Rend. Mat.* **8**, 295–310.

Winfree, A. T. (1972). Spiral waves of chemical activity. *Science* **175**, 634–636.

Winfree, A. T. (1974). Rotating chemical reactions. *Sci. Am.* **230**, 82–95.

Winfree, A. T. (1980). "The Geometry of Biological Time". Springer-Verlag, New York.

Wittenberg, B. A., Wittenberg, J. B. and Caldwell, P. R. B. (1975). Role of myoglobin in the oxygen supply to red skeletal muscle. *J. Biol. Chem.* **250**, 9038–9043.

Wittenberg, J. B. (1959). Oxygen transport: a new function proposed for myoglobin. *Biol. Bull.* **117**, 402–403.

Wittenberg, J. B. (1966). The molecular mechanism of haemoglobin-facilitated oxygen transport. *J. Biol. Chem.* **241**, 104–114.

Wittenberg, J. B. (1970). Myoglobin-facilitated oxygen diffusion: role of myoglobin in oxygen entry into muscle. *Physiol. Rev.* **50**, 559–636.

Wyman, J. (1966). Facilitated diffusion and the possible role of myoglobin as a transport mechanism. *J. Biol. Chem.* **241**, 115–121.

Yanagida, E. (1985). Stability of fast travelling pulse solutions of the Fitzhugh–Nagumo equations. *J. Math. Biol.* **22**, 81–104.

Yu, S. (1985). Asymptotic behaviour of solutions of heterogeneous nonlinear reacting and diffusing systems. *Nonlinear Anal. TMA* **9**, 275–288.

Zhabotinskii, A. M. (1964). Periodic course of oxidation of malonic acid on solution (investigation of the kinetics of the reaction of Belousov). *Biophysics* **9**, 329–335.

Zhou, L. and Pao, C. V. (1982). Asymptotic behaviour of a competition–diffusion system in population dynamics. *Nonlinear Anal. TMA* **6**, 1163–1184.

Index

MORE PRAISE FOR ERIC JEROME DICKEY

Liar's Game

"True to his form, Dickey offers many plot twists in [*Liar's Game*]—he has a wonderful flair."
—*The Cincinnati Enquirer*

"Witty and engrossing." —*Booklist*

"Very simply, Dickey is amazing. . . . *Liar's Game* [is] a sassy story filled with winding drama." —*Chicago Defender*

"Masterful . . . humorous and convincing . . . avoids all the usual clichés." —BookPage

"Dickey hits his stride . . . with wit, energy, and deft sensitivity. . . . [*Liar's Game*] is Dickey's best yet."
—*Heart & Soul*

"Seductive." —*Publishers Weekly*

"[*Liar's Game*] really has the power." —*Kurkus Reviews*

"[Dickey] has his finger on the pulse of contemporary urban romance . . . more twists and turns than the electric slide." —*Pittsburgh Post-Gazette*

"Skillful . . . scandalous . . . a rich gumbo of narrative twists." —*Minneapolis Star Tribune*

"Brimming with steamy romance, stinging betrayal, sweet redemption, and well-placed humor." —*Miami Times*

continued . . .

Cheaters

"A deftly crafted tale about the games people play and the lies they tell on their search for love." —*Ebony*

"Wonderfully written . . . smooth, unique, and genuine." —*The Washington Post Book World*

"Raw, street-savvy humor." —*Publishers Weekly*

"You can't read *Cheaters* without becoming an active participant." —*Los Angeles Times*

"What gives the book a compelling edge is the characters' self-discovery. . . . Thankfully, Dickey often goes beyond the 'men are dogs and women are victims' stereotype." —*USA Today*

"Hot, sexy, and funny . . . *Cheaters* not only makes readers examine their own behavior but keeps them laughing while doing so." —*Library Journal*

"A generous helping of humor and a distinctly male viewpoint." —*The Atlanta Journal-Constitution*